Recent Advances in Lichenology

Dalip Kumar Upreti · Pradeep K. Divakar
Vertika Shukla · Rajesh Bajpai
Editors

Recent Advances in Lichenology

Modern Methods and Approaches in Lichen Systematics and Culture Techniques, Volume 2

Springer

Editors
Dalip Kumar Upreti
Lichenology Laboratory
CSIR-National Botanical Research Institute
Lucknow, Uttar Pradesh
India

Vertika Shukla
Lichenology Laboratory
CSIR-National Botanical Research Institute
Lucknow, Uttar Pradesh
India

Pradeep K. Divakar
Facultad de Farmacia, Departamento de Biologia Vegetal II
Universidad Complutense de Madrid
Madrid, Madrid
Spain

Rajesh Bajpai
Lichenology Laboratory
CSIR-National Botanical Research Institute
Lucknow, Uttar Pradesh
India

ISBN 978-81-322-2234-7 ISBN 978-81-322-2235-4 (eBook)
DOI 10.1007/978-81-322-2235-4

Library of Congress Control Number: 2014957397

Springer New Delhi Heidelberg New York Dordrecht London

Printed on acid-free paper

Springer (India) Pvt. Ltd. is part of Springer Science+Business Media
(www.springer.com)

Dedicated to

Dr. D.D. Awasthi
Father of Indian Lichenology

Foreword

सीएसआईआर-राष्ट्रीय वनस्पति अनुसंधान संस्थान
CSIR-National Botanical Research Institute
(वैज्ञानिक तथा औद्योगिक अनुसंधान परिषद, नई दिल्ली)
राणा प्रताप मार्ग, लखनऊ – 226 001, उ.प्र., भारत
(Council of Scientific & Industrial Research, New Delhi)
Rana Pratap Marg, Lucknow - 226 001, U.P., India

डॉ. चन्द्र शेखर नौटियाल
टांटा इनोवेशन फैलो, एफएनए, एफएनएएससी, एफएनएएएस
निदेशक

Dr. Chandra Shekhar Nautiyal
TATA Innovation Fellow, FNA, FNASc, FNAAS
Director

FOREWORD

Lichens are unique organisms which require multidimensional approach to explore its potential in various fields of environment, botany and chemistry. Modern techniques especially molecular, culture, remote sensing technique has considerably contributed in the field of lichens. Multivariate analyses together with GIS approaches have established lichens as an ideal and reliable indicator of air pollution. Advanced culture techniques have increased the pharmacological application which was earlier restricted due to meager biomass of lichens. Advent of sophisticated analytical instrumentations facilitated isolation and characterization of bioactive constituents of lichens even in lower concentration for their bioprospection.

I am delighted to see that the editors made an effort to compile the advances in the field of lichenology contributed by experts of various fields of lichenology from around the globe, which will not only help to introduce various multidisciplinary approaches and techniques in the field of lichenology but also create interest among researchers to take up research on these unique organisms.

I am glad that this work has been accomplished in the Lichenology Laboratory of CSIR-National Botanical Research Institute, Lucknow.

I congratulate the editors for their endeavour in composing a comprehensive and valuable contribution about advances in the study of lichens worldwide and hope this book will be immensely helpful for environmentalists, botanist and phytochemists.

Date: 15th July 2014

(C.S. Nautiyal)

Phones : (Off.) (0522) 2205848, EPABX Phone : (0522) 2297800 to 2297999 Fax : (0522) 2205839, Post Box No. 436, Lucknow
e-mail : director@nbri.res.in, csnnbri@yahoo.com, Website: http://www.nbri.res.in

Preface

Lichens present an excellent example of symbiotic association. The unique composition of alga and fungi in lichens not only results in conferring differential sensitivity towards a range of environmental factors but also makes them physiologically adapted and chemically diversed to combat various abiotic and biotic environmental stresses.

This book volume covers two very important aspects of lichens, i.e. biomonitoring and bioprospection. In order to understand the role of lichens as biomonitors, it is important to know various factors which influence the growth of lichens in natural conditions.

Lichens have been known for long as bioindicators of air pollution and various studies have explored biomonitoring potential of different lichen species. Recently more standardised protocols have been adopted in Europe and America, which ensures the authenticity and reproducibility of lichen biomonitoring data.

Standardised methods for lichen diversity assessment not only authenticate classical approaches of lichen diversity changes related to ecological studies and air quality, but also provide an opportunity to extend to the studies related to climate change and land use changes. Lichen diversity studies coupled with remote sensing data can be used for mapping lichen species, which provide vital information regarding present scenario and predict future model of climate change or land use change. Applicability of lichen diversity-remote sensing studies further extends to estimating and mapping influence of pollutant in unsampled location based on sampled locations, which help to track the long-range transport of semi-volatile pollutants to high altitude ecosystems.

Lichens also play a major role in functioning of the ecosystem by the process of soil formation (lichens have an ability to grow on barren rock), but this characteristic is disadvantageous for conservation of ancient monuments as lichen growth results in biodeterioration. Studies on restricting lichen growth and preserving the monument are a subject of research and need implementation of strategies for conservation of monuments.

Another aspect of lichens which is gaining importance in recent climate change studies is lichenometric studies, to observe glacier retreat phenomenon. As any change in shape and size of glacier provides vital information regarding climate change, therefore the age of lichens growing on exposed moraines is a possible clue to retreat of glacier.

Apart from being excellent biomonitors, another not much explored aspect of lichens is its chemistry. For lichens their unique chemistry supplements physiological adaptation for sustaining in extreme climates, but for mankind lichens are a treasure house of pharmacologically important bioactive constituents capable of curing simple cough and cold to HIV and cancer.

The therapeutic potential of lichen secondary metabolites is known since medieval times mainly known through folklore and ethnobotanical uses. The main constraint which restricts its commercial utilisation is the slow growth rate and low biomass of lichens. Recent advances in analytical instrumentation techniques have resulted in high throughput techniques for isolation, while highly sensitive detection techniques allow characterisation of bioactive compounds even in very low quantity. Structural characterisation of bioactive compound paves the way for its synthesis in the laboratory and further structural modifications.

Therefore, the book volume intends to introduce researchers to advancements in the field of lichenology with an aim to involve more active participation of multidisciplinary research in the study of lichens.

Contents

Editors and Contributors

About the Editors

Dalip Kumar Upreti FNASc, Head of the Lichenology Laboratory, CSIR-National Botanical Research Institute (CSIR-NBRI), Lucknow, has a Ph.D. (1983) in Botany on *Studies on the lichen family Cladoniaceae from India and Nepal* from Lucknow University under the guidance of Dr. D.D. Awasthi. Dr. Upreti has over 35 years of research experience and has published more than 300 research papers in peer-reviewed journals and co-authored eight books.

Apart from taxonomy, Dr. Upreti has also carried out extensive research on ecology, lichen chemistry, pollution monitoring, in vitro culture and biodeterioration studies and bioprospection of Indian lichens. Dr. Upreti was in Antarctica in 1991–1992. He is the Indian corresponding member to the British Lichen Society and International Association for Lichenology.

Pradeep K. Divakar holds an M.Sc. in Botany and earned his Ph.D. (2002) on *Revisionary studies on the lichen genus Parmelia sensu lato in India* from Lucknow University. Dr. Divakar is Professor at the Department of Plant Sciences, Faculty of Pharmacy, Complutense University of Madrid, Spain. Dr. Divakar has published two books and more than 70 papers in national and international journals. Currently he is editor of *Biodiversity and Conservation*, *Mycokeys*, *Organisms Diversity and Evolution*, and *Lazaroa* (Iberian Journal of Botany).

His research activities centre on molecular systematics, genetic diversity, evolution, speciation events, gene mapping, gene flow, TCS estimation of gene genealogies and phenotypic character mapping, ancestral character state, and genomics of fungi that form mutualistic symbioses with algae or cyanobacteria and mycobiont cultures.

Vertika Shukla holds an M.Sc. in Organic Chemistry and a Ph.D. (2003) on the *Chemical Study of Macrolichens of Garhwal Himalayas* from H.N.B. Garhwal University, Srinagar (Garhwal). She is currently working in the Lichenology Laboratory of CSIR-NBRI as a DST scientist and to date she has published more than 50 scientific articles.

Dr. Shukla has contributed to secondary metabolite chemistry, spatio-temporal behavior of pollutants in high altitude ecosystems, lichen responses to climate change and the role of lichens in bioremediation of atmospheric fallouts.

Rajesh Bajpai holds an M.Sc. in Environmental Science and received his Ph.D. in 2009 from Babasaheb Bhimrao Ambedkar (Central) University, Lucknow on *Studies on lichens of some monuments of Madhya Pradesh with reference to Biodeterioration and Biomonitoring*, is working as a DST scientist in Lichenology Laboratory of CSIR-NBRI, Lucknow. Dr. Bajpai has published 30 research papers in various national and international journals.

Dr. Bajpai has carried out extensive research work on the interaction of lichen with the environment, climate-change modeling, arsenic pollution and biodeterioration.

Contributors

Ólafur S. Andrésson Faculty of Life and Environmental Sciences, University of Iceland, Reykjavik, Iceland

Bhaskar C. Behera Plant Science Division, Mycology and Plant Pathology Group, Agharkar Research Institute, Pune, India

Ana Crespo Departmento de Biologia Vegetal II, Facultad de Farmacia, Universtad Computense de Madrid, Madrid, Spain

Pradeep K. Divakar Departmento de Biologia Vegetal II, Facultad de Farmacia, Universtad Computense de Madrid, Madrid, Spain

Christina Hametner Department of Organismic Biology, University of Salzburg, Salzburg, Austria

Kojiro Hara Faculty of Bioresource Sciences, Akita Prefectural University, Akita, Japan

David L. Hawksworth Departamento de Biología Vegetal II, Facultad de Farmacia, Universidad Complutense de Madrid, Madrid, Spain; Department of Life Sciences, The Natural History Museum, London, UK; Mycology Section, Kew, Surrey, UK

Jae-Seoun Hur Korean Lichen Research Institute, Sunchon National University, Sunchon, Republic of Korea

Zophonías O. Jónsson Faculty of Life and Environmental Sciences, University of Iceland, Reykjavik, Iceland

Yogesh Joshi Lichenology Division, Department of Botany, Kumaun University, Almora, Uttarakhand, India

Hiroko Kawakami Faculty of Bioresource Sciences, Akita Prefectural University, Akita, Japan

Hangun Kim College of Pharmacy and Research Institute of Life and Pharmaceutical Sciences, Sunchon National University, Sunchon, Republic of Korea

Kyung Keun Kim Medical Research Center for Gene Regulation, Chonnam National University Medical School, Gwangju, Republic of Korea

Masashi Komine Faculty of Bioresource Sciences, Akita Prefectural University, Akita, Japan

Steven D. Leavitt Committee on Evolutionary Biology, University of Chicago, Chicago, IL, USA; Department of Science and Education, Field Museum of Natural History, Chicago, IL, USA

Vivian P.W. Miao Department of Microbiology and Immunology, University of British Columbia, Vancouver, Canada

Corrie S. Moreau Department of Science and Education, Field Museum of Natural History, Chicago, IL, USA

Gerhard Rambold Mycology Department, University of Bayreuth, Bayreuth, Germany

Preeti Shukla Lichenology Laboratory, CSIR-National Botanical Research Institute, Lucknow, Uttar Pradesh, India

Elfie Stocker-Wörgötter Department of Organismic Biology, University of Salzburg, Salzburg, Austria

H. Thorsten Lumbsch Department of Science and Education, Field Museum of Natural History, Chicago, IL, USA

Manish Tripathi Lichenology Division, Department of Botany, Kumaun University, Almora, Uttarakhand, India

D.K. Upreti Lichenology Laboratory, CSIR-National Botanical Research Institute, Lucknow, Uttar Pradesh, India

Neeraj Verma Plant Science Division, Mycology and Plant Pathology Group, Agharkar Research Institute, Pune, India

Silke Werth Faculty of Life and Environmental Sciences, University of Iceland, Reykjavik, Iceland; Swiss Federal Research Institute WSL, Birmensdorf, Switzerland; Institute of Plant Sciences, University of Graz, Graz, Austria

Yoshikazu Yamamoto Faculty of Bioresource Sciences, Akita Prefectural University, Akita, Japan

Luciana Zedda BIO-Diverse, Bonn, Germany

Lichenization: The Origins of a Fungal Life-Style

1

David L. Hawksworth

Contents

D.L. Hawksworth (✉)
Departamento de Biología Vegetal II, Facultad de Farmacia, Universidad Complutense de Madrid, Plaza Ramón y Cajal, 28040 Madrid, Spain
e-mail: d.hawksworth@nhm.ac.uk

D.L. Hawksworth
Department of Life Sciences, The Natural History Museum, Cromwell Road, London SW7 5BD, UK

D.L. Hawksworth
Mycology Section, Royal Botanic Gardens, Kew, Kew, Surrey TW9 3DS, UK

D.K. Upreti et al. (eds.), *Recent Advances in Lichenology*,
DOI 10.1007/978-81-322-2235-4_1

Abstract

Following a discussion on the naming of lichens and a definition of "lichen", hypotheses on the origins of lichenization and lichen-forming fungi are reviewed. It is emphasized that lichen associations strictly have no scientific name, while the partners in the symbiosis do. As fungi have a wide range of associations with algae and cyanobacteria, the definition of "lichen" must include the fungal partner enclosing the photosynthetic, and the photosynthetic partner not being incorporated into fungal cells. Hypotheses put forward to explain lichenization are examined in the context of the evidence from the fossil record and molecular biology. There are uncertainties over the interpretation of many of the pre-Devonian fossils, but stratified undisputed lichen-like associations were present in the Lower Devonian, and material referable to modern genera is preserved in Eiocene and Miocene amber. Some early molecular studies suggested that the earliest ascomycetes may have been lichenized, but as more fungi have been sequenced, it has emerged as more likely that there have been repeated lichenization and de-lichenization events in different lineages over time. Some caution is necessary as the molecular trees do not included data from extinct lineages. The possibility that there were early lichen-like fungal associations as far back as the late Pre-Cambrian or early Cambrian cannot be discounted on the basis that they are not recognizable in the fossil record.

Keywords
Algae · *Ascomycota* · Cyanobacteria · Evolution · Lichens · Nomenclature · Palaeomycology · Symbiosis

1.1 Introduction

The origin and nature of lichens was a conundrum for the earliest naturalists, although many species were described and illustrated as plants from the mid-seventeenth century. Even Luyken (1809), in whose dissertation numerous generic names of Acharius were introduced, considered that they were compounded of air and moisture. Acharius (1810) asserted they were distinct from algae, hepatics and fungi. As regards development, Hornschurch (1819) thought that they arose from a vegetable "infusorium" which became green and developed by the action of light and air while, from the observations on the lichenization of *Nostoc* colonies to form a *Collema*, the renowned algologist Agardh (1820) viewed them as transformed algae. By the mid-nineteenth century, it was becoming increasingly clear from microscopical studies that the "gonidia" in lichens were not produced from hyphae inside the lichen thallus but were indeed algal cells, although this interpretation was not universally accepted immediately and was hotly debated (Lorsch 1988; Mitchell 2002, 2005, 2007). The term "symbiosis" was introduced for the lichen association by Frank (1876, as "symbiotismus")[1] as a result of his studies on the anatomy of five[2] crustose lichens (Sapp 1994; Hawksworth 1995). Interestingly, it was Frank who later also coined the word "mycorrhiza" in 1885 for the fungus-root "Pilzensymbiosis" (Sapp 1994).

The issue of how the lichen symbiosis had arisen remained obscure, but the integration of lichenized fungi into the overall system of fungal classification, the advent of molecular systematics, and critical work on newly recognized fossils have all led to new insights which are reviewed here. A prerequisite for a discussion on this topic is, however, an understanding of the status of names given to lichens and how the term "lichen" is defined.

[1] Often cited as published in 1877, but the article is indicated as written in "Marz 1876", and the first two parts of volume 2 of the journal are listed as received in the 11 October 1876 issue of Flora (59: 530).

[2] *Arthonia radiata* (as *A. vulgaris*), *Arthopyrenia cerasi*, *Graphis scripta*, *Lecanora albella* (as *L. pallida*), and *Pertusaria pertusa* (as *Variolaria communis*).

1.2 Names Given to Lichens

While lichens were considered as plants and as single organisms, names were unwittingly applied to the associations rather than to their separate components. When the fungal components were isolated into culture, however, they looked so different from the lichenized thalli that Thomas (1939) opted to give them separate scientific names, such as *Cladoniomyces pyxidatae* for the fungal partner of *Cladonia pyxidata*. This view was not shared by the International Botanical Congress (IBC) in Stockholm in 1950 which ruled that "for nomenclatural purposes names given to lichens shall be considered as applying to their fungal components" (Lanjouw et al. 1952). Notwithstanding that decision, in a series of papers from 1952 to 1957, Cifferi and Tomaselli (e.g. Tomaselli and Ciferri 1952) introduced numerous generic names with the suffix "*-myces*", and also names at higher ranks, for fungal partners. They made proposals to modify the rules to permit this, but they were rejected by the Montreal IBC in 1959, and Tomaselli (1975) subsequently agreed that this was the right decision. The result of this ruling is that lichen associations do not have a separate scientific name from the partners in the symbiosis. Consequently, whole lichens have no names (Hawksworth 1999), and it is more correct to speak of "lichenized fungi", "lichen-forming fungi" or "lichen fungi", rather than "lichens", when using scientific names. A corollary of this ruling is that where the same fungus forms morphologically distinct lichens with different photosynthetic partners, so-called photosymbiodemes or "phototypes", the same fungal name has to be used for both morphs; for example, the shrubby cyanobacteria morph and the green algal foliose morph of *Sticta felix* are both referred to by that name.

As it is the fungal and algal partners of lichens that have scientific names, not the associations, the nomenclature of the fungal partners comes under the auspices of a single Nomenclature Committee established by successive IBCs, at least since 1950. The committee was, however, named as "for Fungi and Lichens" from 1954 to 1993, when it reverted to "for Fungi", a logical change as lichen associations do not have names apart from those of their fungal components. Since the IBC in 1981, lichenized and non-lichenized fungi have been treated together under "Fungi" in the lists of conserved and rejected names adopted by IBCs. Lichenologists are traditionally included amongst the members of the Nomenclature Committee for Fungi (NCF). This provision does not preclude the establishment of a separate international committee to advise on names of lichenized fungi, as proposed by Lendemer et al. (2012), as there is already a precedent with a separate body that considers fungi with a yeast morph, the International Commission on Yeasts (ICY), established in 1964. It is, however, the NCF which reports to the IBC and makes formal recommendations relating to all organisms treated as fungi under the *International Code of Nomenclature for algae, fungi and plants* (McNeill et al. 2012).

1.3 What Is a Lichen?

While the answer to the question "What is a lichen?" may seem obvious when considering cases in which distinctive thalli are formed, the issue becomes more complex when the full range of associations between fungi, algae and cyanobacteria is considered. Early definitions and debates over them have been summarized elsewhere (Hawksworth 1988), and the generally accepted definition currently in use is: "An ecologically obligate, stable mutualism between an exhabitant fungal partner (the mycobiont[3]) and an inhabitant population of extracellularly located unicellular or filamentous algal or cyanobacterial cells (the photobiont[3])" (Hawksworth and Honegger 1994). The emphasis on the fungal partner forming the outer tissues (as "exhabitant") excludes some fungal–algal associations in which the algae forms the outer tissue, as is the case with, for example, *Blodgettia confervoides, Mycophycias ascophylli* (syn. *Stigmidium ascophylli*) and *Phaeospora lemaneae*. The mention of "extracellularly located" is to exclude the glomeromycete *Geosiphon pyriforme* which has cyanobacteria inside bladder-like cells of the fungal partner (Schüßler 2012).

Lichens are essentially "the symbiotic phenotype of lichen-forming fungi" (Honegger 2012), and encompass not only a wide range of morphologies, of which an elegant overview is provided by Honegger (2012), but also differences in the extent to which the associations are obligate. Examples are known of single fungal species able to form either lichen thalli, or survive as saprobes, depending on whether they grow on bark or wood; these were formerly placed in different genera on that basis until proved conspecific by molecular methods (Wedin et al. 2004). Over 20 genera are now known to include both lichen-forming species and ones which are either lichenicolous (i.e. growing on lichens) or saprobic, for example, *Arthonia, Caloplaca, Diploschistes, Diplotomma, Mycomicrothelia, Opegrapha* and *Thelocarpon*. There are also examples of fungal genera, such as *Arthopyrenia, Chaenothecopsis, Cyrtidula, Leptorhaphis* and *Stenocybe,* which have traditionally been studied by lichenologists, but which appear not to be obligately associated with any algal or cyanobacterial partner.

In the last few years, a further complication over a definition of "lichen" has arisen from the discovery that lichen thalli can routinely include specialized non-photosynthetic bacteria which only occur within their tissues, some producing novel compounds (Cardinale et al. 2006; Grube et al. 2012). In addition, the presence of fungi

[3] I regard the terms "mycobiont" and "photobiont" as unnecessary jargon as "fungal partner" and "photosynthetic partner" are more immediately understood by non-specialists

other than the fungal partner, and which do not represent lichenicolous species, but which occur in lichen thalli, so-called endolichenic fungi, have been revealed by molecular methods (U'Ren et al. 2010). It has been suggested from ancestral-state molecular reconstructions that some of these endolichenic fungi may have been the source of endophytic fungi in plants and ancestors of plant pathogens and saprobes (Arnold et al. 2009). Evidence that many of these endolichenic fungi actually live inside the lichen thalli, has, however, yet to be convincingly demonstrated, although hyphae other than of the fungal partner can sometimes be revealed by scanning electron microscopy (Honegger 2012).

A further dimension arises with respect to the photosynthetic partners, as different algal species, and sometimes genera, may form morphologically identical lichens with the same fungal species. Three different *Trebouxia* species, for example, have been isolated from *Parmelia saxatilis* thalli (Friedl 1989). Individual thalli may also contain more than a single algal species. There is also a tendency for lichens in the same habitat to have the same photosynthetic partner; for example, species of four genera in the maritime Antarctic shared the same *Nostoc* strain (Wirtz et al. 2003).

Other issues, not pertinent to expand on here, include the occurrence of different fungal genotypes of the same species, or in some cases different species, in the same lichen thallus, and the probability that some crustose lichens on bark also obtain some nutrients from their hosts and are not totally self-supporting. It is becoming increasingly evident, therefore, that the term "lichen" includes a spectrum of associations of varying degrees of morphological and biological specialization, and that their thalli are not always two-partner systems, but may rather be considered as more complex ecosystems.

Despite the recognition of the partners in lichen associations as different organisms meriting their own scientific names, Margulis (1993) nevertheless, somewhat bizarrely, introduced the phylum name *Mycophycophyta* for the association, ignoring all molecular and morphological evidence. While this has been ignored by lichenologists and other mycologists, it is regrettable that the term has been taken up in numerous introductory texts (e.g. Allen 2012) and on websites (7,200 hits in Google on 18 Aug. 2014). All researchers who work with lichens are urged to discourage this practice.

1.4 Hypotheses on the Origins of the Lichen Life-Style

Lindsay (1856) considered that lichens must have been the first colonizers of land early in Earth history, before any soil was formed, but did not speculate on their origin. Church (1921) hypothesized that seaweeds, trapped on the land as pools dried, lost their chloroplasts, became in effect primitive fungi and were invaded by or captured unicellular algae. Corner (1964) concurred with Church's view that lichens originated during the migration of life from sea to land, and referred to them as the "land seaweeds".

Cain (1972), however, hypothesized that the first ascomycetes arose from autotrophic "ascophytes", on soil in moist tropical sites, before the origin of vascular plants and perhaps even before green algae. He suggested that the ascophytes would have resembled modern lichens, but were not xerophytic and lacked algal partners, later evolving into lichens through acquiring cyanobacteria initially, and eventually green algae. Eriksson (1981) did not accept Cain's views, and argued that heterotrophic fungi evolved first from heterotrophic or parasitic algae in rock pools and lagoons, where they became lichenized through association with cyanobacteria. He suggested that these early lichens had lecanoralean-type asci from which were derived the non-lichenized ascohymenial and ascolocular ascomycetes, with non-fissitunicate and fissitunicate asci, respectively. Independently, a similar conclusion was reached from a study of the numbers of obligately lichenicolous fungi on different families of host lichens, highlighting the *Peltigeraceae* (Hawksworth 1982).

In order to stimulate debate on evolutionary pathways in ascomycetes, Dick and Hawksworth (1985) endeavoured to construct a diagram displaying the then-recognized orders, lichenized and non-lichenized, taking into account ascus type and ecology; in their construction, cyanobacterial *Peltigerales* emerged as the deepest rooted amongst the lichenized orders. In the light of molecular data and recently discovered fossils, Eriksson (2005, 2006) developed his earlier ideas on the origin of filamentous ascomycetes. He considered the different possibilities for the origin of all ascomata-forming ascomycetes (other than *Neolecta*), i.e. subphylum *Pezizomycotina,* and concluded that while origins on early vascular plants, bryophytes or macroalgae were very unlikely, an origin on microalgae and cyanobacteria was very probable. The subphylum was suggested to be derived from a group of hypothetical lichenized ancestors, *Protolichenes,* which were living symbiotically with algae and cyanobacteria long before land plants developed. Eriksson termed this the "Protolichenes Hypothesis", and considered that the many types of asci found in modern lichen-forming fungi indicated that they had evolved on several evolutionary lines over extended periods of time, and that saprobic and parasitic ascomycetes had arisen later through the loss of symbiosis.

In an innovative analysis of fungi in the context of the origin of life and its emergence onto land, Moore (2013) independently concurred that from the beginning, fungi formed lichens with cyanobacteria and noted that these would have been able to colonize terrestrial habitats as they formed abound 1.5 Gyr[4] ago.

1.5 Fossil Evidence

The oldest report of a fossil lichen, named *Thuchomyces lichenoides,* is from 2.8 Gyr-old Pre-Cambrian of South Africa. It was interpreted as a palisade of podetium-like structures, measuring 2–3 × 0.5–0.6 mm, was dismissed as an artefact of the extraction method (Cloud 1976), but later shown to be real from subsequent observations (MacRae 1999). Somewhat similar but more urn-like structures, also from the Pre-Cambrian of South Africa but dated to 2.2 Gyr ago, have recently been described as *Diskagma buttonii* and compared with *Cladonia* and *Geosiphon* (Retallack et al. 2013). In neither case is there incontrovertible evidence that these Pre-Cambrian organisms were lichens, i.e. fungi with an included photosynthetic partner. If these, and some other early enigmatic structures from China and Namibia, commented on by Retallack (1994) and Retallack et al. (2013), were truly fungal, this would have major implications for views on the origins of eukaryotic life; the earliest definite fossil evidence for cyanobacteria and eukaryotes has been considered to date to ~2.5 Gyr and 1.78–1.68 Gyr ago, respectively (Rasmussen et al. 2008).

The last era of the Pre-Cambrian, the Neoproteozoic Ediacaran (*ca* 635–542 Mya[5]), also has fossils interpreted as lichens (Retallack 1994), amongst which are three species from South Australia, described in *Dickinsonia*, one reaching 1.4 m in length (Retallack 2007). However, a dual nature could not be established and that interpretation is questionable (Waggoner 1995). Also in this period, there is evidence of cyanobacterial and closely associated filaments, considered to be fungal, from marine deposits of 635–551 Mya in southern China. These have been interpreted as lichen-like (Yuan et al. 2005), though the actual nature of the relationship is obscure, but the fungi produced structures recalling spores of *Glomeromycota*, rather than of *Ascomycota*.

There are few lichen candidates from the earliest periods of the Palaeozoic, the Cambrian and Ordovician, but spores recovered from the Silurian suggest that ascomycetes were already diverse by that time (Sherwood-Pike and Gray 1985). Whether any of these spores are from lichenized fungi cannot be ascertained.

Structured lichens were, however, definitely present in the mid-Palaeozoic Devonian, and several cases have been critically documented. The earliest are *Chlorolichenomycites salopensis*

[4] Gyr = Giga years (i.e. billions of years)

[5] Mya = Million years ago.

and *Cyanolichenomycites devonicus,* from the Lower Devonian (415 Mya) of Wales, and these had internal stratification and green algal and cyanobacterial partners, respectively (Honegger et al. 2012). Preservation of the structures may have been facilitated by charring. The hypha were septate, ascomata were absent, but pycnidia occurred in *C. devonicus*; the two genera were tentatively referred to the *Pezizomycotina*. *Winfrentia reticulata,* described from the Lower Devonian Rhynie Chert from Scotland (400 Mya), consisted of a reticulum of fungal hyphae with included cyanobacterial cells (Taylor et al. 1997, 2009). It lacked stratified tissues and had a fungal partner that may have belonged to *Zygomycota* rather than *Ascomycota*; it consequently differed from extant lichens, but could represent an extinct lichenized lineage (Honegger 2012). A stronger candidate from this period, however, is the genus *Spongiophyton,* species of which are reported from the Lower to Middle Devonian in Bolivia, Brazil, Canada, Ghana and the USA. They have a structure reminiscent of the cortex of extant foliose lichens (Taylor et al. 2004) but no sporing structures are known. What does seem clear is that lichenized and non-lichenized fungi together with free-living algae and cyanobacteria, may have formed extensive rock and soil-crust communities before the evolution of land plants (Honegger 2012).

The Early to Mid-Devonian (400–350 Mya) enigmatic *Prototaxites* has also been posited as a possible lichen. It is known from many parts of the world and formed pillar-like structures to 9 m tall and 1.5 m diam. It has been considered a basidiomycete, possessing septal pores and unfused clamp connections (Hueber 2001), but no basidia were found. In order to sustain itself, these massive structures must have had a carbon source, and it has been suggested that came from green algae embedded in the thallus, and this would also explain some of the polymers found (Selosse 2002). An alternative subsequent interpretation was that it represented rolled-up mats of liverworts intermixed with fungal and cyanobacterial associates, but that hypothesis does not seem sustainable (Boyce and Hotton 2010). *Nematothallus* species also have a complex stratified anatomy again recalling lichenized fungi (Edwards et al. 2013), and that genus and *Prototaxites* are placed in the extinct order *Nematophytales* which is considered to have affinities with *Mucoromycotina* or *Glomeromycota* (Retallack & Landing 2014).

Honeggeriella complexa is an undoubted stratified foliose or squamulose lichen from the Mesozoic Early Cretaceous of British Columbia dated at *ca* 133 Mya. It lacked rhizines and had a green algal partner with intracellular haustoria (Matsunaga et al. 2013); no ascomata were found, but the sections show a clear upper and lower cortex, a delimited algal layer and a less compacted medulla, similar to modern foliose *Lecanoromycetes*.

More fossils recalling extant lichens survive from the Cenozoic. *Pelichothallus villosus,* described on leaves of *Chrysobalanus* from Eocene (56–34 Mya) deposits in the USA, appears to represent a species of *Strigula,* with a *Cephaleuros* algal partner and pycnidia recalling those of *Strigula* (Sherwood-Pike 1985). A wide and increasing range of lichens trapped in amber dating from the Eocene into the Miocene are being described, especially from the Baltic and Dominican Republic. These finds are listed in Matsunaga et al. (2013), and correspond to modern genera, including *Alectoria, Anzia, Calicium, Chaenotheca, Hypotrachyna, Parmelia* and *Phyllopsora.* There is also an impression recalling *Lobaria pulmonaria* from the Miocene of California (Peterson 2000).

In view of the abundance of lichens today and the antiquity of the symbiosis, there are remarkably few fossils which have been recognized as lichens. This may be partly because they tend to occur in dry habitats, and so fail to preserve, but it is, perhaps, more likely that the crux of the issue "is simply the inability of paleobiologists to recognize them in the fossil record" (Taylor and Osborn 1996). In that connection, it may be significant that the recognition of the earliest undoubted lichens involved an experienced lichenologist (Honegger et al. 2012). There is an extensive early literature on fossil fungi (Pirozynski 1976), and there are 950 named spore types in the fossil record which are considered to be fungal (Kalgutkar and Jansonius 2000). In the light of modern technologies, it may well be that there are reports to be re-discovered and re-interpreted, as well as fresh discoveries to be made from the examination of fossil remains. Further, perhaps indications of the abundance of lichens might be deduced not from recognizable fossils, but rather from their mineralogical effects, such as laminar calcretes which arise from the action of lichens on exposed calcareous rock surfaces (Klappa 1979).

1.6 Molecular Evidence

The first major work assessing the importance of lichenization in the evolution of fungi was that of Lutzoni et al. (2001), who concluded that the major lineages in *Ascomycota* were derived from lichenized ancestors. The rates of loss of lichenization exceeded gains, and where there was loss, the fungi became lichenicolous, plant pathogens or saprobes. As more examples of orders and families were sequenced, a fuller picture emerged through a collaborative effort by numerous researchers (Lutzoni et al. 2004). An earlier origin of the lichen symbiosis was indicated by strong support for a close relationship of the lichen-forming *Lichinomycetes*, *Thelocarpaceae* and *Biatoridium* to the *Eurotiomycetes-Lecanoromycetes* group, indicating a deep transition to lichenization, and supporting the hypothesis of low numbers of lichenization events. In contrast, in *Basidiomycota*, lichenization had been a relatively recent and not an ancestral event.

As more genes and more taxa were sampled, hypotheses were tested and modified. The Lutzoni et al. (2001) hypothesis of lichenized ancestors was not supported (Lücking et al. 2009), but, due to early radiations, assessing the number of lichenization events could not be resolved. The situation is complicated by previously unsampled lineages being found to be independent and deeply rooted, such as the cyanobacterial-partnered *Lichinomycetes* (Reeb et al. 2004) which also have a specialized polysaccharide chemistry distinct from *Lecanoromycetes* (Prieto et al. 2008). The current molecular evidence, based on data from six genes, does, however, suggest that there have been multiple independent origins of the lichen symbiosis, with losses mainly limited to terminal closely related species (Schoch et al. 2009).

There have been various attempts to apply molecular clocks to the dating of divergences, but calibration is a problem as it relies on the interpretation of fossils and this is often unclear. For example, if the fossil non-lichenized *Paleopyrenis devonicus* is considered to belong to (1) *Sordariomycetes,* the lineages of the five fungal phyla would originate at 1.489 Gya. (2) If recognized as at the base of *Pezizomycotina,* it would give an age of 792 Mya; while (3) if treated as a providing a minimum age for *Ascomycota*, a figure of 452 Mya is obtained (Taylor and Berbee 2006). Lücking et al. (2009) revisited this case using uniform calibration points; this placed the origin of fungi between 1.06 Gya and 760 Mya, and the origin of the *Ascomycota* at 650–500 Mya. These results correlate with the fossil records of fungi and plants, and do not require the postulation of hypothetical "protolichens" (see above).

Calibrated molecular phylogenetic trees are increasingly revealing that many extant lichen genera and species can be of considerable antiquity. In the case of *Parmeliaceae*, using two dated fossils referred to *Alectoria* and *Parmelia* (see above), Amo de Pas et al. (2011) demonstrated that the parmelioid lichens diversified around the K/T boundary at the end of the Cretaceous (*ca* 66 Mya) as the dinosaurs became extinct, and the major clades diverged during the Eocene and Oligocene (Tertiary period). The radiation of the genera in this family was dated to the early Oligocene, Miocene and early Pliocene (i.e. from *ca* 33 Mya) and considered to be linked to changing climatic conditions. *Flavoparmelia,* for example, was estimated to have split from *Parmotrema* in South America at the Eocene-Oligocene transition about 33 Mya, with the Australian groups arising much later, at around 5.4–6.5 Mya (Del Prado et al. 2013).

1.7 Origins of the Lichen Life-Style

Early hypotheses regarding the origin of the lichen life-style have now been challenged by new discoveries of fossils and molecular dating techniques. The view now emerging is that the earliest filamentous fungi were probably saprotrophs, and parasites of the earliest land plants, and mutualistic relationships with algae need not be invoked to explain these data (Lücking et al. 2009). Different fungal lineages appear to have established and lost symbiotic associations with algae independently, and there also may be cases of re-lichenization. It, therefore, seems that Aptroot (1998) was prescient when he estimated that there could have been "at least over a 100" lichenization events in the evolution of fungi.

The changing biology of fungi in and out of lichenization events in different phyla, classes, orders and families, mean that it is essential to study the evolution and classification of lichen-forming fungi in the context of the fungi as a whole. Lichenology must therefore be considered a sub-discipline of mycology, exploring a particular fungal life-style, on a par with mycologists and mycological organizations that are devoted to fungi exhibiting other biologies (e.g. medical fungi, mycorrhizas, plant pathogens, yeasts).

A final note of caution may be appropriate, however, as molecular data are derived from the organisms known to be living on Earth today. While regressive extrapolations can be made, these cannot account for fungi of ancient lineages that may have become extinct (see above), or for major lineages of fungi that have only recently been recognized (e.g. *Cryptomycota*). Consequently, while it may not be necessary to postulate "ascophyte" or "protolichen" ancestors of modern fungi to explain their evolution, it cannot be ignored that there may also have been early lichen-like fungal-cyanobacterial or algal associations of which no unequivocal fossils or ancestors remain—or perhaps still await reinterpretation or discovery. With the current revival of interest in palaeomycology, who knows what exciting discoveries lie ahead….

Acknowledgment This contribution was prepared while in receipt of funding from the Spanish Ministerio de Ciencia e Innovación project CGL2011-25003.

References

Acharius E (1810) Lichenographia Universalis. Dankwerts, Göttingen.

Agardh CA (1820) Dissertatio de Metamorphosi Algarum. Berling, Lund

Allen LZ (2012) Living things (Science works for kids). Evan-Moore Educational Publishers, Monterey

Amo de Paz G, Cubas P, Divakar PK, Lumbsch HT, Crespo A (2011) Origin and diversification of major clades in parmelioid lichens (*Parmeliaceae*, *Ascomycota*) during the Paleogene inferred by Bayesian analysis. PLoS ONE 6:e28161

Aptroot A (1998) Aspects of the integration of the taxonomy of lichenized and non-lichenized pyrenocarpous ascomycetes. Lichenologist 30:501–514

Arnold AE, Miadlikowska J, Higgins KL, Sarvaye SD, Gugger P, Way A, Hofstettrer V, Kauff F, Lutzoni F (2009) A phylogenetic estimation of trophic transition networks for ascomycetous fungi: are lichens cradles of symbiotrophic fungal diversification? Systematic Biol 58:283–297

Boyce CK, Hotton CL (2010) *Prototaxites* was not a taphonomic artifact. Am J Bot 97:1073

Cain RF (1972) Evolution of the fungi. Mycologia 64:1–14

Cardinale M, Puglia AM, Grube M (2006) Molecular analysis of lichen-associated bacterial communities. FEMS Microb Ecol 57:484–495

Church AH (1921) The lichen as transmigrant. J Bot 59:40–46

Cloud PE (1976) The beginnings of biospheric evolution and their biochemical consequences. Palaeobiology 2:351–387

Corner EJH (1964) The life of plants. Weidenfeld & Nicolson, London

Del-Prado R, Blanco O, Lumbsch HT, Divakar PK, Elix JA, Molina MC, Crespo A (2013) Molecular phylogeny and historical biogeography of the lichen-forming fungal genus *Flavoparmelia* (*Ascomycota*: *Parmeliaceae*). Taxon 62:928–939

Dick MW, Hawksworth DL (1985) A synopsis of the biology of the *Ascomycotina*. Bot J Linn Soc London 91:175–179

Edwards D, Axe A, Honegger R (2103) Contributions to the diversity in cryptogamic covers in the mid-Palaeozoic: *Nematothallus* revisited. Bot J Linn Soc London 173:505–534

Eriksson OE (1981) The families of bitunicate ascomycetes. Oper Bot 60:1–220

Eriksson OE (2005) Ascomyceternas ursprung och evolution—protolichenes hypotesen. Svensk Mykologische Tidskrift 26:22–29

Eriksson OE (2006) Notes on ascomycete systematics. Nos 2299–4407. Myconet 12:83–101

Frank AB (1876) Ueber die biologischen Verhältnisse des Thallus einiger Krustenflechten. Beiträge zur Biologie der Planzen 2:123–200

Friedl T (1989) Systematik und Biologie von *Trebouxia* (*Microthalmniales*, *Chlorophyta*) als Phycobiont der *Parmeliaceae* (lichenisierte Ascomyceten). Dissertations Druck, Darmstadt

Grube M, Cardinale M, Berg G (2012) Bacteria and the lichen symbiosis. In: Hock B (ed) The Mycota, vol 9, Fungal associations 2nd edition. Springer, Berlin, pp 363–372

Hawksworth DL (1982) Co-evolution and the detection of ancestry in lichens. J Hattori Bot Lab 52:323–329

Hawksworth DL (1988) The variety of fungal-algal symbioses, their evolutionary significance, and the nature of lichens. Bot J Linn Soc London 96:3–20

Hawksworth DL (1995) Symbiosis evolving. Nature 374:841–842

Hawksworth DL (1999) ["1997"] Raffaele Ciferri, the crisis precipitated in the naming of lichen-forming fungi, and why whole lichens have no names. Archivio Geobotanico 3:3–9

Hawksworth DL, Honegger R (1994) The lichen thallus: a symbiotic phenotype of nutritionally specialized fungi and its response to gall producers. In: Williams MAJ (ed) Plant galls: organisms, interactions, populations. Clarendon Press, Oxford, pp 77–98

Honegger R (2012) The symbiotic phenotype of lichen-forming ascomycetes and their endo- and epibionts. In: Hock B (ed) The Mycota, vol 9, Fungal associations 2nd edition. Springer, Berlin, pp 287–339

Honegger R, Edwards D, Axe E (2012) The earliest records of internally stratified cyanobacterial and algal lichens from the Lower Devonian of the Welsh borderland. New Phytol 197:264–275

Hornschurch F (1819) Einige Beobachtungen über das Entstehen der Algen, Flechten und Laubmoose. Flora 2:140–144

Hueber FM (2001) Rotted wood-alga-fungus: the history and life of *Prototaxites* Dawson 1859. Rev Palaeobot Palynol 116:123–158

Kalgutkar RM, Jansonius J (2000) Synopsis of fossil fungal spores, mycelia and fructifications (AASP contributions series no. 39). American Association of Stratigraphic Palynologists, Dallas

Klappa CF (1979) Lichen stromatolites: criterion for subaerial exposure and a mechanism for the formation of laminar calcretes (caliche). J Sediment Petrol 49:387–400

Lanjouw J, Baehni C, Merrill ED, Rickett HW, Robyns W, Sprague TA, Stafleu FA (eds) (1952) International code of botanical nomenclature adopted by the seventh international botanical congress, Stockholm, July 1950. (Regnum vegetabile no. 3). International Bureau for Plant Taxonomy and Nomenclature, Utrecht

Lendemer JC, Benatti N, Esslinger TL, Hafellner J, Hodkinson BP, Knudsen K, Kocourkova J, Spribille T (2012) Notice of the formation of the International Committee for the Nomenclature of Lichens and Allied Fungi (ICNLAF). Opuscula Philolichenum 11:1–3

Lindsay WL (1856) A popular history of British lichens. Lovell Reeve, London

Lorsch J (1988) The true nature of lichens—a historical survey. In: Galun M (ed) CRC handbook of lichenology. CRC Press, Boca Raton, pp 3–32

Lücking R, Huhndorf S, Pfister DH, Rivas Plata E, Lumbsch HT (2009) Fungi evolved right on track. Mycologia 101:810–822

Lutzoni F, Pagel M, Reeb V (2001) Major fungal lineages are derived from lichen symbiotic ancestors. Nature 411:937–940

Lutzoni F, Kauff F, Cox CJ, McLaughin D, Celio G et al (2004) Assembling the fungal tree of life: progress, classification, and evolution of subcellular traits. Am J Bot 91:1446–1480

Luyken JA (1809) Tentamen Historiae Lichenum. H. Dieterich, Göttingen

MacRae C (1999) Life etched in stone. Geological Society of South Africa, Johannesburg

Margulis L (1993) Symbiosis in cell evolution: microbial communities in the Archean and Proterozoic eons, 2nd edn. W.H. Freeman, New York

Matsunaga KKS, Stockley RA, Tomescu MF (2013) *Honeggeriella complexa* gen. et sp. nov., a heteromerous lichen from the Lower Cretaceous of Vancouver Island (British Columbia, Canada). Am J Bot 100:450–459

McNeill J, Barrie FR. Buck WR, Demoulin V, Greuter W, Hawksworth DL et al (eds) (2012) International code of nomenclature for algae, fungi, and plants (Melbourne Code) adopted by the eighteenth international botanical congress Melbourne, Australia, July 2011 (Regnum vegetabile no. 154). Koeltz Scientific Books, Königsten

Mitchell ME (2002) "Such a strange theory": anglophone attitudes to the discovery that lichens are composite organisms, 1871–1890. Huntia 11:193–207

Mitchell ME (2005) The inside story: a commentary on the study of lichen structure in the 18th and 19th centuries. Huntia 12:13–29

Mitchell ME (2007) Signposts to symbiosis: a review of early attempts to establish the constitution of lichens. Huntia 13:101–120

Moore D (2013) Fungal biology in the origin and emergence of life. Cambridge University Press, Cambridge

Peterson EB (2000) An overlooked fossil lichen (*Lobariaceae*). Lichenologist 32:298–300

Pirozynski KA (1976) Fungal spores in fossil record. Biol Mem 1:104–120

Prieto A, Leal JA, Bernabe M, Hawksworth DL (2008) A polysaccharide from *Lichina pygmaea* and *L. confinis* supports the recognition of *Lichinomycetes*. Mycol Res 112:381–388

Rasmussen B, Fletcher IR, Brocks JJ, Kilburn MR (2008) Reassessing the first appearance of eukaryotes and cyanobacteria. Nature 455:1101–1104

Reeb V, Lutzoni F, Roux C (2004) Contribution of *RPB2* to multilocus phylogenetic studies of the euascomycetes (*Pezizomycotina*, *Fungi*) with special emphasis on the lichen-forming *Acarosporaceae* and evolution of polyspory. Mol Phylogenet Evol 32:1036–1060

Retallack GJ (1994) Were the Ediacaran fossils lichens? Paleobiology 20:523–544

Retallack GJ (2007) Growth, decay and burial compaction of *Dickinsonia*, an iconic Ediacaran fossil. Alcheringa 31:215–240

Retallack GJ, Krull ES, Thackray GD, Parkinson D (2013) Problematic urn-shaped fossils from a Palaeoproterozoic (2.2. Ga) paleosol in South Africa. Precambr Res 235:71–87

Retallack GJ, Landing E (2014) Affinities and architecture of Devonian trunks of *Prototaxites loganii*. Mycologia 106:1143–1158

Sapp J (1994) Evolution by association: a history of symbiosis. Oxford University Press, New York

Schoch CL, Sung G-H, Lopez-Giráldez F, Townsend JP, Mialikowska J, Hofstetter V, Robbertse B et al (2009) The *Ascomycota* tree of life: a phylum-wide phylogeny clarifies the origin and evolution of fundamental reproductive and ecological traits. Syst Biol 58:224–239

Schüβler A (2012) The *Geosiphon-Nostoc* endosymbiosis and its role as a model for arbuscular mycorrhiza research. In: Hock B (ed) The mycota. Fungal associations, vol 9, 2nd edn. Springer, Berlin, pp 77–91

Selosse MA (2002) *Prototaxites*: a 400 Myr old giant fossil, a saprophytic basidiomycetes, or a lichen. Mycol Res 106:642–644

Sherwood-Pike MA (1985) *Pelichothallus* Dilcher, an overlooked fossil lichen. Lichenologist 17:114–115

Sherwood-Pike MA, Gray J (1985) Silurian fungal remains: probable records of ascomycetes. Lethaia 19:1–20

Taylor JW, Berbee ML (2006) Dating divergence in the fungal tree of life: review and new analysis. Mycologia 98:838–849

Taylor TN, Osborn JM (1996) The importance of fungi in shaping the paleoecosystem. Rev Palaeobot Palynol 90:249–262

Taylor TN, Hass H, Kerp H (1997) A cyanolichen from the Lower Devonian Rhynie Chert. Am J Bot 84:992–1004

Taylor TN, Taylor EL, Krings M (2009) Paleobotany: the biology and evolution of fossil plants, 2nd edn. Elsevier Academic Press, Amsterdam

Taylor WA, Free C, Boyce C, Helgemo R, Ochoada J (2004) SEM analysis of *Spongiophyton* interpreted as a fossil lichen. Int J Plant Sci 165:875–881

Thomas EA (1939) Über die Biologie von Flechtenbildnern. Buchdruckerei Büchler, Bern

Tomaselli R (1975) The systemic position of lichens. Arch Botanico 51:3–10

Tomaselli R, Cifferi R (1952) Scissioni di generi di Licheni sulla base delle caratteristiche del fungo. Arch Botanico 28:1–15

U'Ren J, Lutzoni F, Miadlikowska J, Arnold AE (2010) Community analysis reveals close affinities between endophytic and endolichenic fungi in mosses and lichens. Microb Ecol 60:340–353

Waggoner BM (1995) Edicaran lichens: a critique. Paleobiol 21:393–397

Wedin M, Döring H, Gilenstam G (2004) Saprotrophy and lichenization as options for the same fungal species on different substrata: environmental plasticity and fungal lifestyles in the *Stictis-Conotrema* complex. New Phytol 164:459–465

Wirtz N, Lumbsch HT, Green TGA, Türk R, Pintado A, Sancho L, Schroeter B (2003) Lichen fungi have low cyanobiont selectivity in maritime Antarctica. New Phytol 160:177–183

Yuan X, Xiao S, Taylor TN (2005) Lichen-like symbiosis 600 million years ago. Science 308:1017–1020

The Dynamic Discipline of Species Delimitation: Progress Toward Effectively Recognizing Species Boundaries in Natural Populations

2

Steven D. Leavitt, Corrie S. Moreau and H. Thorsten Lumbsch

Contents

Abstract

Species represent a fundamental unit in evolutionary biology and provide a valuable context for organizing, evaluating, and communicating important biological concepts and principles. Empirical species delimitation is a dynamic discipline, with ongoing methodological and bioinformatical developments. Novel analytical methods, increasing availability of genetic/genomic data, increasing computational power, reassessments of morphological and chemical characters, and improved availability of distributional and ecological records offer exciting avenues for empirically exploring species delimitation and evolutionary relationships among species-level lineages. In this chapter, we aim to contribute a contemporary perspective on delimiting species, including a brief discussion on species concepts and practical direction for empirical species delimitation studies. Using lichen-forming fungi as an example, we illustrate the importance and difficulties in documenting and describing species-level biodiversity.

Keywords

Barcoding · Coalescence · DNA taxonomy · Fungi · Gene tree · Genomics · Lichens · Species circumscription · Species concept · Species tree

S.D. Leavitt (✉)
Committee on Evolutionary Biology, University of Chicago, Chicago, IL, USA
e-mail: sleavitt@fieldmuseum.org

S.D. Leavitt · C.S. Moreau · H. Thorsten Lumbsch
Department of Science and Education Field Museum of Natural History, Chicago, IL, USA

D.K. Upreti et al. (eds.), *Recent Advances in Lichenology*,
DOI 10.1007/978-81-322-2235-4_2

2.1 Introduction: What's in a Name? The Importance of Accurate Species Delimitations

Although there are over 1.5 million species formally named by scientists, current estimates of the number of species alive on the planet today range from approximately two million to over one hundred million (Caley et al. 2014). Documenting, describing, and naming this diversity is paramount for conservation, human health, food security, and recreation (Tewksbury et al. 2014). In a broad sense, species delimitation is the process of identifying how individuals and populations fit into natural, species-level clusters, and not simply constructs of classification (Carstens et al. 2013). Empirical species delimitation is a dynamic discipline, with ongoing methodological and bioinformatical developments due to a growing interest in empirical species delimitations. Novel analytical methods, increasing availability of genetic/genomic data, increasing computation power, reassessments of morphological and chemical characters, and improved availability of distributional and ecological records offer exciting avenues for empirically exploring species delimitation and evolutionary relationships in species all over the world. In this chapter, we aim to contribute a contemporary perspective on delimiting species and offer practical direction for empirical species delimitation studies.

To illustrate the importance and difficulties in documenting and describing biodiversity, we will use the lichens as an example. Lichens describe a mutualism between a fungus (mycobiont) and a photosynthetic partner (photobiont), which can be either a green algae and/or cyanobacterium. Lichens are ubiquitous components of most terrestrial ecosystems, playing important ecological roles and contributing to global biogeochemical cycles (Porada et al. 2014; Bonan and Shugart 1989; Longton 1997). Due to the fact that many lichens live and grow continuously for decades, or even hundreds of years, showing cumulative responses to changes in atmospheric pollution levels, land management practices, and fluctuation climate, these are commonly used as bioindicators to monitor the impacts of air pollution, forest age, soil quality, and climate change (McCune 2000). As iconic examples of symbiosis, lichens also provide crucial insight into general patterns and processes in symbiotic systems. Central to understanding the dynamic roles of lichens is our ability to accurately delimit and recognize species boundaries. Increased accuracy in recognizing species boundaries in lichenized fungi has major implications for enhancing our perspective on biological diversity, evolution, ecology, symbiotic interactions, biomonitoring research, and conservation policy.

2.1.1 Species: Concepts and Criteria

Species serve as a central unit for categorizing biological diversity. Humans, including birdwatchers, doctors, fisherman, gardeners, politicians, scientists, and others, rely to varying degrees on recognizing species for distinguishing different kinds of organisms and effective communication. In the biological sciences, species represent one of the most fundamental units and provide a valuable context for organizing, evaluating, and communicating important biological concepts and principles (Coyne and Orr 2004; Mayr 1963; Darwin 1859). Due to the fact that biological information is commonly provided with reference to a species unit, accurate species circumscriptions are integral to interpreting biological patterns and processes across a wide range of subdisciplines in biology (e.g., anatomy, behavior, ecology, evolution, physiology, etc.).

In spite of the underlying importance of species in biology, the conceptualization of the term "species" remains somewhat contentious (de Queiroz 2007; Hausdorf 2011; Coyne and Orr 2004; Mayden 1997; Simpson 1951; Mayr 1963). Most biologists agree that biological nature is aggregated into discrete, evolutionarily independent entities, i.e., "species" (Coyne and Orr 2004), although theorists and empiricists alike continue to debate over an all-encompassing species concept and appropriate operational criteria for delimiting species (Hausdorf 2011;

de Queiroz 2007; Hey 2006; Donoghue and Gauthier 2004; Cracraft 1983; Mishler and Brandon 1987; Mayr 1970). Over two-dozen different and at least partially incompatible species concepts have been proposed, each based on distinct biological properties, e.g., differences in genetic or morphological features, adaptive zones or ecological niches, mate recognition systems, reproductive compatibility, monophyly, etc. (de Queiroz 2007; Mayden 1997). Hausdorf (2011) argues that most species concepts are useable, but acceptance of a specific concept requires an appropriate adaptation of the term "species" and of species delimitation. In contrast, de Queiroz (1998) and Mayden (1997) argue that distinct species concepts emphasize different properties of species rather than fundamental conceptual differences, and all modern species concepts share an important commonality—equating species with segments of metapopulation lineages. This "general lineage concept" (GLC; de Queiroz 1999) highlights that no single property should be regarded as defining for the recognition of species, apart from forming lineages (Simpson 1951), and segments of metapopulation lineages (i.e., "species") may exist regardless of our ability to empirically delimit them (Camargo and Sites 2013).

We concur that the GLC provides a practical solution to the species concept impasse, and our discussion of species delimitation is based on the GLC. Arguably, the major implication of the GLC is that most of the earlier species concepts should be regarded as secondary species "criteria," rather than "concepts," that can provide evidence of lineages separation (Sites and Marshall 2003; Camargo and Sites 2013; Mayden 1997; de Queiroz 2007). This pivotal distinction disentangles the conceptual issues of defining "species" from methodological issues of delimiting species boundaries (Camargo and Sites 2013). The GLC allows researchers to delimit species using different empirical properties and facilitates the development of new methods to test hypotheses of lineage separation (de Queiroz 2007). Although different datasets and operational criteria may give conflicting or ambiguous results due to multiple evolutionary processes occurring within and between populations (e.g., Miralles and Vences 2013; Satler et al. 2013), the use of several independent suites of characters, such as genetic data, morphology, geographic range, host preference, chemistry, and cross-validation using inferences from multiple empirical operational criteria, can provide robust hypotheses of species boundaries (Carstens et al. 2013; Fujita et al. 2012).

2.1.2 Species in Lichenized Fungi: Cases of Cryptic Diversity, Polymorphic Lineages, and Striking Biogeographic Patterns

Similar to most major biological groups, including birds (McKay et al. 2013), fish (Wagner et al. 2013), plants (Griffin and Hoffmann 2014), arthropods (Schlick-Steiner et al. 2010; Moreau 2009), and many others, finding and applying the appropriate character sets and analytical tools is one of the greatest challenges with empirical species delimitation in lichen-forming fungi (Lumbsch and Leavitt 2011). Understanding the differences between morphological variation within a species and among closely related groups is central to identifying diagnostic characters required for establishing accurate phenotype-based taxonomic boundaries. However, in practice, a clear demarcation between intraspecific and interspecific variation is commonly subject to observational bias and individual interpretation.

Traditionally, differences in morphological, chemical, and ecological features have been the predominant source of diagnostic taxonomic characters for circumscribing lichen-forming fungal species (Printzen 2009). However, lichens generally display few taxonomically useful characters, relative to other groups (e.g., vascular plants, vertebrates, and arthropods) (Printzen 2009), and varying levels of intraspecific variation among different species groups may confound accurate taxonomic circumscriptions. While some species may have little variation, high levels of intraspecific phenotypic variation are well documented in some lichen-forming fungi (e.g., *Xanthoparmelia*; Hale 1990). Therefore, molecular genetic data now play a prominent role in delimiting fungal species and understanding evolutionary relationships in lichens (Lumbsch and Leavitt 2011; Printzen 2009).

Arguably, the use of molecular data has led to an improved perspective on the taxonomic value of many phenotypic characters in lichenized fungi and species delimitation in general. Cryptic species-level lineages are commonly identified using molecular data, and in some cases, these previously unrecognized lineages are corroborated by formerly overlooked phenotypic characters (Pino-Bodas et al. 2012a; Spribille et al. 2011; Divakar et al. 2010; Argüello et al. 2007). Other studies have revealed the fact that some species-level lineages are likely comprised of chemically and morphologically polymorphic individuals, which are conventionally considered as separate species (e.g., Leavitt et al. 2011a; Pino-Bodas et al. 2011; Velmala et al. 2009). While these studies highlight the limitations of using traditional taxonomic characters for distinguishing natural groups in lichen-forming fungi, they also provide a valuable perspective on the importance of ongoing research in even the best-studied lichen groups. Furthermore, an improved perspective of species boundaries has led to a striking improvement in understanding diversification and distribution in many groups of lichenized fungi.

Traditional phenotype-based approaches to species recognition appear to vastly underestimate diversity in lichen-forming fungi. While molecular research has corroborated traditional, phenotype-based hypotheses of species boundaries in a number of cases, studies repeatedly demonstrate that our current interpretation of morphological and chemical characters often fails to accurately characterize species-level diversity (reviewed in Lumbsch and Leavitt 2011). A growing body of evidence reveals that a significant proportion of unknown diversity estimated for fungi, including lichen-forming fungi, is hidden under names of supposedly common and widespread species. For example, approximately 80 unrecognized species-level lineages are estimated to occur in Parmeliaceae (Crespo and Lumbsch 2010). Even higher levels of unrecognized species diversity are estimated to occur in other families, such as Graphidaceae (Rivas Plata and Lücking 2013; Rivas Plata and Lumbsch 2011; Lücking 2012).

The topic of cryptic species, cases where two or more distinct species-level lineages are erroneously classified (and hidden) under one nominal taxon (Bickford et al. 2007), has been frequently reviewed for lichen-forming fungi (Lumbsch and Leavitt 2011; Crespo and Lumbsch 2010; Crespo and Pérez-Ortega 2009; Printzen 2009). Although novel diagnostic phenotypic characters may potentially be identified corroborating "cryptic" species, these previously unrecognized lineages generally remain difficult to classify within a traditional phenotype-based taxonomy (Leavitt et al. 2013c, d). In some cases, it appears that similar phenotypic characters may arise in parallel at local or regional scales, but may not be correlated with natural groups or genetic isolation (Muggia et al. 2014; Rivas Plata and Lumbsch 2011; Grube and Hawksworth 2007). For example, in the cosmopolitan species *Tephromela atra* (Fig. 2.1), up to 15 independent lineages were identified using phylogenetic analyses of molecular sequence data. However, the continuum of morphological and chemical variability in the *T. atra* complex currently prevents the description of new species using traditional phenotype-based characters (Muggia et al. 2014).

Recent research on the genus *Cladonia* (Cladoniaceae) highlights a fitting example of the complexities associated within using phenotypic characters for delimiting species in lichen-forming fungi (Fig. 2.1; Pino-Bodas et al. 2011, 2012a, b, 2013a, b). In the *Cladonia gracilis* group, most currently accepted species were not recovered as monophyletic clades and traditional diagnostic morphological characters were shown to be highly homoplasious (Pino-Bodas et al. 2011). Similarly, *C. iberica* and *C. subturgida* have been shown to constitute a single morphologically and chemically polymorphic species (Pino-Bodas et al. 2012b). Similar patterns of high degrees of morphological and chemical polymorphisms have also been observed in the

Fig. 2.1 Examples of common lichens in which traditional morphology-based species circumscriptions fail to reflect natural species-level fungal lineages. **a** *Cladonia gracilis* photographed in the Clearwater Valley, British Columbia, Canada (see Pino-Bodas et al. 2011) (*photograph credit* Curtis Björk). **b** *R. shushanii*, a member of the *Rhizoplaca melanophthalma* species group, from the Aquarius Plateau, Utah, USA (see Leavitt et al. 2011a) (*photograph credit* S. Leavitt). **c** *Xanthoparmelia* aff. *wyomingica* occurring in Colorado's Front Range, USA (see Leavitt et al. 2011b, c; Leavitt et al. 2013e) (*photograph credit* S. Leavitt). **d** *Tephromela atra* sensu lato found in the Santa Monica Mountains, California, USA (see Muggia et al. 2014) (*photograph credit* J. Hollinger)

C. cariosa group (Pino-Bodas et al. 2012a) and *C. humilis* species complex (Pino-Bodas et al. 2013a). High levels of intraspecific morphological and chemical polymorphisms are not restricted to *Cladonia*. A clear demarcation between intraspecific and interspecific morphological and/or chemical variation does not exist for a large proportion of *Xanthoparmelia* species in western North America, and high intraspecific morphological and chemical variation are common for a number of species-level genetic groups (Fig. 2.1; Leavitt et al. 2011b, c, 2013e). In some cases, as many as eight traditionally circumscribed *Xanthoparmelia* species were recovered in a single species-level genetic group (Leavitt et al. 2011c). High levels of intraspecific phenotypic variation have been observed in a number of other genera in Parmeliaceae, including *Bryoria* (Velmala et al. 2009), *Cetraria* (Pérez-Ortega et al. 2012), *Vulpicida* (Mark et al. 2012), and others.

The importance of biogeography in lichen-forming fungal evolution has remained somewhat ambiguous due to the occurrence of phenotypically similar lichens occurring across broad, intercontinental distributions and uncertainty of appropriate species circumscriptions. While a number of lichen-forming fungal species have been found to be truly widespread (e.g., Lindblom and Søchting 2008; Fernández-Mendoza et al. 2011; Ahti and Hawksworth 2005; Del-Prado et al. 2013), improved recognition of species boundaries has provided insight into important biogeographical patterns in lichens previously assumed to have cosmopolitan distribution patterns (Leavitt et al. 2013d; Del-Prado et al. 2013; Amo de Paz et al. 2012; Sérusiaux et al. 2011; Otálora et al. 2010; Elix et al. 2009). Cryptic diversity and complex biogeographic patterns are highlighted in the *Rhizoplaca melanophthalma* species group (Fig. 2.1; Leavitt et al. 2013d). Analyses of *R. melanophthalma* sensu lato collected from five continents supported the presence of cryptic species within this complex. Two of these lineages were found to have broad intercontinental distributions, while the other four were limited to western North America (Leavitt et al. 2013d). Most strikingly, of the six lineages, five were found on a single mountain in the western USA and the sixth occurred no more than 200 km away from this mountain. A recent study of *Pseudocyphellaria* (Lobariaceae) sensu lato in Hawaii revealed a surprising number of previously unrecognized species hidden within nominal taxa with putative broad geographic distributions, suggesting a high degree of endemism in Hawaii (Moncada et al. 2014). These studies provide crucial impetus to reevaluate species boundaries in order to improve our understanding of diversity, distributions, and evolution in lichenized fungi.

2.2 A Practical Guide to Contemporary Species Delimitation

As it has become clear that conventional phenotype-based criteria for species circumscriptions are often problematic (Bickford et al. 2007), molecular data are thereby particularly valuable for assessing traditional species boundaries and for species delimitation in general (Lumbsch and Leavitt 2011; Fujita et al. 2012). Below, we provide an overview of relevant operational species delimitation methods, the majority of which are reliant on molecular data coupled with molecular phylogenetic and population genetic analyses. Assuming that species do in fact represent "segments of metapopulation lineages" (de Queiroz 1998), direct genetic evidence of lineage status is particularly relevant to species delimitation studies when analyzed within a rigorous statistical framework, regardless of whether lineages differ in phenotypic characters that are apparent to human observers (Fujita et al. 2012). This perspective should not be taken as support for disregarding phenotypic characters in species delimitation studies (Edwards and Knowles 2014; Fujita et al. 2012; Yeates et al. 2011). Rather, hypotheses of species boundaries should be considered more robust with increasing corroboration from independent data sources (i.e., molecular, chemistry, morphology, ecology, etc.), and the integration of independent data for empirical species delimitation studies should be a major focus of species delimitation research (Fujita et al. 2012).

The majority of molecular phylogenetic studies of lichenized fungi focus on generating sequence data from a number of specimens representing the focal group, inferring a gene tree from the genetic data matrix, and assessing the monophyly of the sampled taxa. This approach has provided valuable acumen into evolutionary relationships, the value of morphological characters for taxonomy, and insight into diversity of lichenized fungi (Thell et al. 2009; Reese Næsborg et al. 2007; Westberg et al. 2007; Martín et al. 2003; Arup and Grube 2000; Lohtander et al. 2000; Stenroos and DePriest 1998). However, simply assessing the monophyly of traditional phenotype-based species often offers an incomplete perspective on species boundaries. Studies explicitly designed for empirically delimiting species are pivotal to advancing our understanding of speciation and species diversity in lichens. For example, although a nominal species may be recovered as monophyletic in a gene tree, intraspecific phylogenetic substructure may correspond to evolutionarily independent lineages (e.g., Leavitt et al. 2011a). As a construct of taxonomy, recognizing a monophyletic clade comprised of multiple morphological indistinguishable species-level lineages as a single species may hold some appeal; however, failing to formally recognize this diversity can have far-reaching implications in our biological (e.g., ecology, evolution, reproduction) interpretation of the group. Alternatively, well-supported intraspecific phylogenetic substructure may be the result of stochastic evolutionary processes, uniparental inheritance, etc., rather than evolutionary independence. Interpreting this type of phylogenetic substructure as species-level diversity can introduce detrimental bias.

Empirical species delimitation methods have broadly focused on four main areas: detecting putative species, individual specimen assignment to a species group or operational taxonomic unit (OTU), validation of candidate species or OTUs as evolutionarily distinct lineages, and inferring species relationships (i.e., species tree inference). Ideally, operational species delimitation criteria should be based on explicit statistical protocols in order to objectively assess species boundaries and minimize the need of subjective interpretations or taxonomic expertise. In this section, we provide a brief overview on a number of empirical species delimitation methods that to varying degrees fit these criteria. For convenience, we divide these methods into three general categories: (i) species discovery methods for assigning samples to populations without a priori information; (ii) species validation approaches and coestimating individual assignments and species trees; and (iii) species delimitation using genomic data (Table 2.1).

Empirical species delimitation has received increasing attention, including ongoing development of bioinformatical approaches, and the methods and programs provided in this chapter are by no means intended to be a comprehensive list of all available analytical approaches. Rather, the methods provided here have been shown to be useful in a number of previous studies or show particular promise for future species delimitation studies. Our aim is that these can serve as a starting point when designing studies to assess species limits. Not surprisingly, various approaches to species delimitation may yield different estimates of species boundaries, and the researcher may be required to make some degree of subjective interpretation of the most biologically appropriate species boundaries. Complementarily, recently developed metrics for quantifying the congruence between two taxonomies (C_{tax}) and the potential for an approach to capture a high number of species boundaries (R_{tax}) provide a means to objectively assess discrepancies among species delimitation methods (Miralles and Vences 2013). More sophisticated approaches, including selection of species delimitation models using approximate Bayesian computing (Camargo et al. 2012; Fan and Kubatko 2011; Beaumont et al. 2010) and designing and conducting a simulation study that matches the characteristics of the empirical study (Carstens et al. 2013; Camargo et al. 2012), can be used to more objectively evaluate competing hypotheses of species boundaries. In most contexts, it is likely better to fail to delimit species than it is to falsely circumscribe entities that do not represent actual species, and therefore, the inferences drawn from species delimitation studies should generally be conservative (Carstens et al. 2013; Miralles and Vences 2013).

2.2.1 Corroborating Traditional Taxonomy and Discovering Cryptic Species Using Single-Locus Data

Objectively defining a threshold separating intraspecific population substructure from interspecific divergence is the general aim of species delimitation studies using single-marker datasets. Most species delimitation methods based on single-locus sequence data fall under two general categories: either genetic distance or tree-based approaches (Sites and Marshall 2004). Distance-based approaches attempt to detect a difference between intra- and interspecific distances (i.e., "barcode gap") where the pairwise genetic

Table 2.1 Some methods used for species delimitation, including species discovery methods for assigning samples to populations without a *priori* information (BAPS, Gaussian clustering, Guillot's Unified Model, STRUCTURE, STRUCTURAMA), genetic distance-based method for sorting sequences into hypothetical species (ABGD), tree-based species discovery methods (bGMYC, GMYC, bPTP, PTP, "Species Delimitation" plug-in for Geneious), and joint discovery and validation methods (BP&P, Brownie, DISSECT, SpeDeStem)

Method	Description	Input data
BAPS—population assignment using Bayesian clustering	A program for Bayesian inference of the genetic structure in a population clustering molecular data and performing admixture analyses. Genetic mixture analyses can be performed at both group and individual levels using either a non-spatial or spatial model. BAPS treats both the allele frequencies of the molecular markers (or nucleotide frequencies for DNA sequence data) and the number of genetically diverged groups in population as random variables. In the "clustering with linked loci" model, a genetic mixture analysis can be done using haploid sequence data or other linked genetic markers. Analyses and model comparisons can also be performed using a fixed number of genetically diverged groups or prespecified population structures *Limitations* Temporal divergence and relationships among putative groups are not explicitly estimated. Equivalence to genetic clusters to species-level groups is uncertain, and validation approaches can be used to assess evolutionary independence of clusters *Source* Available from http://www.helsinki.fi/bsg/software/BAPS/; described in Corander et al. (2004, 2006, 2008); Corander and Marttinen 2006)	Genotypic data, haploid sequence data, or linked markers (AFLP or SNPs)
Gaussian clustering—population assignment using Gaussian clustering	A program for Bayesian inference of the genetic structure in a population. Model groups sample into populations using genotypic data by searching for clusters that can be attributed to being mixtures of normal allele frequency distributions. Gaussian clustering requires a dataset where the cases are defined by variable of metric scale and has been used with genetic, environmental, and morphological datasets individually, in addition to integrated datasets *Limitations* Temporal divergence and relationships among putative groups are not explicitly estimated. Equivalence to genetic clusters to species-level groups is uncertain, and validation approaches can be used to assess evolutionary independence of clusters *Source* Implemented in R using the prabclus (Hausdorf and Hennig 2010) and mclust packages (Fraley and Raftery 2007)	Genotypic data (flexible)

(continued)

Table 2.1 (continued)

Method	Description	Input data
Guillot's Unified Model—population assignment using Bayesian clustering	This approach provides a statistical model that allows one to analyze genetic and phenotyptic data within a unified model and inference framework and optionally incorporate information about the spatial distribution of samples. A Bayesian clustering algorithm assumes that each cluster in a geographical domain can be approximated by polygons that are centered around points generated by a Poisson process. Guillot's Unified Model is flexible in terms of the genetic data that it can utilize and capable of accurately delimiting species *Limitations* Genetic and phenotypic data can trace different evolutionary histories, for instance, phylogenetic divergence for neutral genetic markers and adaptation for a morphological structure *Source* Available as an extension of the R GENELAND package (Guillot et al. 2005, 2012) (http://www2.imm.dtu.dk/~gigu/Geneland/)	Genotypic and non-genetic (e.g., phenotypical, ecological, geographical, behavioral) data
STRUCTURE—population assignment using Bayesian clustering	A model-based clustering method using multilocus genotype data to infer population structure and assign individuals to populations. Individuals in the sample are assigned probabilistically to populations, or jointly to two or more populations if their genotypes indicate that they are admixed. The model does not assume a particular mutation process, and it can be applied to most of the commonly used genetic markers, provided that they are not closely linked. The method can produce highly accurate assignments using modest numbers of loci (Pritchard et al. 2000). The most appropriate level of population structure can be inferred by assessing likelihood scores or the ad hoc ΔK statistic (Evanno et al. 2005) *Limitations* Identifying the most appropriate number of genetic clusters is challenging; clusters produced by STRUCTURE can be strongly influenced by variation in sample size; clusters created by STRUCTURE may not be consistent with the evolutionary history of the populations when there are relatively long divergence times within evolutionary lineages. Temporal divergence and relationships among putative groups is not explicitly estimated. Equivalence to genetic clusters to species-level groups is uncertain, and validation approaches can be used to assess evolutionary independence of clusters *Source* http://pritchardlab.stanford.edu/structure.html; described in Falush et al. (2003) and Pritchard et al. (2000)	Genotypic data

(continued)

Table 2.1 (continued)

Method	Description	Input data
Structurama—population assignment using Bayesian clustering	Implements the clustering algorithm used in STRUCTURE (see above) that clusters samples into populations by minimizing Hardy–Weinberg disequilibrium for a given partitioning level. However, Structurama also includes the addition of reversible-jump MCMC to identify the optimal partitioning level. Nearly any type of genetic data can be input into Structurama, and the program can assign individuals to population with or without the admixture *Limitations* Temporal divergence and relationships among putative groups is not explicitly estimated. Equivalence to genetic clusters to species-level groups is uncertain, and validation approaches can be used to assess evolutionary independence of clusters *Source* http://cteg.berkeley.edu/~structurama/index.html; described in Huelsenbeck et al. (2011)	Genotypic data
ABGD—barcode gap using genetic distances	"Automatic Barcode Gap Discovery" sorts sequences into hypothetical species based on the barcode gap. The method uses a recursive approach to partition the data and test for significant gaps. ABGD is fast, simple method to split a sequence alignment dataset into candidate species that should be complemented with other evidence in an integrative taxonomic approach *Limitations* Single-locus data alone should only be used to provide a preliminary perspective of species boundaries and not as the sole evidence in species circumscriptions. *Source* http://wwwabi.snv.jussieu.fr/public/abgd/; described in Puillandre et al. (2012)	Single-locus sequence alignment
GMYC&bGMYC—gene tree	The GMYC approach combines a coalescent model of intraspecific branching with a Yule model for interspecific branching, which is then fit to an inferred single-gene topology to estimate species boundaries and a statistical measure of confidence for the inferred boundaries. As an input, the GMYC approach requires an ultrametric gene tree, and recent refinements can account for uncertainty in phylogenetic relationships and parameters of the GMYC model. The GMYC is generally stable across a wide range of circumstances, including various methods of phylogenetic reconstruction, the presence of a high number singletons, high numbers of sampled species, and gaps in intraspecific sampling; the accuracy of the GMYC is most significantly affected by the mean population size relative to divergence times between them *Limitations* The GMYC may delimit well-supported clades of haplotypes as independent lineages and as such may be prone to over-delimitation. Single-locus data alone should only be used to provide a preliminary perspective of species boundaries and not as the sole evidence in species circumscriptions *Source* (http://r-forge.r-project.org/projects/splits, https://sites.google.com/site/noahmreid/home/software); described in Monaghan et al. (2009), Fujisawa and Barraclough (2013) and Pons et al. (2006)	Single-locus ultrametric gene tree

(continued)

Table 2.1 (continued)

Method	Description	Input data
PTP&bPTP—gene tree	The Poisson tree processes (PTP) model can be used to infer putative species boundaries on a given phylogenetic input tree. PTP can infer putative species boundaries that are consistent with the PSC. An important advantage of this method is that it models speciation in terms of the number of substitutions, thereby circumventing the potentially error-prone and compute-intensive process of generating ultrametric trees, which are required as an input for GMYC model (see above). Furthermore, it appears that the PTP model may outperform the GMYC and other OTU-picking methods when evolutionary distances are small *Limitations* Single-locus data alone should only be used to provide a preliminary perspective of species boundaries and not as the sole evidence in species circumscriptions *Source* http://sco.h-its.org/exelixis/web/software/PTP/index.html; described in Zhang et al. (2013)	Single-locus gene tree
Geneious Species Delimitation plug-in—gene tree	A plug-in to the Geneious software provides an exploratory tool allowing the user to assess phylogenetic support and diagnosability of species defined as user-selected collections of taxa on user-supplied trees. The plug-in computes statistics relating to the probability of the observed monophyly or exclusivity having occurred by chance in a coalescent process and assesses the within- and between-species genetic distances to infer the probability with which members of a putative species might be identified successfully with tree-based methods *Limitations* The plug-in summarizes measures of phylogenetic support and diagnosability of species defined as user-selected collections of taxa, but it does not provide definitive support for species groups. It assumes species are monophyletic *Source* Implemented as a plug-in to Geneious (geneious.com); described in Masters et al. (2011)	Single-locus gene tree
BP&P—multispecies coalescent model for species validation	This approach to species delimitation uses a Bayesian modeling approach to generate the posterior probabilities of species assignments taking account of uncertainties due to unknown gene trees and the ancestral coalescent process. The method relies on a user-specified guide tree, implementing a reversible-jump Markov chain Monte Carlo search of parameter space that includes θ, population divergence, and estimated distributions of gene trees from multiple loci *Limitations* Misspecifications of priors and the guide tree can result in inflated speciation probabilities, it assumes no recombination, and computational limitations restrict its utility with next-generation datasets with 100s of loci *Source* http://abacus.gene.ucl.ac.uk/software.html; described in Yang and Rannala (2010) and Rannala and Yang (2003)	Multilocus sequence alignments and group membership

(continued)

Table 2.1 (continued)

Method	Description	Input data
DISSECT—multispecies coalescent model for species delimitation	DISSECT explores the full space of possible clusterings of individuals and species tree topologies in a Bayesian framework. To avoid the need for reversible-jump MCMC, it uses an approximation in the form of a prior that is a modification of the birth–death prior for the species tree. It is implemented as part of BEAST and requires only a few changes from a standard *BEAST analysis. Analyses of simulated and empirical data suggest that the method is shown to be insensitive to the degree of approximation, but quite sensitive to other parameters *Limitations* Recently described method lacking a thorough theoretical and empirical evaluation. It appears that a large number of sequences are required to draw firm conclusions *Source* http://code.google.com/p/beast-mcmc/&http://www.indriid.com/dissectinbeast.html; described in Jones and Oxelman (2014)	Multilocus sequence data
SpeDeSTEM—multispecies coalescent model for discovery, validation, and joint estimation	This maximum likelihoodand/or information theory-based method was developed to test species boundaries in a system with existing subspecies taxonomy (Carstens and Dewey 2010), and computes the probability of the gene trees given the species tree for all hierarchical permutations of lineage grouping. Species boundaries are compared using Akaike information criteria, and phylogenetic uncertainty in the species tree topologies does not affect species delimitations *Limitations* Accuracy is dependent on quality of the gene tree estimates *Source* (http://carstenslab.org.ohio-state.edu/software.html).criteria; described in Ence and Carstens (2011)	Multilocus sequence alignments and group membership
Brownie—multispecies coalescent model for species delimitation	The nonparametric heuristic species delimitation approach implemented in the program Brownie (O'Meara 2010) jointly sorts anonymous samples into species and infers a species tree from input gene trees from different loci, assuming that for a speciation even the corresponding nodes on gene trees will be more consistent with each than the divergences within species *Limitations* Finding both the optimum species tree and species boundaries remains computationally challenging, and Brownie has been shown to frequently yield incorrect results. The accuracy of the method is likely correlated with nodal support values in the individual gene topologies *Source* http://www.brianomeara.info/brownie; described in O'Meara (2010)	Individual gene trees

(continued)

Table 2.1 (continued)

Method	Description	Input data
BFD*—multispecies coalescent model for species delimitation	The recently developed method, Bayes factor delimitation (*with genomic data; BFD*), combines a dynamic programming algorithm for estimating species trees that bypasses the computationally intensive MCMC integration over gene trees to provide a rigorous technique for species delimitation studies using genome-wide SNP data. Competing species delimitation models are compared using Bayes factors, and it appears that this approach is robust to sample sizes (i.e., few loci and limited samples per species) and misspecification of the prior for population size (θ) *Limitations* Recently described method lacking a thorough theoretical and empirical evaluation *Source* http://www.beast2.org/wiki/index.php/BFD*; described in Leaché et al. (2014)	Genome-wide SNP data

distances among organisms belonging to the same species are smaller than distances among organisms from different species (Puillandre et al. 2012; Hebert et al. 2003). Genetic distance approaches hold particular promise as an identification tool shortcutting the difficulties of morphology-based identification (Hebert et al. 2004), although in practice a barcode gap may not exist for many groups (Wiemers and Fiedler 2007). Furthermore, the role of distance-based approaches using a single genetic marker for species discovery remains more controversial (Rubinoff 2006), and without other corroborating evidence, OTUs inferred from single-locus dataset should only be considered "candidate" species. Tree-based methods aim to detect monophyletic clades corresponding to species-level diversity by detecting discontinuities associated within inter- and intraspecific branching patterns (Fujisawa and Barraclough 2013; Zhang et al. 2013; Monaghan et al. 2009; Pons et al. 2006). Tree-based species delimitation methods can also be used on the basis of other properties related to phylogenetic tree topologies (monophyly, concordance with geography, etc. (Sites and Marshall 2003, 2004). Both distance- and tree-based methods have been applied for assessing species boundaries in lichen-forming fungi (Leavitt et al. 2012c, 2014; Parnmen et al. 2012; Del-Prado et al. 2010, 2011; Molina et al. 2011).

A number of tree-based methods partially automate the species delimitation process with specific bioinformatical analyses (Table 2.1), including the general mixed Yule coalescent (GMYC) approach (Fujisawa and Barraclough 2013; Monaghan et al. 2009; Pons et al. 2006) and the Poisson tree processes (PTP) model (Zhang et al. 2013). These methods provide relatively straightforward and objective pipelines for delimiting putative species-level lineages from inferred gene trees by fitting within- and between-species branching models to an inferred single-locus topology. A number of other tree-based methods for species delimitation are effectively summarized in Sites and Marshall (2004).

When only single-locus data are available, the GMYC has been shown to be a relatively robust tool for species delimitation. The GMYC approach combines a coalescent model of intra-specific branching with a Yule model for inter-specific branching, which is then fit to an inferred single-gene topology to estimate species boundaries and a statistical measure of confidence for the inferred boundaries. As an input, the GMYC approach requires an ultrametric gene tree, and recent refinements can account for uncertainty in phylogenetic relationships and parameters of the GMYC model (Fujisawa and Barraclough 2013; Reid and Carstens 2012). Regardless of these improvements, it may be difficult to accurately infer an adequate ultrametric tree for large datasets. Although it appears that the GMYC is generally stable across a wide range of circumstances, including various methods of phylogenetic reconstruction, the presence of a high number singletons, high numbers of sampled species, and gaps in intraspecific sampling, the accuracy of the GMYC is most significantly affected by the mean population size relative to divergence times between them (Fujisawa and Barraclough 2013; Talavera et al. 2013). Furthermore, research suggests that the so-called single-threshold version of the GMYC method likely outperforms the multiple-threshold approach (Fujisawa and Barraclough 2013; Monaghan et al. 2009). However, other studies suggest that the GMYC method may often provide higher estimates for the total number of OTUs than other molecular species delimitation methods (Hamilton et al. 2014; Miralles and Vences 2013; Talavera et al. 2013), warranting a cautious interpretation of results from GMYC analyses.

Compared to the GMYC approach, the recently introduced PTP method for species delimitation has been suggested to be more accurate for preliminary species delimitation (Zhang et al. 2013). Relative to the GMYC, PTP offers a more straightforward implementation, requiring a simple phylogenetic tree, rather than an ultrametric chronogram. However, more research is required to assess the general performance of PTP across wide range of empirical and simulated species delimitation studies. At this point, data support the use of the GMYC and PTP methods as objective and reasonable starting points for species delimitation using single-gene topologies.

The "Species Delimitation" plug-in to the Geneious software provides statistical approaches for exploring species boundaries in single-gene topologies (Masters et al. 2011). Using a priori specimen assignments to putative species, the "Species Delimitation" plug-in computes statistics relating to the probability of the observed monophyly or exclusivity having occurred by chance in a coalescent process and assesses the within- and between-species genetic distances in order to infer the probability with which members of a putative species might be identified successfully with tree-based methods. An important contribution of the "Species Delimitation" plug-in is that it provides an objective means for users to assess putative species within an empirical, statistic-based framework, rather than qualitative evaluations of what level of hierarchy constitutes a species in a phylogeny.

In contrast to simple sequence similarity thresholds (OTU-picking) for delimiting putative evolutionarily significant units, the Automatic Barcode Gap Discovery (ABGD) method is an automated procedure that sorts sequences into hypothetical species based on the existence of a barcode gap, observed whenever intraspecific genetic distances are less than those among organisms from different species (Puillandre et al. 2012). Ultimately, ABGD is a fast, simple method that can be used to group individuals represented in a single-locus sequence alignment into candidate species that should be complemented with other lines of evidence in an integrative taxonomic approach (Kekkonen and Hebert 2014; Puillandre et al. 2012).

In general, the ABGD, GMYC, and PTP analytical protocols using single-locus data are repeatable and computationally relatively fast, providing a valuable starting point for a preliminary perspective into species boundaries in understudied groups that can be validated with subsequent studies (Kekkonen and Hebert 2014). Similarly, analyses of single-locus data can be used to corroborate traditional phenotype-based species boundaries and identify candidate species that have previously been hidden within nominal species. However, single-locus species delimitation methods may be particularly prone to failure in recognizing significant proportions of species-level biodiversity due to the strict criterion for reciprocal monophyly or, alternatively, provide spurious inflations of estimated species diversity based on genetic differences that do not correspond to species-level lineages. For example, recent estimates suggest that the incidence of species-level gene tree paraphyly is approximately 20 % (Ross 2014), suggesting that analyses of single-locus datasets would likely fail to accurately delimit species in one in every five cases. In contrast, recent empirical studies suggest that in some cases the GMYC provides a striking overestimate of species diversity (Miralles and Vences 2013).

The internal transcribed spacer region (ITS) has played a central role in molecular phylogenetic studies of lichenized fungi and has recently been adopted as the official barcoding marker for fungi (Schoch et al. 2012). In many cases, the ITS is sufficiently variable to resolve species boundaries for lichenized fungi (Schoch et al. 2012; Kelly et al. 2011), although accurate specimen identification using sequence-based identification approaches requires a thoroughly curated reference database (Leavitt et al. 2014; Orock et al. 2012; Kelly et al. 2011). In spite of the general utility of the ITS marker, a number of issues may potentially limit its effective use in species delimitation studies, including the potential for intragenomic variation of the nuclear ribosomal tandem repeat and difficulties in aligning ITS sequences from divergent taxa (Kiss 2012). Because the ITS has been formally adopted as the official barcoding marker for fungi, we recommend that species delimitation studies attempt to include this region for consistency across studies. However, we recognize that in some taxonomic groups the inclusion of the ITS may be problematic and therefore suggest a careful screening of other markers to identify appropriate loci for resolving species-level relationships.

Ultimately, the success of any single-locus species delimitation method largely depends on the evolutionary history of the species group under study and the variability of the selected marker (Fig. 2.2). Species delimitation will be

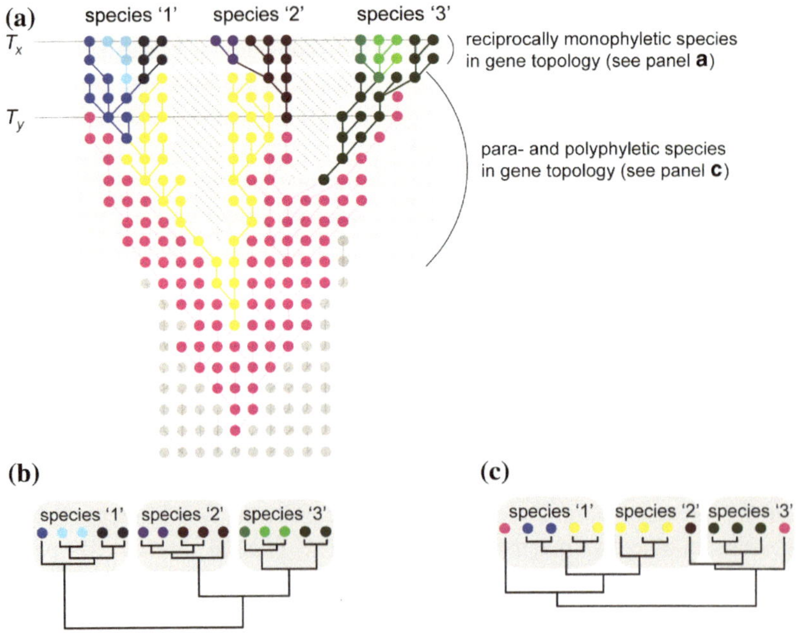

Fig. 2.2 A simplified diagram illustrating the process of speciation through time in a single gene history and resulting gene topologies sampled at two points in the speciation history (modified from Leliaert et al. 2014). **a** Each *dot* represents a distinct gene copy and each row one non-overlapping generation, with lines connecting gene copies to their ancestors in the previous generation (*one row below*); *dashed diagonal lines* represent reproductive barriers; two hypothetical sampling intervals at different points in the speciation history are shown, T_x and T_y; species delimitation methods using single-locus data are effective only when species are reciprocally monophyletic in the sampled gene tree; genetic clustering and coalescent-based species delimitation methods can circumscribe species when species may not be monophyletic in sampled genetic loci. **b** Gene topology representing sampled haplotypes at time T_x (shown in panel **a**); hypothetical species are reciprocally monophyletic. **c** Gene topology representing sampled haplotypes at time T_y (shown in panel **a**); hypothetical species are para- and polyphyletic

more difficult in more recently diverged lineages and in cases with some level of interspecific gene flow, relative to older, well-diverged lineages (e.g., Leavitt et al. 2012c, 2013e). In the end, we reemphasize that single-locus data alone should only be used to provide a preliminary perspective of species boundaries and not as the sole evidence in species circumscriptions.

2.2.2 Sampling Across the Genome: Multilocus Sequence Data and Genome-Wide Markers

Although single-locus methods can provide an efficient approach for preliminary large-scale assessments of species diversity, there are significant limitations, particularly in recently diverged species groups where retention of ancestral polymorphisms and incomplete lineage sorting will likely result in different neutral loci having unique gene topologies that do not mirror the speciation process (Knowles and Carstens 2007; Heled and Drummond 2010; Taylor et al. 2000). In contrast to single-locus and strictly phenotype-based approaches for species delimitation, analyses of genetic data collected from independent genomic regions can provide robust hypotheses of species boundaries with increasing confidence (Satler et al. 2013; Zhang et al. 2011; Yang and Rannala 2010). Sequences from multiple independent loci provide an important source of data for species delimitation studies, including recently developed models that combine individual gene genealogies and species phylogenies via modeling the coalescent history of markers (Yang and Rannala 2010; Edwards 2009; Knowles and Carstens 2007). As a response to advances in sequencing technologies, bioinformatical approaches for multimarker species delimitation analyses continue to be developed (Table 2.1; Camargo and Sites 2013).

Long-term reproductive isolation of candidate species can be assessed with multilocus sequence data by evaluating genealogical concordance of unlinked markers (Baum and Shaw 1995; Avise and Ball 1990). Within species, the mixing effects of recombination cause unlinked loci to have distinct genealogical histories, but genetic drift and long-term divergence leads to concordant genealogical histories at loci across the genome. Relationships of individuals belonging to distinct candidate species can be evaluated using gene genealogies (e.g., haplotype networks or single-gene topologies) to identify lineages that exhibit genealogical exclusivity across unlinked neutral loci (Hudson and Coyne 2002; Dettman et al. 2003b; Avise and Ball 1990). The presence of the same clades in the majority of single-locus genealogies is taken as evidence that the clades represent reproductively isolated lineages (Dettman et al. 2003a; Pringle et al. 2005). In practice, the criteria of reciprocal monophyly and genealogical concordance of unlinked loci provide a conservative approach for assessing species boundaries due to the fact that a substantial amount of time is required after the initial divergence of species until ancestral polymorphisms have fully sorted (Knowles and Carstens 2007; Hudson and Coyne 2002). Consequently, groups with recent divergence histories will likely go undiscovered under a genealogical concordance criterion due to the fact that species boundaries likely are not reflected in gene genealogies. For example, it would take more than 1 million years after speciation before species would be delimited if 15 loci were sampled in species with an effective population size (N_e) of 100,000, assuming one generation a year under a strict reciprocal monophyly criterion (Knowles and Carstens 2007). The amount of time required for a species to be recognized using a criterion of reciprocal monophyly increases proportionally with increasing population sizes (Hudson and Coyne 2002). In a number of studies of lichen-forming fungi, a genealogical concordance criterion has been used to both delimit previously unrecognized species-level lineages and validate some conventional phenotype-based species (Leavitt et al. 2012a, c, 2013b; Molina et al. 2011; Kroken and Taylor 2001).

Although analyses of molecular sequence data relying on reciprocal monophyly or fixed character differences for species delimitation can serve as important criteria for identifying species, these criteria are not commonly met across multiple loci, particularly in recent speciation histories (Fig. 2.2). To accommodate the observed conflict among genealogies from multiple loci with the underlying speciation history, the recent merge of coalescent theory with phylogenetics has driven a major paradigm shift in species delimitation and molecular systematics in general (Fujita et al. 2012; Edwards 2009). The multispecies coalescent model can be applied to genealogical histories from multiple independent loci to assemble separate coalescent processes occurring in populations into a species tree (Degnan and Rosenberg 2009; Rannala and Yang 2003). Within this coalescent-based framework, multiple individuals can be assigned to a single species/population and the speciation history of ancestral and descendant species-level lineages can be inferred as a "species tree," in contrast to estimating gene genealogies from individual samples (Degnan and Rosenberg 2006, 2009; Rannala and Yang 2003).

A number of multispecies coalescent-based species delimitation methods have recently been introduced, offering an exciting framework for empirically assessing species boundaries by selecting the best species tree model from a set of alternative models representing different hypotheses of species boundaries (Table 2.1). For example, SpeDeSTEM (Ence and Carstens 2011) finds the maximum likelihood values for a species tree assuming all putative species are separate lineages and for alternative species trees where two or more species are collapsed into a single lineage. The best fitting model is then selected using the Akaike information criterion, under the assumption of constant population sizes and fixed gene topologies across the species tree. The species delimitation program Bayesian Phylogenetic and Phylogeography (BP&P;Yang and Rannala 2010) accommodates gene tree uncertainty and variable population sizes and samples from the Bayesian posterior distribution of species delimitation models using reversible-jump Markov chain Monte Carlo (rjMCMC) sampling. BP&P requires a user-provided guide tree with resolution finer than the species level and evaluates alternative modes derived from all possible subtrees that are generated by collapsing or splitting nodes on the guide tree (Yang and Rannala 2010; Rannala and Yang 2003). The proportion of time spent on each model is proportional to the posterior probability of the model, i.e., "speciation probabilities." While BP&P ranks among the most popular coalescent-based species delimitation programs (Fujita et al. 2012), misspecifications of the guide tree and/or the prior distribution for population size (θ) can result in strong support for models containing an inflated number of species (Leaché and Fujita 2010; Zhang et al. 2011). The nonparametric heuristic species delimitation approach implemented in the program Brownie (O'Meara 2010) jointly sorts anonymous samples into species and infers a species tree from input gene trees from different loci, assuming that for a speciation event, the corresponding nodes on gene trees will be more consistent with each than the divergences within species. However, finding both the optimum species tree and species boundaries remains computationally challenging, and Brownie has been shown to frequently yield incorrect results (O'Meara 2010).

Most recently, available coalescent-based species delimitation methods require individual assignments to a species or population a priori. In a number of scenarios, correct assignments of samples to species may be difficult, including presence of cryptic species, incongruence between conventional species and molecular data, or simply the fact that accurate specimen identification in complex groups with subtle or difficult to discern diagnostic characters is a nontrivial task (Leavitt et al. 2013e). Statistical methods for assessing individual assignment and species detection prior to coalescent-based species delimitation and species tree reconstruction and species provide a more objective approach for understanding species boundaries.

Population assignment tests using a variety of clustering algorithms using genetic data offer a useful approach for grouping individuals into putative reproductively isolated groups (Table 2.1; Hausdorf and Hennig 2010; Corander et al. 2004, 2008; Falush et al. 2003; Pritchard et al. 2000), particularly for species delimitation problems that exist at the interface of traditional population genetic and phylogenetic analyses (Carstens et al. 2013; Weisrock et al. 2010; Knowles and Carstens 2007).

The programs STRUCTURE (Falush et al. 2003; Pritchard et al. 2000) and STRUCTURAMA (Huelsenbeck et al. 2011) cluster samples into populations by minimizing Hardy–Weinberg disequilibrium for a given number of population clusters using unlinked genetic markers. In general, unlinked markers are not available for most groups of lichen-forming fungi and a number of studies have used information from multilocus sequence data for STRUCTURE analyses (Leavitt et al. 2011a, b, c, 2013e; Fernández-Mendoza and Printzen 2013; Fernández-Mendoza et al. 2011). Varying approaches have been implemented to convert DNA sequence data to allelic data for Bayesian clustering (see O'Neill et al. 2013). STRUCTURE is expected to perform well when there is sufficient independence across regions such that linkage disequilibrium within regions does not dominate the data (STRUCTURE manual), but can also be effective using multilocus sequence data and treating all SNPs as independent loci regardless of physical linkage within each locus (O'Neill et al. 2013; Falush et al. 2003). A recent study of the lichen-forming genus *Letharia* showed that Bayesian clustering implemented in the program STRUCTURE was generally able to recover the same putative *Letharia* lineages circumscribed employing a gene genealogical approach in Kroken and Taylor's iconic species delimitation study (Altermann et al. 2014; Kroken and Taylor 2001). Altermann et al. (2014) show that population assignments were largely consistent across a range of scenarios, including extensive amounts of missing data, the exclusion of SNPs from variable markers, and inferences based on SNPs from as few as three gene regions. In contrast to STRUCTURE and STRUCTURAMA, the program BAPS (Corander et al. 2008) can explicitly infer genetic structure using haploid sequence data or other linked genetic markers in the "clustering with linked loci" model. Another advantage of BAPS is that the program is much more computationally efficient than STRUCTURE or STRUCTURAMA. Simulation studies indicate that both BAPS and STRUCTURE perform well at low levels of population differentiation and when clusters are not well differentiated (Latch et al. 2006).

In practice, inferring the most appropriate level of population structure using Bayesian clustering algorithms remains challenging (Latch et al. 2006; Evanno et al. 2005). BAPS and STRUCTURAMA can infer both individual assignments and the most likely number of genetic clusters (Corander et al. 2008; Huelsenbeck et al. 2011), and the ad hoc ΔK statistic proposed by Evanno et al. (2005) provides an objective approach for identifying the uppermost hierarchical level of structure. However, careful consideration of other levels of population structure may ultimately reveal more biologically meaning results and researchers should examine individual assignments across a range of genetic groupings. Some markers generated from independent genomic regions, such as SNPs and fast-evolving microsatellites, can be used to distinguish fine-scale population structuring on the basis of allele frequencies, and species delimitation analyses based on these markers may imply the risk of severe taxonomic oversplitting. In these cases, validation approaches, such as BP&P and SpeDeSTEM, and corroboration among different species delimitation approaches can provide perspective between intraspecific population structure and species-level clusters.

Another limitation of clustering approaches is that they do not assess or take into account evolutionarily divergences and relationships among population clusters. Coalescent-based species tree methods (discussed below) provide a means to assess the diversification histories of populations inferred from clustering analyses. Therefore, a potential working protocol for an informed species delimitation study that takes into account population structure could consist of first applying a genetic clustering analysis under a population genetics criterion (e.g., BAPS, STRUCTURE, STRUCTURAMA) to identify genetically distinct population clusters that can be considered "candidate species." From these candidate species, a species tree can be inferred for focal group using coalescent-based species tree reconstruction methods (e.g., *BEAST). Subsequently, a coalescent-based validation method can be applied to assess whether the distinct population clusters represent independent evolutionary lineages (e.g., BP&P). This multistep approach would provide a consistent and standard criterion for distinguishing between population- and species-level lineages (Camargo and Sites 2013), and has been applied to a number of cases, including lichen-forming fungi (Leavitt et al. 2011a, b, c, 2013e; Leaché and Fujita 2010).

High-throughput sequencing methods provide the means to effectively sample hundreds to thousands of loci from across a species genome for a large number of species and continue to revolutionize studies that can be performed even in non-model organisms. Targeted high-throughput sequencing approaches, such as anchored phylogenomics, transcriptome sequencing, reduced-representation genomic library sequencing (restriction-site-associated DNA sequencing: RAD-Seq and genotype-by-sequencing: GBS), and high-throughput amplicon sequencing, provide important insight into species boundaries for a number groups, although none of these approaches has yet been applied to studies of lichenized fungi.

For example, restriction-site-associated (RAD-Seq) DNA can simultaneously detect and genotype thousands of genome-wide SNPs by focusing the sequencing effort on a reduced representation of the entire genome (Baird et al. 2008) and has been successfully applied to intraspecies (Lewis et al. 2007; Miller et al. 2007; Emerson et al. 2010) and interspecies studies (Rubin et al. 2012; Eaton and Ree 2013; Wagner et al. 2013). This approach has provided striking insight into the recent adaption radiation of Lake Victoria cichlids (Wagner et al. 2013). An alternative approach targeting the sequencing of specific loci using next-generation sequencing platforms provides an efficient means of generating data for loci of known genomic location, orthology, size, and expected level of variation (O'Neill et al. 2013). Markers can be targeted using well-established PCR techniques or newer hybridization techniques and subsequently pooled for high-throughput sequencing using parallel-tagged sequencing of multiple individual samples within a single sequencing run (O'Neill et al. 2013). A recent study of North American tiger salamanders (*Ambystoma tigrinum*) demonstrated the potential for amplicon-based parallel-tagged sequencing to rapidly generate large-scale genomic data for species delimitation and species tree research (O'Neill et al. 2013).

Although these methods provide an enormous amount of data and insight into diversification at an unprecedented scale, computational limitations restrict the applicable analytical methods for large datasets, although computational and analytical advances are happening rapidly. Most recent coalescent-based species delimitation and species tree models using gene trees have been limited to handle tens of loci for multiple individuals, and combining hundreds or thousands of gene trees into a single species delimitation framework presents considerable computational challenges (Camargo and Sites 2013). The recently developed method, Bayes factor delimitation (*with genomic data; BFD*), combines a dynamic programming algorithm for estimating species trees that bypasses the computationally intensive MCMC integration over gene trees to provide a rigorous technique for species delimitation studies using genome-wide SNP data (Leaché et al. 2014). Competing species delimitation models are compared using Bayes factors, and it appears that this approach is robust to sample sizes (i.e., few loci and limited samples per species) and misspecification of the prior for population size (θ) (Leaché et al. 2014).

Fungi are an ideal model for assessing divergence in eukaryotes due to their simple morphologies, small genomes, broad ecological roles, and diverse lifestyles (Gladieux et al. 2014). However, the use of genomic data from high-throughput sequencing methods have not yet been included in species delimitation studies of lichen-forming fungi. This is due, in part, to challenges in dealing with metagenomic data containing genomic information from a plethora of symbionts associated with a lichen thallus, scant genomic resources, and discipline-specific inertia. A number of lichen-forming fungal genomes are currently available, including *Endocarponpusillum* (Wang et al. 2014), *Cladonia* spp. (Park et al. 2013a, 2014), *Caloplaca flavorubescens* (Park et al. 2013b), and *Cladonia greyi* and *Xanthoria parientina* through the Fungal Genomics Program at Joint Genome Institute of the United States Department of Energy (http://www.jgi.doe.gov). The foreseeable ongoing expansion of genomic resources for lichen-forming fungi will be central to developing approaches for delimiting species using high-throughput sequencing. Specifically, genomic resources will be crucial in identifying novel markers (variable genes regions, SNPs, microsatellites, etc.), identifying conserved fungal markers for targeted high-throughput sequencing approaches, and references for entire genome comparisons (Devkota et al. 2014; Werth et al. 2013).

2.3 Can We Make Species Delimitation in Lichen-Forming Fungi Truly Integrative?

The widespread availability of genetic data has created a biased viewpoint that only genetic data should be used for statistical species delimitation. However, an ongoing appeal to researchers to assess species boundaries from multiple and complementary perspectives (phylogenetics, population genetics, comparative morphology, development, ecology, etc.) has resulted in an integrative taxonomic framework bringing these conceptual and methodological perspectives together (Hamilton et al. 2014; Fujita et al. 2012; Salicini et al. 2011; Yeates et al. 2011; Padial et al. 2010; Dayrat 2005; Will et al. 2005; Wiens and Penkrot 2002). In practice, any study linking different kinds of data to support hypotheses of species boundaries, including mapping morphological characters onto a molecular phylogeny, can be considered integrative. Integrative methods for species delimitation fall across a broad spectrum, ranging from verbal and qualitative assessments of data classes to quantitative methods that allow different data types to contribute to statistical species delimitation (Yeates et al. 2011). In most taxonomic studies utilizing both molecular and morphological data, expert opinion is eventually used to some degree to evaluate the final status of a candidate species (e.g., Bond and Stockman 2008). In this sense, many studies using evidence for independent data sources for delimiting species boundaries use an iterative approach (Yeates et al. 2011) where species boundaries can be tested within a *hypothetico-deductive* framework with diverse datasets.

From a practical perspective, we advocate a process of iterative taxonomy (sensu Yeates et al. 2011) to circumscribe and refine species limits using multiple lines of evidence. This iterative process involves comparisons of morphological data with a phylogenetic hypothesis (i.e., single-locus gene tree or concatenated multilocus phylogeny) to identify the least inclusive monophyletic clade in the topology characterized by at least one unambiguously diagnostic morphological character (Miralles and Vences 2013). This phylogenetic tree-informed approach to assessing species boundaries represents a common practice in studies of lichen-forming fungi. Previously unrecognized species-level clades with corresponding subtle, or overlooked, phenotypic characters have been commonly observed in both crustose lichens [e.g., Acarospraceae (Wedin et al. 2009), Graphidaceae (Rivas Plata and Lumbsch 2011; Papong et al. 2009), Lecanoraceae (Leavitt et al. 2011a), Lecideaceae (Ruprecht et al. 2010), Mycoblastaceae (Spribille et al. 2011), and Teloschistaceae (Vondrák et al. 2009; Muggia et al. 2008)] and foliose and fruticose lichens [e.g., Lobariaceae (Moncada et al. 2014; McDonald et al. 2003), Parmeliaceae (Leavitt et al. 2013a; Wirtz et al. 2012; Divakar et al. 2005; Molina et al. 2004; Kroken and Taylor 2001), Peltigeraceae (O'Brien et al. 2009; Goffinet et al. 2003), Physciaceae (Elix et al. 2009; Divakar et al. 2007), and Sphaerophoraceae (Högnabba and Wedin 2003)]. As a specific example, Lücking et al. (2008) used a combination of medullary chemistry and underside pigmentation in specimens from the *Heterodermia obscurata* group in Costa Rica to corroborate monophyletic clades in an ITS gene tree as species-level lineages.

While this iterative approach to species delimitation and taxonomy has proven valuable for understanding species boundaries and describing new taxa in some groups of lichenized fungi, a posteriori examination of morphological and chemical features has failed to reveal diagnostic phenotypic characters in a number of studies (Muggia et al. 2014; Leavitt et al. 2011a; Otálora et al. 2010; O'Brien et al. 2009). Furthermore, a study of widespread species in the genus *Melanelixia* (Parmeliaceae) indicated that phenotypically cryptic lichen-forming fungal species-level lineages may be relatively ancient and diagnosable phenotypic differences may be absent even millions of years after the initial divergence (Leavitt et al. 2012b). The latter study highlights the fact that species-level lineages may commonly exist without any observable diagnostic phenotypic characters, calling into question the impetus for a universal application of integrative taxonomy.

In keeping with the principle that as many lines of evidence as available should be combined to delimit species (Dayrat 2005), the formalized integrative taxonomic approach (ITAX; Miralles and Vences 2013) uses a mtDNA guide tree and observations from different types of data that might provide conclusive evidence for the independence of lineages and thus their identity as different species. Miralles and Vences (2013) provide a non-exhaustive list of species delimitation criteria to be integrated in the ITAX approach, including (a) sympatric occurrence without admixture as revealed by consistent differences in morphological or molecular characters at the same geographic location; (b) strong differences in a behavioral, morphological, or genetic character known to mediate premating isolation; (c) unviability or infertility of hybrids; (d) lack of gene flow across a geographical hybrid zone; (e) congruent diagnostic differences between sister lineages in various unlinked morphological character; (f) absence of haplotype sharing in several unlinked nuclear loci; and (g) a combination of criteria e–f. Species boundaries are based on seeking the least inclusive monophyletic group in the mtDNA tree which fulfills at least one of the criteria listed above. Clearly, the ITAX approach is sensitive to sample size, and in order to reliably support the distinctiveness of a given species, it has been recommended that the sampling strategy includes at least five individuals per species (Miralles and Vences 2013). A similar approach could be adopted for lichenized fungi using an ITS gene topology as guide tree, rather than mtDNA.

A total evidence approach, including concatenation, has been a common approach for integrating information from different sources, including independent genetic markers in phylogenetic reconstructions (Kluge 1989; Wiens 1998; de Queiroz et al. 1995). Proponents for concatenation of independent data in phylogenetic analyses argue that when combining all data, the underlying signal of speciation may emerge even if weakly contradictory signal is contained in the individual data partitions (Gatesy et al. 1999). For example, establishing a preliminary perspective of species boundaries from multilocus sequence data using concatenation may provide a reasonable starting point for screening for cryptic species and species tree inference (Šlapeta et al. 2006; Leavitt et al. 2011a; Le Gac et al. 2007; Leaché 2009). However, concatenation and consensus methods imply a risk of obtaining inflated support for incorrect relationships and information about variance in gene coalescence is lost (Degnan and Rosenberg 2006, 2009; Kubatko and Degnan 2007; Edwards 2009). In spite of the impetus for integrating evidence from independent data sources into empirical species delimitation studies and taxonomy (Fujita et al. 2012; Padial et al. 2009), most currently available species delimitation methods are unable to accommodate non-genetic data sources in a statistical framework.

2.3.1 Selecting the Appropriate Data

In the face of increasing availability of genetic data and associated bioinformatical approaches for delimiting species, researchers should carefully consider what information is being sacrificed by the failure to consider non-genetic data in species delimitation studies and whether accuracy could be improved by the addition of multiple data types. Morphological data have historically served as a proxy to identify reproductively isolated groups (i.e., "species") (Ray 1686; Fujita et al. 2012). Current methods for delimiting species using non-genetic data (e.g., chemistry, morphology, and ecology) remain woefully understudied. For example, morphology-based species circumscriptions are generally based on one or more qualitative (or quantitative) morphological characters that do not appear to overlap with other species. However, ascertaining that a given trait is truly fixed within a population with statistical confidence requires unrealistic sample sizes, even when allowing for some level of polymorphism in the diagnostic character (Wiens and Servedio 2000). Now various combinations of data—from morphology, genetics, geography, and ecology—are accepted as standard information for species delimitation studies (Ruiz-Sanchez and Sosa 2010; Ross et al. 2010; Edwards and Knowles 2014).

Below, we briefly discuss appropriate data sources for species delimitation studies of lichen-forming fungi. However, homoplastic characters (similar traits that are not derived from a common ancestor) are common among many traits commonly used to circumscribe fungal taxa, the biological significance of secondary metabolite variation remains largely unknown (Lawrey 1986), and ecological niches may be difficult to adequately characterize and model due to microhabitat requirements and data resolution. Therefore, we advocate a cautious approach to selecting appropriate and relevant data for assessing species boundaries.

In spite of the potential challenges and limitations of using phenotypic data, these traits have provided a plethora of valuable information for understanding species boundaries. For example, in the *Melanelixia fuliginosa* group (Parmeliaceae), a morphometric analysis, using color, isidia, and marginal zone free of isidia as characters, revealed a general pattern of differentiation between material formerly recognized as subspecies *fuliginosa* and *glabratula,* in spite of the fact that considerable overlap between groups occurs in some characters individually (Arup and Berlin 2011). Furthermore, the distinction of the two groups was supported by a phylogenetic analysis of the ITS marker and ecological differences, with *M. fuliginosa* occurring predominantly on rock and *M. glabratula* on bark (Arup and Berlin 2011). This study of the *M. fuliginosa* group provides a fitting example of using multiple independent suites of data, including ecology, morphology, and genetic information, to establish a robust hypothesis of species boundaries.

In lichen systematics, phenotypic data, including thallus organization, secondary metabolites, mode of reproduction, ascoma-type and ontogeny, ascus and ascospore characters, have historically played a prominent role (Printzen 2009). Ascomatal characters have traditionally held a major role in higher-level classification (Printzen 2009), in contrast to species-level classification which also tends to include a wide array of vegetative and chemical characters. Commonly assessed morphological characters include thallus form and size, cortical features (e.g., maculae and pseudocyphellae), presence/form/color of attachment structure (e.g., rhizines), and reproductive mode (ascomata vs. vegetative diaspores) (Printzen 2009). Different morphological types of vegetative diaspores—corticated isidia and ecorticated soredia—and their location are commonly used to distinguish species. Ascomatal characters, including morphology, location (laminal vs. marginal), position (sessile vs. immersed), presence of thalline margins, color of an apothecial disc, and presence or color of pruina, are also commonly used in species delimitations of lichen-forming fungi (Printzen 2009). Other important characters may include thalline characters, form, color, size and septation of ascospores, size and form and structure of asci, the hamathecium, type of epihymenium and hypothecium and the type of excipulum or peridium, conidiomatal characters, etc. (Printzen 2009).

Assessments of secondary metabolites has played an important role in lichen taxonomy, beginning with the introduction of simple spot tests by Nylander (1866a, b). The use of chemistry in lichen taxonomy has been discussed in detail in numerous reviews (Lumbsch 1998a, b; Rogers 1989; Brodo 1986; Egan 1986; Leuckert 1985; Brodo 1978; Hawksworth 1976; Culberson 1969, 1970), and we refer readers to these valuable sources for a more comprehensive perspective on lichen chemistry. In short, extrolites (secondary metabolites) belong to various classes; the most common and diverse include depsides, depsidones, chlorinated xanthones, and anthraquinones (Lumbsch 2002; Culberson 1969). The presence or absence of specific extrolites, or their replacements by another substance, is widely used to distinguish species, particularly when correlated with differences in geographic distributions. However, if morphological or geographical differences between populations containing different extrolites are not apparent, the taxonomic significance has been disputed, with some authors distinguishing them as species and others preferring to regard them as chemical races within a species. In addition to simply using the presence or absence of extrolites, Culberson and Culberson (1976) proposed to arrange lichen substances into chemosyndromes of closely related substances. The presence or absence of these chemosyndromes may potentially be used as characters to delimit species, regarding differences involving the same chemosyndromes as intraspecific variation and distinct chemosyndromes as evidence for interspecific populations (Lumbsch 1994; Gowan 1986).

Due to the fact that species delimitation studies often incorporate a substantial biogeographical or ecological component, ecological niche modeling plays an increasing role in phylogenetic and taxonomic research. Niche modeling can provide evidence for ecological isolation between populations based on either conserved or divergent ecological niches and therefore can provide additional evidence supporting lineage independence between putative species. The application of ecological niche modeling has been applied to species delimitation studies in a number of cases (Ruiz-Sanchez and Sosa 2010; Leaché et al. 2009; Raxworthy et al. 2007; Rissler and Apodaca 2007), although it has not been explicitly applied to assess species boundaries in lichen-forming fungi. By mapping the spatial distribution of environmental suitability of climatic variables for candidate species, the application of ecological niche modeling can be particularly important in cases where species have allopatric distributions (Raxworthy et al. 2007).

Ecological niche modeling utilizes known associations between a species' occurrence, localities, and environmental variables to define abiotic conditions within which populations can be maintained (Guisan and Thuiller 2005). The methodological approach for modeling is based on four general properties: (i) The current known species' localities is the dependent variable, (ii) the distribution is modeled as a map composed of grid cells at a specified resolution, (iii) a range of environmental variables (e.g., temperature, precipitation, and solar exposure) are collected to describe the characteristics of each cell, and (iv) classifying the degree to which each cell is either suitable or unsuitable for each species under a range of models (Guisan and Thuiller 2005).

Ecological data also have the potential to play an important role in understanding species boundaries in lichen-forming fungi. For example, Nash and Zavada (1977) demonstrated that *Xanthoparmelia* populations with distinct chemistries occurring in the northern portion of the Sonoran Desert exhibit habitat selection among different rock substrates within a region with relatively uniform climate and topography. In another case, the parmelioid species, *Parmelia mayi*, is morphologically indistinguishable from *P. saxatilis*, but can be separated by bioclimatic features and genetic and chemical characters (Molina et al. 2011). McCune and Printzen(2011) assess distributions and climatic niches of species in the *Lecanora varia* group in western USA and provide a model that uses continental influence and annual temperature as the major factors predicting species distributions. The distribution of *Usnea hirta*, a lichen commonly used in air quality biomonitoring research, was modeled for a section of the White River National Forest in central Colorado, based on the presence of *U. hirta* at 72 biomonitoring reference sites distributed in the intermountain western United States (Shrestha et al. 2012). The best model for predicting *U. hirta* distribution included four variables—solar radiation, average monthly precipitation, and average monthly minimum and maximum temperatures (Shrestha et al. 2012). These studies support the potential use of ecological niche modeling methods in species delimitation studies of lichen-forming fungi.

2.4 Conclusions: What About Taxonomy?

In most cases, species circumscription and taxonomy requires some degree of qualitative judgment and individual interpretation. Integrating multiple types of data into an empirical framework for delimiting species boundaries where species boundaries can be tested within a *hypothetico*-deductive framework with diverse datasets can provide robust hypotheses of species boundaries and taxonomic stability (Yeates et al. 2011). However, it has long been known to evolutionary biologists that distinct species do not need to have diagnosable morphological differences (Mayr 1963), and increasing availability of genetic data has allowed researchers to identify species and to rigorously test species boundaries with a level of precision that was unimaginable a decade ago. While analytical advances in statistical species delimitation have been largely based on genetic data, the utility of these approaches to formal taxonomy remains elusive. Due to these challenges, an eclectic approach to delimiting species and caution against the reliance on any single dataset or method is required when delimiting species.

While results from many recent studies may contradict traditional species boundaries across many groups of lichen-forming fungi, we are optimistic that this research represents substantial progress toward a more accurate perspective on species boundaries and diversity in fungi. As a result of ongoing research of species boundaries in lichen-forming fungi, the taxonomic value of many phenotypes is now better understood; our understanding of ecological, evolutionary, and biogeographic patterns has improved, and we can begin to better understand patterns of symbiont interactions in lichens. Integrating new data (including novel morphological characters and genetic data) will be essential to accurately represent species-level diversity across all groups of lichenized fungi. Hopefully, an improved perspective on lichen diversity also increases our appreciation of these incredible symbiotic systems. While we are strong advocates for the application of independent data types in developing an integrative taxonomy, there is an increasing need to formally recognize the existence of phenotypically cryptic species-level lineages in lichen-forming fungi (see Hibbett et al. 2011). In some cases, a molecular taxonomy may provide the most practical approach to consistent treatment of mycobiont species within lichen groups where diagnostic morphological characters are unidentifiable or practically not feasible (Leavitt et al. 2013c, d).

These are exciting times for taxonomists and phylogeneticists. A closer look at lichen taxonomy, with the inclusion of new data, will help us to better understand the diversity of these fascinating organisms, accurately interpret distribution patterns, and play a more important role in meaningful conservation practices. However, some level of uncertainty will accompany progress. In many taxonomic groups, our traditional approach for species identification will likely need to be substantially modified. The search for corroborating morphological support for cryptic species identified using molecular data will require meticulous and creative approaches to assess phenotypic variation in potentially unorthodox ways. We are hopeful that lichenologists, who traditionally have been eager to include new methods, such as chromatography, in their routine identifications, will be amenable to include molecular techniques to their routine examination of specimens for identification and classification. Although this may prove difficult to achieve by single individuals, especially citizen scientists that traditionally play an important role in lichen taxonomy, the increasing number of collaborative projects in the discipline (e.g., Lumbsch et al. 2011; Crespo et al. 2010; Gueidan et al. 2009) make us optimistic that broad-scale collaborative approaches will facilitate the inclusion of molecular data in lichen research at all levels.

Acknowledgments We are indebted to various colleagues for valuable, thought-provoking discussion, notably Matthew Nelsen (University of Chicago), Ana Crespo (Universidad Complutense de Madrid), Pradeep Divakar (Universidad Complutense de Madrid), Beckett Sterner (The Field Museum), and Joyce Havstad (The Field Museum). We also thank anonymous reviewers who provided valuable comments that improved this chapter. Support by the US National Science Foundation is gratefully acknowledged ("Hidden diversity in parmelioid lichens," DEB-0949147).

References

Ahti T, Hawksworth DL (2005) *Xanthoparmelia stenophylla*, the correct name for *X. somloënsis*, one of the most widespread usnic acid containing species of the genus. The Lichenologist 37(4):363–366. doi:10.1017/S0024282905015197

Altermann S, Leavitt SD, Goward T, Nelsen MP, Lumbsch HT (2014) How do you solve a problem like *Letharia*? A new look at cryptic species in lichen-forming fungi using Bayesian clustering and SNPSs from multilocus sequence data. PLoS ONE 9(5): e97556. doi:10.1371/journal.pone.0097556

Amo de Paz G, Cubas P, Crespo A, Elix JA, Lumbsch HT (2012) Transoceanic dispersal and subsequent diversification on separate continents shaped diversity of the *Xanthoparmelia pulla* group (Ascomycota). PLoS ONE 7(6):e39683. doi:10.1371/journal.pone.0039683

Argüello A, Del Prado R, Cubas P, Crespo A (2007) *Parmelina quercina* (Parmeliaceae, Lecanorales) includes four phylogenetically supported morphospecies. Biol J Linn Soc 91(3):455–467. doi:10.1111/j.1095-8312.2007.00810.x

Arup U, Berlin ES (2011) A taxonomic study of *Melanelixia fuliginosa* in Europe. The Lichenologist 43(02):89–97. doi:10.1017/S0024282910000678

Arup U, Grube M (2000) Is *Rhizoplaca* (Lecanorales, lichenized Ascomycota) a monophyletic genus? Can J Bot 78(3):318–327. doi:10.1139/b00-006

Avise J, Ball R (1990) Principles of genealogical concordance in species concepts and biological taxonomy. Oxf Surv Evol Biol 7:45–67

Baird NA, Etter PD, Atwood TS, Currey MC, Shiver AL, Lewis ZA, Selker EU, Cresko WA, Johnson EA (2008) Rapid SNP discovery and genetic mapping using sequenced RAD markers. PLoS ONE 3(10): e3376. doi:10.1371/journal.pone.0003376

Baum DA, Shaw KL (1995) Genealogical perspectives on the species problem. In: Hoch PC, Stephenson AG (eds) Experimental and molecular approaches to plant biosystematics. Missouri Botanical Garden, St. Louis, pp 289–303

Beaumont MA, Nielsen R, Robert C, Hey J, Gaggiotti O, Knowles L, Estoup A, Panchal M, Corander J, Hickerson M, Sisson SA, Fagundes N, Chikhi L, Beerli P, Vitalis R, Cornuet J-M, Huelsenbeck J, Foll M, Yang Z, Rousset F, Balding D, Excoffier L (2010) In defence of model-based inference in phylogeography. Mol Ecol 19(3):436–446. doi:10.1111/j.1365-294X.2009.04515.x

Bickford D, Lohman DJ, Sodhi NS, Ng PKL, Meier R, Winker K, Ingram KK, Das I (2007) Cryptic species as a window on diversity and conservation. Trends Ecol Evol 22(3):148–155. doi:10.1016/j.tree.2006.11.004

Bonan GB, Shugart HH (1989) Environmental factors and ecological processes in boreal forests. Annu Rev Ecol Syst 20(1989):1–28

Bond J, Stockman A (2008) An integrative method for delimiting cohesion species: finding the population-species interface in a group of Californian trapdoor spiders with extreme genetic divergence and geographic structuring. Syst Biol 57(4):628–646. doi:10.1080/10635150802302443

Brodo IM (1978) Changing concepts regarding chemical diversity in lichens. The Lichenologist 10(1):1–11. doi:10.1017/S0024282978000031

Brodo IM (1986) Interpreting chemical variation in lichens for systematic purposes. The Bryologist 89(2):132–138

Caley MJ, Fisher R, Mengersen K (2014) Global species richness estimates have not converged. Trends Ecol Evol 29(4):187–188. doi:10.1016/j.tree.2014.02.002

Camargo A, Sites JW (2013) Species delimitation: a decade after the Renaissance. In: Pavlinov I (ed) The species problem—ongoing issues. InTech. doi:10.5772/52664

Camargo A, Morando M, Avila LJ, Sites JW (2012) Species delimitation with ABC and other coalescent-based methods: a test of accuracy with simulations and an empirical example with lizards of the *Liolaemus darwinii* complex (Squamata: Liolaemidae). Evolution 66(9):2834–2849. doi:10.1111/j.1558-5646.2012.01640.x

Carstens BC, Dewey TA (2010) Species delimitation using a combined coalescent and information-theoretic approach: an example from North American *Myotis* bats. Syst Biol 59(4):400–414. doi:10.1093/sysbio/syq024

Carstens BC, Pelletier TA, Reid NM, Satler JD (2013) How to fail at species delimitation. Mol Ecol 22(17):4369–4383. doi:10.1111/mec.12413

Corander J, Marttinen P (2006) Bayesian identification of admixture events using multi-locus molecular markers. Mol Ecol 15(10):2833–2843. doi:10.1111/j.1365-294X.2006.02994.x

Corander J, Waldmann P, Marttinen P, Sillanpaa M (2004) BAPS 2: enhanced possibilities for the analysis of genetic population structure. Bioinformatics 20(15):2363–2369. doi:10.1093/bioinformatics/bth250

Corander J, Marttinen P, Mantyniemi S (2006) Bayesian identification of stock mixtures from molecular marker data. Fish Bull 104:550–558

Corander J, Marttinen P, Siren J, Tang J (2008) Enhanced Bayesian modelling in BAPS software for learning genetic structures of populations. BMC Bioinf 9(1):539. doi:10.1186/1471-2105-9-539

Coyne JA, Orr HA (2004) Speciation. Sinauer Associates, Sunderland

Cracraft J (1983) Species concepts and speciation analysis. Curr Ornithol 1:159–187

Crespo A, Lumbsch HT (2010) Cryptic species in lichen-forming fungi. IMA Fungus 1:167–170. doi:10.5598/imafungus.2010.01.02.09

Crespo A, Pérez-Ortega S (2009) Cryptic species and species pairs in lichens: a discussion on the relationship between molecular phylogenies and morphological characters. Anales del Jardin Botanico de Madrid 66(S1):71–81. doi:10.3989/ajbm.2225

Crespo A, Kauff F, Divakar PK, del Prado R, Perez-Ortega S, Amo de Paz G, Ferencova Z, Blanco O, Roca-Valiente B, Nunez-Zapata J, Cubas P, Arguello A, Elix JA, Esslinger TL, Hawksworth DL, Millanes A, Molina MC, Wedin M, Ahti T, Aptroot A, et al (2010) Phylogenetic generic classification of parmelioid lichens (Parmeliaceae, Ascomycota) based on molecular, morphological and chemical evidence. Taxon 59(6):1735–1753

Culberson WL (1969) The use of chemistry in the systematics of the lichens. Taxon 18:498–505

Culberson WL (1970) Chemosystematics and ecology of lichen-forming fungi. Annu Rev Ecol Syst 1:153–170

Culberson CF, Culberson WL (1976) Chemosyndromic variation in lichens. Syst Bot 1:325–339

Darwin C (1859) On the origin of species by means of natural selection or the preservation of favoured races in the struggle for life. J. Murray, London

Dayrat B (2005) Towards integrative taxonomy. Biol J Linn Soc 85(3):407–415. doi:10.1111/j.1095-8312.2005.00503.x de Queiroz K (1998) The general lineage concept of species, species criteria, and the process of speciation: a conceptual unification and terminological recommendations. In: Howard DJ, Berlocher SH (eds) Endless forms: species and speciation. Oxford University Press, Oxford, pp 57–75

de Queiroz K (1999) The general lineage concept of species and the defining properties of the species category. In: Wilson RA (ed) Species: new interdisciplinary essays. MIT Press, Cambridge, pp 49–89

de Queiroz K (2007) Species concepts and species delimitation. Syst Biol 56(6):879–886. doi:10.1080/10635150701701083

de Queiroz A, Donoghue MJ, Kim J (1995) Separate versus combined analysis of phylogenetic evidence. Annu Rev Ecol Syst 26(1):657–681. doi:10.1146/annurev.es.26.110195.003301

Degnan J, Rosenberg N (2006) Discordance of species trees with their most likely gene trees. PLoS Genet 2 (5):e68. doi:10.1371/journal.pgen.0020068

Degnan JH, Rosenberg NA (2009) Gene tree discordance, phylogenetic inference and the multispecies coalescent. Trends Ecol Evol 24(6):332–340. doi:http://dx.doi.org/10.1016/j.tree.2009.01.009

Del-Prado R, Cubas P, Lumbsch HT, Divakar PK, Blanco O, de Paz GA, Molina MC, Crespo A (2010) Genetic distances within and among species in monophyletic lineages of Parmeliaceae (Ascomycota) as a tool for taxon delimitation. Mol Phylogenet Evol 56(1):125–133. doi:10.1016/j.ympev.2010.04.014

Del-Prado R, Divakar PK, Crespo A (2011) Using genetic distances in addition to ITS molecular phylogeny to identify potential species in the *Parmotrema reticulatum* complex: a case study. The Lichenologist 43 (06):569–583. doi:10.1017/S0024282911000582

Del-Prado R, Blanco O, Lumbsch HT, Divakar PK, Elix JA, Molina MC, Crespo A (2013) Molecular phylogeny and historical biogeography of the lichen-forming fungal genus *Flavoparmelia* (Ascomycota: Parmeliaceae). Taxon 62(5):928–939. doi:10.12705/625.22

Dettman J, Jacobson D, Taylor J (2003a) A multilocus genealogical approach to phylogenetic species recognition in the model eukaryote *Neurospora*. Evolution 57(12):2703–2720. doi:10.1111/j.0014-3820.2003.tb01514.x

Dettman JR, Jacobson DJ, Turner E, Pringle A, Taylor JW (2003b) Reproductive isolation and phylogenetic divergence in *Neurospora*: comparing methods of species recognition in a model eukaryote. Evolution 57(12):2721. doi:10.1554/03-074

Devkota S, Cornejo C, Werth S, Chaudhary RP, Scheidegger C (2014) Characterization of microsatellite loci in the Himalayan lichen fungus *Lobaria pindarensis* (Lobariaceae). Appl Plant Sci 2(5):1300101. doi:10.3732/apps.1300101

Divakar PK, Molina MC, Lumbsch HT, Crespo A (2005) *Parmelia barrenoae*, a new lichen species related to *Parmelia sulcata* (Parmeliaceae) based on molecular and morphological data. The Lichenologist 37 (01):37–46. doi:10.1017/S0024282904014641

Divakar PK, Amo De paz G, del Prado R, Esslinger TL, Crespo A (2007) Upper cortex anatomy corroborates phylogenetic hypothesis in species of *Physconia* (Ascomycota, Lecanoromycetes). Mycol Res 111 (11):1311–1320. doi:10.1016/j.mycres.2007.08.009

Divakar PK, Figueras G, Hladun N, Crespo A (2010) Molecular phylogenetic studies reveal an undescribed species within the North American concept of *Melanelixia glabra* (Parmeliaceae). Fungal Divers 42 (1):47–55. doi:10.1007/s13225-010-0027-3

Donoghue MJ, Gauthier A (2004) Implementing the phylocode. Trends Ecol Evol 19(6):281–282. doi:10.1016/j.tree.2004.04.004

Eaton DAR, Ree RH (2013) Inferring phylogeny and introgression using RADseq data: an example from flowering plants (*Pedicularis*: Orobanchaceae). Syst Biol 62(5):689–706. doi:10.1093/sysbio/syt032

Edwards SV (2009) Is a new and general theory of molecular systematics emerging? Evolution 63(1):1–19. doi:10.1111/j.1558-5646.2008.00549.x

Edwards DL, Knowles LL (2014) Species detection and individual assignment in species delimitation: can integrative data increase efficacy? Proc R Soc B: Biol Sci 281(1777). doi:10.1098/rspb.2013.2765

Egan RS (1986) Correlations and non-correlations of chemical variation patterns with lichen morphology and geography. The Bryologist 89:99–110

Elix JA, Corush J, Lumbsch HT (2009) Triterpene chemosyndromes and subtle morphological characters characterise lineages in the *Physcia aipolia* group in Australia (Ascomycota). Syst Biodivers 7(04):479–487. doi:10.1017/S1477200009990223

Emerson KJ, Merz CR, Catchen JM, Hohenlohe PA, Cresko WA, Bradshaw WE, Holzapfel CM (2010) Resolving postglacial phylogeography using high-throughput sequencing. Proc Natl Acad Sci 107 (37):16196–16200. doi:10.1073/pnas.1006538107

Ence DD, Carstens BC (2011) SpedeSTEM: a rapid and accurate method for species delimitation. Mol Ecol Resour 11(3):473–480. doi:10.1111/j.1755-0998.2010.02947.x

Evanno G, Regnaut S, Goudet J (2005) Detecting the number of clusters of individuals using the software STRUCTURE: a simulation study. Mol Ecol 14(8):2611–2620. doi:10.1111/j.1365-294X.2005.02553.x

Falush D, Stephens M, Pritchard JK (2003) Inference of population structure using multilocus genotype data: linked loci and correlated allele frequencies. Genetics 164(4):1567–1587. doi:10.1111/j.1471-8286.2007.01758.x

Fan HH, Kubatko LS (2011) Estimating species trees using approximate Bayesian computation. Mol Phylogenet Evol 59(2):354–363. doi:10.1016/j.ympev.2011.02.019

Fernández-Mendoza F, Printzen C (2013) Pleistocene expansion of the bipolar lichen *Cetraria aculeata* into the Southern hemisphere. Mol Ecol 22(7):1961–1983. doi:10.1111/mec.12210

Fernández-Mendoza F, Domaschke S, García MA, Jordan P, Martín MP, Printzen C (2011) Population structure of mycobionts and photobionts of the widespread lichen *Cetraria aculeata*. Mol Ecol 20(6):1208–1232. doi:10.1111/j.1365-294X.2010.04993.x

Fraley C, Raftery A (2007) Model-based methods of classification: Using the mclust software in chemometrics. J Stat Softw 18:i06 [Available online at http://www.doaj.org/doaj?func=abstract&id=218544]

Fujisawa T, Barraclough TG (2013) Delimiting species using single-locus data and the generalized mixed yule coalescent (GMYC) approach: a revised method and evaluation on simulated datasets. Syst Biol 62(5):707–724. doi:10.1093/sysbio/syt033

Fujita MK, Leaché AD, Burbrink FT, McGuire JA, Moritz C (2012) Coalescent-based species delimitation in an integrative taxonomy. Trends Ecol Evol 27(9):480–488. doi:10.1016/j.tree.2012.04.012

Gatesy J, O'Grady P, Baker RH (1999) Corroboration among data sets in simultaneous analysis: hidden support for phylogenetic relationships among higher level Artiodactyl taxa. Cladistics 15(3):271–313. doi:10.1111/j.1096-0031.1999.tb00268.x

Gladieux P, Ropars J, Badouin H, Branca A, Aguileta G, de Vienne DM, Rodríguez de la Vega RC, Branco S, Giraud T (2014) Fungal evolutionary genomics provides insight into the mechanisms of adaptive divergence in eukaryotes. Mol Ecol 23(4):753–773. doi:10.1111/mec.12631

Goffinet B, Miadlikowska J, Goward T (2003) Phylogenetic inferences based on nrDNA sequences support five morphospecies within the *Peltigera didactyla* complex (Lichenized Ascomycota). The Bryologist 106(3):349–364. doi:10.1639/01

Gowan SP (1986) Evolution of secondary natural products in the genus *Porpidia* (Ascomycata, Porpidiaceae). Am J Bot 73:606

Griffin PC, Hoffmann AA (2014) Limited genetic divergence among Australian alpine *Poa* tussock grasses coupled with regional structuring points to ongoing gene flow and taxonomic challenges. Ann Bot. doi:10.1093/aob/mcu017

Grube M, Hawksworth DL (2007) Trouble with lichen: the re-evaluation and re-interpretation of thallus form and fruit body types in the molecular era. Mycol Res 111(9):1116–1132. doi:10.1016/j.mycres.2007.04.008

Gueidan C, Savi S, Thues H, Roux C, Keller C, Tibell L, Prieto M, Heimarsson S, Breuss O, Orange A, Froberg L, Wynns AA, Navarro-Rosines P, Krzewicka B, Pykaelae J, Grube M, Lutzoni F (2009) Generic classification of the Verrucariaceae (Ascomycota) based on molecular and morphological evidence: recent progress and remaining challenges. Taxon 58 (1):184–208

Guillot G, Mortier F, Estoup A (2005) Geneland: a computer package for landscape genetics. Mol Ecol Notes 5(3):712–715. doi:10.1111/j.1471-8286.2005.01031.x

Guillot G, Renaud S, Ledevin R, Michaux J, Claude J (2012) A unifying model for the analysis of phenotypic, genetic, and geographic data. Syst Biol 61 (6):897–911. doi:10.1093/sysbio/sys038

Guisan A, Thuiller W (2005) Predicting species distribution: offering more than simple habitat models. Ecol Lett 8(9):993–1009. doi:10.1111/j.1461-0248.2005.00792.x

Hale ME (1990) A synopsis of the lichen genus *Xanthoparmelia* (Vainio) Hale (Ascomycotina, Parmeliaceae). Smithsonian Institution Press, Washington, DC

Hamilton CA, Hendrixson BE, Brewer MS, Bond JE (2014) An evaluation of sampling effects on multiple DNA barcoding methods leads to an integrative approach for delimiting species: a case study of the North American tarantula genus Aphonopelma (Araneae, Mygalomorphae, Theraphosidae). Mol Phylogenet Evol 71:79–93. doi:10.1016/j.ympev.2013.11.007

Hausdorf B (2011) Progress toward a general species concept. Evolution 65(4):923–931. doi:10.1111/j.1558-5646.2011.01231.x

Hausdorf B, Hennig C (2010) Species delimitation using dominant and codominant multilocus markers. Syst Biol 59(5):491–503. doi:10.1093/sysbio/syq039

Hawksworth DL (1976) Lichen chemotaxonomy. In: Bailey RH (ed) Lichenology: progress and problems. Academic, London, pp 139–184

Hebert PDN, Cywinska A, Ball SL, deWaard JR (2003) Biological identifications through DNA barcodes. Proc R Soc Lond B Biol Sci 270(1512):313–321. doi:10.1098/rspb.2002.2218

Hebert PDN, Stoeckle MY, Zemlak TS, Francis CM (2004) Identification of birds through DNA barcodes. PLoS Biol 2(10):e312. doi:10.1371/journal.pbio.0020312

Heled J, Drummond AJ (2010) Bayesian inference of species trees from multilocus data. Mol Biol Evol 27 (3):570–580. doi:10.1093/molbev/msp274

Hey J (2006) On the failure of modern species concepts. Trends Ecol Evol 21(8):447–450. doi:10.1016/j.tree.2006.05.011

Hibbett DS, Ohman A, Glotzer D, Nuhn M, Kirk P, Nilsson RH (2011) Progress in molecular and morphological taxon discovery in Fungi and options for formal classification on environmental sequences. Fungal Biol Rev 25:38–47. doi:10.1016/j.fbr.2011.01.001

Högnabba F, Wedin M (2003) Molecular phylogeny of the *Sphaerophorus globosus* species complex. Cladistics 19(3):224–232. doi:10.1111/j.1096-0031.2003.tb00365.x

Hudson RR, Coyne JA (2002) Mathematical consequences of the genealogical species concept. Evolution 56(8):1557. doi:10.1111/j.0014-3820.2002.tb01467.x

Huelsenbeck JP, Andolfatto P, Huelsenbeck ET (2011) Structurama: Bayesian inference of population structure. Evol Bioinf Online 7(2011):55–59. doi:10.4137/EBO.S6761

Jones G, Oxelman B (2014) DISSECT: an assignment-free Bayesian discovery method for species delimitation under the multispecies coalescent. Bioinformatics. doi:10.1101/003178

Kekkonen M, Hebert PD (2014) DNA barcode-based delineation of putative species: efficient start for taxonomic workflows. Mol Ecol Res. doi:10.1111/1755-0998.12233

Kelly LJ, Hollingsworth PM, Coppins BJ, Ellis CJ, Harrold P, Tosh J, Yahr R (2011) DNA barcoding of lichenized fungi demonstrates high identification success in a floristic context. New Phytol 191(1):288–300. doi:10.1111/j.1469-8137.2011.03677.x

Kiss L (2012) Limits of nuclear ribosomal DNA internal transcribed spacer (ITS) sequences as species barcodes for fungi. Proc Natl Acad Sci 109(27):E1811–E1811. doi:10.1073/pnas.1207143109

Kluge AG (1989) A concern for evidence and a phylogenetic hypothesis for relationships among *Epicrates* (Boidae, Serpentes). Syst Zool 38(1):7–25. doi:10.1093/sysbio/38.1.7

Knowles LL, Carstens BC (2007) Delimiting species without monophyletic gene trees. Syst Biol 56 (6):887–895. doi:10.1080/10635150701701091

Kroken S, Taylor JW (2001) A gene genealogical approach to recognize phylogenetic species boundaries in the lichenized fungus *Letharia*. Mycologia 93(1):38–53

Kubatko LS, Degnan JH (2007) Inconsistency of phylogenetic estimates from concatenated data under coalescence. Syst Biol 56(1):17–24. doi:10.1080/10635150601146041

Latch EK, Dharmarajan G, Glaubitz JC, Rhodes OE Jr (2006) Relative performance of Bayesian clustering software for inferring population substructure and individual assignments at low levels of population differentiation. Conserv Genet 7(2):295–302. doi:10.1007/s10592-005-9098-1

Lawrey JD (1986) Biological role of lichen substances. The Bryologist 89(2):111–122

Le Gac M, Hood ME, Fournier E, Giraud T (2007) Phylogenetic evidence of host-specific cryptic species in the anther smut fungus. Evolution 61(1):15–26. doi:10.1111/j.1558-5646.2007.00002.x

Leaché AD (2009) Species tree discordance traces to phylogeographic clade boundaries in North American fence lizards (*Sceloporus*). Syst Biol 58(6):547–559. doi:10.1093/sysbio/syp057

Leaché AD, Fujita MK (2010) Bayesian species delimitation in West African forest geckos (*Hemidactylus fasciatus*). Proc R Soc B: Biol Sci 277:3071–3077. doi:10.1098/rspb.2010.0662

Leaché AD, Koo M, Spencer C, Papenfuss T, Fisher R, McGuire J (2009) Quantifying ecological, morphological, and genetic variation to delimit species in the coast horned lizard species complex (*Phrynosoma*). Proc Natl Acad Sci USA 106:12418–12423. doi:10.1073/pnas.0906380106

Leaché AD, Fujita MK, Minin VN, Bouckaert RR (2014) Species delimitation using genome-wide SNP data. Systematic Biology. doi:10.1093/sysbio/syu018

Leavitt SD, Fankhauser JD, Leavitt DH, Porter LD, Johnson LA, St. Clair LL (2011a) Complex patterns of speciation in cosmopolitan "rock posy" lichens—discovering and delimiting cryptic fungal species in the lichen-forming *Rhizoplaca melanophthalma* species-complex (Lecanoraceae, Ascomycota). Mol Phylogenet Evolu 59(3):587–602. doi:10.1016/j.ympev.2011.03.020

Leavitt SD, Johnson L, St. Clair LL (2011b) Species delimitation and evolution in morphologically and chemically diverse communities of the lichen-forming genus *Xanthoparmelia* (Parmeliaceae, Ascomycota) in western North America. Am J Bot 98 (2):175–188. doi:10.3732/ajb.1000230

Leavitt SD, Johnson LA, Goward T, St. Clair LL (2011c) Species delimitation in taxonomically difficult lichen-forming fungi: an example from morphologically and chemically diverse Xanthoparmelia (Parmeliaceae) in North America. Mol Phylogen Evol 60(3):317–332. doi:10.1016/j.ympev.2011.05.012

Leavitt S, Esslinger T, Divakar P, Lumbsch H (2012a) Miocene and Pliocene dominated diversification of the lichen-forming fungal genus *Melanohalea* (Parmeliaceae, Ascomycota) and Pleistocene population expansions. BMC Evol Biol 12(1):176. doi:10.1186/1471-2148-12-176

Leavitt SD, Esslinger TL, Divakar PK, Lumbsch HT (2012b) Miocene divergence, phenotypically cryptic lineages, and contrasting distribution patterns in common lichen-forming fungi (Ascomycota: Parmeliaceae). Biol J Linn Soc 1007:920–937. doi:10.1111/j.1095-8312.2012.01978.x

Leavitt SD, Esslinger TL, Lumbsch HT (2012c) Neogene-dominated diversification in neotropical montane lichens: dating divergence events in the lichen-forming fungal genus *Oropogon* (Parmeliaceae). Am J Bot 99(11):1764–1777. doi:10.3732/ajb.1200146

Leavitt SD, Esslinger TL, Nelsen MP, Lumbsch HT (2013a) Further species diversity in Neotropical Oropogon (Lecanoromycetes: Parmeliaceae) in Central America. The Lichenologist 45(04):553–564. doi:10.1017/S0024282913000212

Leavitt SD, Esslinger TL, Spribille T, Divakar PK, Lumbsch HT (2013b) Multilocus phylogeny of the lichen-forming fungal genus *Melanohalea* (Parmeliaceae, Ascomycota): insights on diversity, distributions, and a comparison of species tree and concatenated topologies. Mol Phylogenet Evol 66(2013):138–152. doi:10.1016/j.ympev.2012.09.013

Leavitt SD, Fernández-Mendoza F, Pérez-Ortega S, Sohrabi M, Divakar PK, Lumbsch HT, St. Clair LL (2013c) DNA barcode identification of lichen-forming fungal species in the *Rhizoplaca melanophthalma* species-complex (Lecanorales, Lecanoraceae), including five new species. MycoKeys 7:1–22. doi:10.3897/mycokeys.7.4508

Leavitt SD, Fernández-Mendoza F, Pérez-Ortega S, Sohrabi M, Divakar PK, Vondrák J, Thorsten Lumbsch H, Clair LLS (2013d) Local representation of global diversity in a cosmopolitan lichen-forming fungal species complex (*Rhizoplaca*, Ascomycota). J Biogeogr 40(9):1792–1806. doi:10.1111/jbi.12118

Leavitt SD, Lumbsch HT, Stenroos S, St. Clair LL (2013e) Pleistocene speciation in North American lichenized fungi and the impact of alternative species circumscriptions and rates of molecular evolution on divergence estimates. PLoS ONE 8(12):e85240. doi:10.1371/journal.pone.0085240

Leavitt SD, Esslinger TL, Hansen ES, Divakar PK, Crespo A, Loomis BF, Lumbsch HT (2014) DNA barcoding of brown Parmeliae (Parmeliaceae) species: a molecular approach for accurate specimen identification, emphasizing species in Greenland. Organ Divers Evol 14 (1):11–20. doi:10.1007/s13127-013-0147-1

Leliaert F, Verbruggen H, Vanormelingen P, Steen F, Lopez-Bautista JM, Zuccarello GC, De Clerck O (2014) DNA-based species delimitation in algae. Eur J Phycol 49(2). doi:10.1080/09670262.2014.904524

Leuckert C (1985) Probleme der Flechten-Chemotaxonomie—Stoffkombinationen und ihre taxonomische Wertung. Ber Deut Bot Ges 98:401–408

Lewis ZA, Shiver AL, Stiffler N, Miller MR, Johnson EA, Selker EU (2007) High-density detection of restriction-site-associated DNA markers for rapid mapping of mutated loci in *Neurospora*. Genetics 177(2):1163–1171. doi:10.1534/genetics.107.078147

Lindblom L, Søchting U (2008) Taxonomic revision of *Xanthomendoza borealis* and *Xanthoria mawsonii* (Lecanoromycetes, Ascomycota). The Lichenologist 40(05):399–409. doi:10.1017/S0024282908007937

Lohtander K, Källersjö M, Roland M, Tehler A (2000) The family physciaceae in fennoscandia: phylogeny inferred from its sequences. Mycologia 92(4):728–735

Longton RE (1997) The role of bryophytes and lichens in polar ecosystems. In: Woodin SJ, Marquiss M (eds) Ecology of Arctic Environments. Blackwell Science, Oxford, pp 69–96 (Special publication No. 13)

Lücking R (2012) Predicting species richness in tropical lichenized fungi with 'modular' combinations of character states. Biodiver Conserv 21(9):2341–2360. doi:10.1007/s10531-011-0217-7

Lücking R, del Prado R, Lumbsch HT, Will-Wolf S, Aptroot A, Sipman HJM, Umana L, Chaves JL (2008) Phytogenetic patterns of morphological and chemical characters and reproductive mode in the Heterodermia obscurata group in Costa Rica (Ascomycota, Physciaceae). Syst Biodivers 6(1):31–41. doi:10.1017/S1477200007002629

Lumbsch HT (1994) Die *Lecanora subfusca*-Gruppe in Australasien. J Hattori Bot Lab 77:1–175

Lumbsch HT (1998a) Taxonomic use of metabolic data in lichen-forming fungi. In: Frisvad JC, Bridge PD, Arora DK (eds) Chemical Fungal Taxonomy. Marcel Dekker, New York, pp 345–387

Lumbsch HT (1998b) The use of metabolic data in lichenology at the species and subspecific levels. The Lichenologist 30(4–5):357–367. doi:10.1017/S0024282992000380

Lumbsch HT (2002) Analysis of phenolic products in lichens for identification and taxonomy. Protocols in lichenology. Culturing, biochemistry, ecophysiology and use in biomonitoring. Springer, Berlin

Lumbsch HT, Leavitt SD (2011) Goodbye morphology? A paradigm shift in the delimitation of species in lichenized fungi. Fungal Divers 50(1):59–72. doi:10.1007/s13225-011-0123-z

Lumbsch HT, Ahti T, Altermann S, Amo de Paz G, Aptroot A, Arup U, Barcenas Peña A, Bawingan PA, Benatti MN, Betancourt L, Björk CR, Boonpragob K, Brand M, Bungartz F, Caceres MES, Candan M, Chaves JL, Clerc P, Common R, Coppins BJ, Crespo A, Dal Forno M, Divakar PK, Duya MV, Elix JA, Elvebakk A, Fankhauser J, Farkas E, Ferraro LI, Fischer E, Galloway DJ, Gaya E, Giralt M, Goward T, Grube M, Hafellner J, Hernandez JE, Herrera-Campos MA, Kalb K, Kärnefelt I, Kantvilas G, Killmann D, Kirika P, Knudesn K, Komposch H, Kondratyuk S, Lawrey JD, Mangold A, Marcelli MP, McCune BP, Michlig A, Miranda Gonzalez R, Moncada B, Naikatini A, Nelsen MP, Øvstedal DO, Palice Z, Papong K, Parnmen S, Pérez-Ortega S, Printzen C, Rico VJ, Rivas Plata E, Robayo J, Rosabal D, Ruprecht U, Salazar Allen N, Sancho L, Santos de Jesus L, Santos Vieira T, Schultz M, Seaward MRD, Sérusiaux E, Schmitt I, Sipman HJM, Sohrabi M, Søchting U, Søgaard MZ, Sparrius LB, Spielmann A, Spribille T, Sutjaritturakan J, Thammathaworn A, Thell A, Thor G, Thüs H, Timdal E, Truong C, Türk R, Umaña Tenorio L, Upreti D, van den Boom P, Vivas Rebuelta M, Wedin M, Will-Wolf S, Wirth V, Wirtz N, Yahr R, Yeshitela K, Ziemmeck F, Wheeler T, Lücking R (2011) One hundred new species of lichenized fungi: a signature of undiscovered global diversity. Phytotaxa 18:1–127

Mark K, Saag L, Saag A, Thell A, Randlane T (2012) Testing morphology-based delimitation of *Vulpicida juniperinus* and *V. tubulosus* (Parmeliaceae) using three molecular markers. The Lichenologist 44 (06):757–772. doi:10.1017/S0024282912000448

Martín MP, LaGreca S, Lumbsch HT (2003) Molecular phylogeny of *Diploschistes* inferred from ITS sequence data. The Lichenologist 35(01):27–32. doi:10.1006/lich.2002.0427

Masters BC, Fan V, Ross HA (2011) Species delimitation—a geneious plugin for the exploration of species boundaries. Mol Ecol Res 11(1):154–157. doi:10.1111/j.1755-0998.2010.02896.x

Mayden RL (1997) A hierarchy of species concepts: the denouement in the saga of the species problem. In: Claridge MF, Dawah HA, Wilson MR (eds) Species: the units of biodiversity. Chapman & Hall, London, pp 381–424

Mayr E (1963) Animal species and evolution. Harvard University Press, Cambridge

Mayr E (1970) Populations, species, and evolution. Belknap Press of Harvard University Press, Cambridge

McCune B (2000) Lichen communities as indicators of forest health. The Bryologist 103(2):353–356. doi:10.1639/0007-2745(2000)103[0353:LCAIOF]2.0.CO;2

McCune B, Printzen C (2011) Distribution and climatic niches of the *Lecanora varia* group in western U.S.A. Bibliotheca Lichenologica 106:225–234

McDonald T, Miadlikowska J, Lutzoni F (2003) The lichen genus *Sticta* in the Great Smoky Mountains: a phylogenetic study of morphological, chemical, and molecular data. The Bryologist 106(1):61–79. doi:10.1639/0007-2745(2003)106[0061:TLGSIT]2.0.CO;2

McKay BD, Mays HL, Wu Y, Li H, Yao C-T, Nishiumi I, Zou F (2013) An empirical comparison of character-based and coalescent-based approaches to species delimitation in a young avian complex. Mol Ecol 22 (19):4943–4957. doi:10.1111/mec.12446

Miller MR, Dunham JP, Amores A, Cresko WA, Johnson EA (2007) Rapid and cost-effective polymorphism identification and genotyping using restriction site associated DNA (RAD) markers. Genome Res 17 (2):240–248. doi:10.1101/gr.5681207

Miralles A, Vences M (2013) New metrics for comparison of taxonomies reveal striking discrepancies among species delimitation methods in *Madascincus* lizards. PLoS ONE 8(7):e68242. doi:10.1371/journal.pone.0068242

Mishler BD, Brandon RN (1987) Individuality, pluralism, and the phylogenetic species concept. Biol Philos 2:397

Molina M, Crespo A, Blanco O, Lumbsch HT, Hawksworth DL (2004) Phylogenetic relationships and species concepts in *Parmelia* s. str. (Parmeliaceae) inferred from nuclear ITS rDNA and β-tubulin sequences. The Lichenologist 36(01):37–54. doi:10.1017/S0024282904013933

Molina M, Del-Prado R, Divakar P, Sánchez-Mata D, Crespo A (2011) Another example of cryptic diversity in lichen-forming fungi: the new species *Parmelia mayi* (Ascomycota: Parmeliaceae). Organ Divers Evol 11(5):331–342. doi:10.1007/s13127-011-0060-4

Monaghan MT, Wild R, Elliot M, Fujisawa T, Balke M, Inward DJG, Lees DC, Ranaivosolo R, Eggleton P, Barraclough TG, Vogler AP (2009) Accelerated species inventory on Madagascar using coalescent-based models of species delineation. Syst Biol 58 (3):298–311. doi:10.1093/sysbio/syp027

Moncada B, Reidy B, Lücking R (2014) A phylogenetic revision of Hawaiian *Pseudocyphellaria* (lichenized Ascomycota: Lobariaceae) reveals eight new species and a high degree of inferred endemism. The Bryologist 117(2):119–160. doi: http://dx.doi.org/10.1639/0007-2745-117.2.119

Moreau CS (2009) Inferring ant evolution in the age of molecular data. Myrmecological New 12:201

Muggia L, Grube M, Tretiach M (2008) A combined molecular and morphological approach to species delimitation in black-fruited, endolithic *Caloplaca*: high genetic and low morphological diversity. Mycol Res 112(1):36–49. doi:10.1016/j.mycres.2007.02.001

Muggia L, Pérez-Ortega S, Fryday A, Spribille T, Grube M (2014) Global assessment of genetic variation and phenotypic plasticity in the lichen-forming species *Tephromela atra*. Fungal Divers 64(1):233–251. doi:10.1007/s13225-013-0271-4

Nash TH, Zavada M (1977) Population studies among Sonoran Desert species of *Parmelia* subg. *Xanthoparmelia* (Parmeliaceae). Am J Bot 64(6):664–669

Nylander W (1866a) Circa novum in studio Lichenum criterium chemicum. Flora 49:198 201

Nylander W (1866b) Hypochlorite of lime and hydrate of potash. Two new criteria for the study of lichens (Tanslated and communicated by the Rev. W. A. Leighton). Bot J Linn Soc 9:358–365

O'Brien HE, Miadlikowska J, Lutzoni F (2009) Assessing reproductive isolation in highly diverse communities of the lichen-forming fugnal genus *Peltigera*. Evolution 63(8):2076–2086. doi:10.1111/j.1558-5646.2009.00685.x

O'Meara BC (2010) New heuristic methods for joint species delimitation and species tree inference. Syst Biol 59(1):59–73. doi:10.1093/sysbio/syp077

O'Neill EM, Schwartz R, Bullock CT, Williams JS, Shaffer HB, Aguilar-Miguel X, Parra-Olea G, Weisrock DW (2013) Parallel tagged amplicon sequencing reveals major lineages and phylogenetic structure in the North American tiger salamander (*Ambystoma tigrinum*) species complex. Mol Ecol 22(1):111–129. doi:10.1111/mec.12049

Orock EA, Leavitt SD, Fonge BA, St. Clair LL, Lumbsch HT (2012) DNA-based identification of lichen-forming fungi: Can publicly available sequence databases aid in lichen diversity inventories of Mount Cameroon (West Africa)? The Lichenologist 44(6):833–839. doi:10.1017/S0024282912000424

Otálora MAG, Martínez I, Aragón G, Molina MC (2010) Phylogeography and divergence date estimates of a lichen species complex with a disjunct distribution pattern. Am J Bot 97(2):216–223. doi:10.3732/ajb.0900064

Padial J, Castroviejo-Fisher S, Kohler J, Vila C, Chaparro J, De la Riva I (2009) Deciphering the products of evolution at the species level: the need for an integrative taxonomy. Zoolog Scr 38(4):431–447. doi:10.1111/j.1463-6409.2008.00381.x

Padial J, Miralles A, De la Riva I, Vences M (2010) The integrative future of taxonomy. Front Zoo 7(1):16. doi:10.1186/1742-9994-7-16

Papong K, Lücking R, Thammathaworn A, Boonpragob K (2009) Four new taxa of *Chroodiscus* (thelotremoid Graphidaceae) from Southeast Asia. The Bryologist 112(1):152–163. doi:10.1639/0007-2745-112.1.152

Park S-Y, Choi J, Kim JA, Jeong M-H, Kim S, Lee Y-H, Hur J-S (2013a) Draft genome sequence of *Cladonia macilenta* KoLRI003786, a lichen-forming fungus producing biruloquinone. Genome Announcements 1 (5):e00695-13. doi:10.1128/genomeA.00695-13

Park S-Y, Choi J, Kim JA, Yu N-H, Kim S, Kondratyuk SY, Lee Y-H, Hur J-S (2013b) Draft genome sequence of lichen-forming fungus *Caloplaca flavorubescens* Strain KoLRI002931. Genome Announcements 1(4): e00678-13. doi:10.1128/genomeA.00678-13

Park S-Y, Choi J, Lee G-W, Kim JA, Oh S-O, Jeong M-H, Yu N-H, Kim S, Lee Y-H, Hur J-S (2014) Draft genome sequence of lichen-forming fungus *Cladonia metacorallifera* Strain KoLRI002260. Genome Announcements 2(1):e01065-13. doi:10.1128/genomeA.01065-13

Parnmen S, Rangsiruji A, Mongkolsuk P, Boonpragob K, Nutakki A, Lumbsch HT (2012) Using phylogenetic and coalescent methods to understand the species diversity in the *Cladia aggregata* complex (Ascomycota, Lecanorales). PLoS ONE 7(12):e52245. doi:10.1371/journal.pone.0052245

Pérez-Ortega S, Fernández-Mendoza F, Raggio J, Vivas M, Ascaso C, Sancho LG, Printzen C, de los Ríos A (2012) Extreme phenotypic variation in *Cetraria aculeata* (lichenized Ascomycota): adaptation or incidental modification? Ann Bot 109(6):1133–1148. doi:10.1093/aob/mcs042

Pino-Bodas R, Burgaz A, Martín M, Lumbsch HT (2011) Phenotypical plasticity and homoplasy complicate species delimitation in the *Cladonia gracilis* group (Cladoniaceae, Ascomycota). Organ Divers Evol 11 (5):343–355. doi:10.1007/s13127-011-0062-2

Pino-Bodas R, Burgaz AR, Martin MP, Lumbsch HT (2012a) Species delimitations in the *Cladonia cariosa* group (Cladoniaceae, Ascomycota). The Lichenologist 44(01):121–135. doi:10.1017/S002428291100065X

Pino-Bodas R, Martín M, Burgaz A (2012b) *Cladonia subturgida* and *C. iberica* (Cladoniaceae) form a single, morphologically and chemically polymorphic species. Mycol Prog 11(1):269–278. doi:10.1007/s11557-011-0746-1

Pino-Bodas R, Ahti T, Stenroos S, Martín MP, Burgaz AR (2013a) Multilocus approach to species recognition in the *Cladonia humilis* complex (Cladoniaceae, Ascomycota). Am J Bot 100(4):664–678. doi:10.3732/ajb.1200162

Pino-Bodas R, Martín AP, Burgaz AR, Lumbsch HT (2013b) Species delimitation in *Cladonia* (Ascomycota): a challenge to the DNA barcoding philosophy. Mol Ecol Resour 13(6):1058–1068. doi:10.1111/1755-0998.12086

Pons J, Barraclough TG, Gomez-Zurita J, Cardoso A, Duran DP, Hazell S, Kamoun S, Sumlin WD, Vogler AP (2006) Sequence-based species delimitation for the DNA taxonomy of undescribed insects. Syst Biol 55 (4):595–609. doi:10.1080/10635150600852011

Porada P, Weber B, Elbert W, Pöschl U, Kleidon A (2014) Estimating impacts of lichens and bryophytes on global biogeochemical cycles. Global Biogeochem Cycles 28(2):71–85. doi:10.1002/2013GB004705

Pringle A, Baker DM, Platt JL, Wares JP, Latgé JP, Taylor JW (2005) Cryptic speciation in the cosmopolitan and clonal human pathogenic fungus *Aspergillus fumigatus*. Evolution 59(9):1886–1899. doi:10.1554/04-241.1

Printzen C (2009) Lichen systematics: the role of morphological and molecular data to reconstruct phylogenetic relationships. Progress in botany, vol 71. Springer, Berlin, pp 233–275

Pritchard JK, Stephens M, Donnelly P (2000) Inference of population structure using multilocus genotype data. Genetics 155(2):945–959

Puillandre N, Lambert A, Brouillet S, Achaz G (2012) ABGD, automatic barcode gap discovery for primary species delimitation. Mol Ecol 21(8):1864–1877. doi:10.1111/j.1365-294X.2011.05239.x

Rannala B, Yang Z (2003) Bayes estimation of species divergence times and ancestral population sizes using DNA sequences from multiple loci. Genetics 164:1645–1656

Raxworthy CJ, Ingram CM, Rabibisoa N, Pearson RG (2007) Applications of ecological niche modeling for species delimitation: a review and empirical evaluation using day geckos (*Phelsuma*) from Madagascar. Syst Biol 56(6):907–923. doi:10.1080/10635150701775111

Ray J (1686) Historia planarum, vol 1. Clark, London

Reese Næsborg R, Ekman S, Tibell L (2007) Molecular phylogeny of the genus *Lecania* (Ramalinaceae, lichenized Ascomycota). Mycol Res 111(5):581–591. doi:10.1016/j.mycres.2007.03.001

Reid N, Carstens B (2012) Phylogenetic estimation error can decrease the accuracy of species delimitation: a Bayesian implementation of the general mixed Yule-coalescent model. BMC Evol Biol 12(1):196. doi:10.1186/1471-2148-12-196

Rissler L, Apodaca J (2007) Adding more ecology into species delimitation: ecological niche models and phylogeography help define cryptic species in the Black Salamander (*Aneides flavipunctatus*). Syst Biol 56:924–942. doi:10.1080/10635150701703063

Rivas Plata E, Lücking R (2013) High diversity of Graphidaceae (lichenized Ascomycota: Ostropales) in Amazonian Perú. Fungal Divers 58(1):13–32. doi:10.1007/s13225-012-0172-y

Rivas Plata E, Lumbsch HT (2011) Parallel evolution and phenotypic divergence in lichenized fungi: a case study in the lichen-forming fungal family Graphidaceae (Ascomycota: Lecanoromycetes: Ostropales). Mol Phylogenet Evol 61(1):45–63. doi:10.1016/j.ympev.2011.04.025

Rogers RW (1989) Chemical variation and the species concept in lichenized ascomycetes. Bot J Linn Soc 101:229–239

Ross HA (2014) The incidence of species-level paraphyly in animals: a re-assessment. Mol Phylogenet Evol 76:10–17. doi:10.1016/j.ympev.2014.02.021

Ross KG, Gotzek D, Ascunce MS, Shoemaker DD (2010) Species delimitation: a case study in a problematic ant taxon. Syst Biol 59(2):162–184. doi:10.1093/sysbio/syp089

Rubin BER, Ree RH, Moreau CS (2012) Inferring phylogenies from RAD sequence data. PLoS ONE 7 (4):e33394. doi:10.1371/journal.pone.0033394

Rubinoff D (2006) Utility of mitochondrial DNA barcodes in species conservation. Conserv Biol 20(4):1026–1033. doi:10.1111/j.1523-1739.2006.00372.x

Ruiz-Sanchez E, Sosa V (2010) Delimiting species boundaries within the Neotropical bamboo *Otatea* (Poaceae: Bambusoideae) using molecular, morphological and ecological data. Mol Phylogenet Evol 54 (2):344–356. doi:10.1016/j.ympev.2009.10.035

Ruprecht U, Lumbsch HT, Brunauer G, Green TGA, Türk R (2010) Diversity of *Lecidea* (Lecideaceae, Ascomycota) species revealed by molecular data and morphological characters. Antarctic Sci 22 (Special Issue 06):727–741. doi:10.1017/S0954102010000477

Salicini I, Ibáñez C, Juste J (2011) Multilocus phylogeny and species delimitation within the Natterer's bat species complex in the Western Palearctic. Mol Phylogenet Evol 61(3):888–898. doi:10.1016/j.ympev.2011.08.010

Satler JD, Carstens BC, Hedin M (2013) Multilocus species delimitation in a complex of morphologically conserved trapdoor spiders (Mygalomorphae, Antrodiaetidae, *Aliatypus*). Syst Biol 62(6):805–823. doi:10.1093/sysbio/syt041

Schlick-Steiner BC, Steiner FM, Seifert B, Stauffer C, Christian E, Crozier RH (2010) Integrative taxonomy: a multisource approach to exploring biodiversity. Annu Rev Entomol 55(1):421–438. doi:10.1146/annurev-ento-112408-085432

Schoch CL, Seifert KA, Huhndorf S, Robert V, Spouge JL, Levesque CA, Chen W, Fungal Barcoding Consortium (2012) Nuclear ribosomal internal transcribed spacer (ITS) region as a universal DNA barcode marker for Fungi. Proc Natl Acad Sci 109(16):6241–6246. doi:10.1073/pnas.1117018109

Sérusiaux E, Villarreal AJC, Wheeler T, Goffinet B (2011) Recent origin, active speciation and dispersal for the lichen genus *Nephroma* (Peltigerales) in Macaronesia. J Biogeogr 38(6):1138–1151. doi:10.1111/j.1365-2699.2010.02469.x

Shrestha G, Peterson SL, St. Clair LL (2012) Predicting the distribution of the air pollution sensitive lichen species *Usnea hirta*. The Lichenologist 44(04):511–521. doi:10.1017/S0024282912000060

Simpson GG (1951) The species concept. Evolution 5:285–298

Sites JW, Marshall JC (2003) Delimiting species: a renaissance issue in systematic biology. Trends Ecol Evol 18:462–470. doi:10.1016/S0169-5347(03)00184-8

Sites JW, Marshall JC (2004) Operational criteria for delimiting species. Annu Rev Ecol Evol Syst 35 (1):199–227. doi:10.1146/annurev.ecolsys.35.112202.130128

Šlapeta J, López-García P, Moreira D (2006) Global dispersal and ancient cryptic species in the smallest marine eukaryotes. Mol Biol Evol 23(1):23–29. doi:10.1093/molbev/msj001

Spribille T, Klug B, Mayrhofer H (2011) A phylogenetic analysis of the boreal lichen *Mycoblastus sanguinarius* (Mycoblastaceae, lichenized Ascomycota) reveals cryptic clades correlated with fatty acid profiles. Mol Phylogenet Evol 59(3):603–614. doi:10.1016/j.ympev.2011.03.021

Stenroos SK, DePriest PT (1998) SSU rDNA phylogeny of cladoniiform lichens. Am J Bot 85(11):1548–1559

Talavera G, Dincă V, Vila R (2013) Factors affecting species delimitations with the GMYC model: insights from a butterfly survey. Methods Ecol Evol 4 (12):1101–1110. doi:10.1111/2041-210X.12107

Taylor JW, Jacobson DJ, Kroken S, Kasuga T, Geiser DM, Hibbett DS, Fisher MC (2000) Phylogenetic species recognition and species concepts in fungi. Fungal Genet Biol 31(1):21–32. doi:10.1006/fgbi.2000.1228

Tewksbury J, Anderson JGT, Bakker JD, Billo TJ, Dunwiddie PW, Groom MJ, Hampton SE, Herman SG, Levey DJ, Machnicki NJ, del Rio CM, Power ME, Rowell K, Salomon AK, Stacey L, Trombulak SC, Wheeler TA (2014) Natural history's place in science and society. BioScience. doi:10.1093/biosci/biu032

Thell A, Högnabba F, Elix JA, Feuerer T, Kärnefelt I, Myllys L, Randlane T, Saag A, Stenroos S, Ahti T, Seaward MRD (2009) Phylogeny of the cetrarioid core (Parmeliaceae) based on five genetic markers. The Lichenologist 41(05):489–511. doi:10.1017/S0024282909990090

Velmala S, Myllys L, Halonen P, Goward T, Ahti T (2009) Molecular data show that *Bryoria fremontii* and *B. tortuosa* (Parmeliaceae) are conspecific. The Lichenologist 41(03):231–242. doi:10.1017/S0024282909008573

Vondrák J, Říha P, Arup U, Søchting U (2009) The taxonomy of the *Caloplaca citrina* group (Teloschistaceae) in the Black Sea region; with contributions to the cryptic species concept in lichenology. The Lichenologist 41(06):571–604. doi:10.1017/S0024282909008317

Wagner CE, Keller I, Wittwer S, Selz OM, Mwaiko S, Greuter L, Sivasundar A, Seehausen O (2013) Genome-wide RAD sequence data provide unprecedented resolution of species boundaries and relationships in the Lake Victoria cichlid adaptive radiation. Mol Ecol 22(3):787–798. doi:10.1111/mec.12023

Wang Y-Y, Liu B, Zhang X-Y, Zhou Q-M, Zhang T, Li H, Yu Y-F, Zhang X-L, Hao X-Y, Wang M, Wang L, Wei J-C (2014) Genome characteristics reveal the impact of lichenization on lichen-forming fungus *Endocarpon pusillum* Hedwig (Verrucariales, Ascomycota). BMC Genom 15(1):34. doi:10.1186/1471-2164-15-34

Wedin M, Westberg M, Crewe AT, Tehler A, Purvis OW (2009) Species delimitation and evolution of metal bioaccumulation in the lichenized *Acarospora smaragdula* (Ascomycota, Fungi) complex. Cladistics 25 (2):161–172. doi:10.1111/j.1096-0031.2009.00240.x

Weisrock DW, Rasoloarison RM, Fiorentino I, Ralison JM, Goodman SM, Kappeler PM, Yoder AD (2010) Delimiting species without nuclear monophyly in Madagascar's mouse lemurs. PLoS ONE 5(3):e9883. doi:10.1371/journal.pone.0009883

Werth S, Cornejo C, Scheidegger C (2013) Characterization of microsatellite loci in the lichen fungus *Lobaria pulmonaria* (Lobariaceae). Appl Plant Sci 1 (2):1200290. doi:10.3732/apps.1200290

Westberg M, Arup U, Kärnefelt I (2007) Phylogenetic studies in the *Candelariaceae* (lichenized Ascomycota) based on nuclear ITS DNA sequence data. Mycol Res 111(11):1277–1284. doi:10.1017/S0953756204002102

Wiemers M, Fiedler K (2007) Does the DNA barcoding gap exist?—A case study in blue butterflies

(Lepidoptera: Lycaenidae). Front Zoo 4(1):8. doi:10.1186/1742-9994-4-8

Wiens JJ (1998) Combining data sets with different phylogenetic histories. Syst Biol 4:568–581. doi:10.1080/106351598260581

Wiens JJ, Penkrot T (2002) Delimiting species using DNA and morphological variation and discordant species limits in spiny lizards (*Sceloporus*). Syst Biol 51:69–91. doi:10.1080/106351502753475880

Wiens JJ, Servedio MR (2000) Species delimitation in systematics: inferring diagnostic differences between species. Proc R Soc Lond B Biol Sci 267(1444):631–636. doi:10.1098/rspb.2000.1049

Will KW, Mishler BD, Wheeler QD (2005) The perils of DNA barcoding and the need for integrative taxonomy. Syst Biol 54(5):844–851. doi:10.1080/10635150500354878

Wirtz N, Printzen C, Lumbsch HT (2012) Using haplotype networks, estimation of gene flow and phenotypic characters to understand species delimitation in fungi of a predominantly Antarctic *Usnea* group (Ascomycota, Parmeliaceae). Organ Divers Evol 12(1):17–37. doi:10.1007/s13127-011-0066-y

Yang Z, Rannala B (2010) Bayesian species delimitation using multilocus sequence data. Proc Natl Acad Sci 107(20):9264–9269. doi:10.1073/pnas.0913022107

Yeates DK, Seago A, Nelson L, Cameron SL, Joseph LEO, Trueman JWH (2011) Integrative taxonomy, or iterative taxonomy? Syst Entomol 36(2):209–217. doi:10.1111/j.1365-3113.2010.00558.x

Zhang C, Zhang D-X, Zhu T, Yang Z (2011) Evaluation of a Bayesian coalescent method of species delimitation. Syst Biol 60(6):747–761. doi:10.1093/sysbio/syr071

Zhang J, Kapli P, Pavlidis P, Stamatakis A (2013) A general species delimitation method with applications to phylogenetic placements. Bioinformatics 29(22):2869–2876. doi:10.1093/bioinformatics/btt499

3 Molecular Phylogenetic and Phylogenomic Approaches in Studies of Lichen Systematics and Evolution

Pradeep K. Divakar and Ana Crespo

Contents

Abstract

Molecular phylogenies and phylogenomics are important for addressing several biological questions such as evolutionary relationships among organisms or genes or genomes, trait evolution, the demographic changes, distribution patterns, and diversification of species, DNA barcoding, prediction of gene function, prediction, and retracing gene transfer. The advancement of DNA sequencing technologies such as next generation sequencing has taken phylogenetic analysis to a new summit. Molecular phylogenies have encompassed almost every branch of biology including lichens, and various phylogenetic methods and software packages are now available. Here, we review the main methods for phylogenetic and phylogenomic analyses, including distance, maximum parsimony, maximum likelihood, and Bayesian approaches; and assembly of phylogenomic dataset and approaches based on whole genome features. We provide a practical guide for their uses and discuss their strengths and weaknesses.

Keywords

Ancestral area reconstruction · Ancestral state reconstruction · Bayesian inference · Maximum likelihood · Maximum parsimony · Molecular dating · Phylogenomics · Orthologous genes

P.K. Divakar (✉) · A. Crespo
Departmento de Biologia Vegetal II, Facultad de Farmacia, Universtad Computense de Madrid, Plaza de Ramon y Cajal, 28040 Madrid, Spain
e-mail: pdivakar@farm.ucm.es

D.K. Upreti et al. (eds.), *Recent Advances in Lichenology*,
DOI 10.1007/978-81-322-2235-4_3

3.1 Introduction

Traditionally, the phylogenetic trees have been used almost exclusively to describe the relationship among species in systematic studies. However, the advent of DNA sequencing technologies so-called molecular phylogeny has helped to understand the evolutionary relationships among organisms and allowed us to address various biological questions such as trait evolution, speciation, prediction of gene functions and gene transfer, relationships between paralogues in a gene family, and population histories. Moreover, molecular phylogeny has become an essential tool for genome comparisons. It is widely used to classify metagenomic sequences, reconstruct ancestral genomes, and identify complete genes (Paten et al. 2008; Brady and Salzberg 2011).

Lichenized fungi exhibit few taxonomically useful characters, of which numerous are homoplasious. Sometimes, the interpretation of morphological features is difficult to evaluate especially within and between groups (Stenroos et al. 2002; Crespo et al. 2007; Lumbsch and Huhndorf 2010; Printzen 2010). In the lack of taxonomically significant morphological features (especially at lower taxonomic level), molecular data have, therefore, gained great magnitude in lichen systematics. Molecular data have helped in identifying the phylogenetic position of lichen-forming fungi, and studies have shown multiple origins of these organisms within Ascomycota (Gargas et al. 1995; Schoch et al. 2009a). Given that the mycobiont is recognized only rarely in its non-symbiotic state, traditional taxonomy treated lichens as a separate group. The integration of lichenized fungi into the fungal system was accepted before the arrival of molecular approaches. However, the molecular phylogeny has largely changed our view on the classification of lichens at lower hierarchical levels, mainly the delimitation of genera and families, but also orders (Hibbett et al. 2007; Crespo et al. 2010; Lumbsch and Huhndorf 2010). Since 1995, phylogenetic analyses carried out on the SSU of nuclear ribosomal DNA of lichenized Ascomycota (Gargas et al. 1995), and the huge number of taxonomical changes both at higher and lower taxonomic level have been made within last 20 years based on DNA sequence data. These molecular studies are already compiled in comprehensive reviews (see e.g., DePriest 2004; Lumbsch 2007; Printzen 2010; Crespo et al. 2011; Thell et al. 2012). We here aim to evaluate the major phylogenetic approaches used in lichen systematic and evolutionary studies. In this review, we describe the suit of current methodologies for phylogenetic inference using sequence data. Further, we also discuss next generation sequencing technologies for generating huge datasets.

3.2 Phylogenetic Analysis

The reconstruction of phylogenetic relationships is a statistical estimation procedure, and we have to infer the phylogenetic tree. A phylogeny is a model of genealogical history in which the lengths of branches are unknown parameters. Each branch represents the determination of a genetic lineage through time, and each node represents the origin of a new lineage. In a tree elucidating the relationship among a group of species, the nodes represent speciation events. However, in a tree of paralogous gene families, the nodes may represent gene duplication events.

A number of methods for phylogenetic tree reconstruction have been developed, which are listed in Table 3.1. The results from these approaches can vary, and therefore, it is crucial to determine which method could be most appropriate for the data and question under study. A summary of strengths and weakness of commonly used methods for phylogeny reconstruction are listed in Table 3.2. Given that all of the available methods of phylogenetic inference generalize the true complexity of evolutionary processes, it is generally advisable to use more than one method to analyze the data and evaluate consistency of results.

Phylogeny reconstruction approaches are either distance-based or character-based. In distance matrix methods, the distance between every

Table 3.1 Some commonly used programs and software packages for phylogenetic and phylogenomic analyses

Program/ Package	Brief description	Source	References
MrBayes	A Bayesian inference program for analysis molecular sequences using MCMC. It includes substitution models of nucleotide, amino acid, and codon for likelihood analysis	http://mrbayes.sourceforge.net	Ronquist et al. (2012)
BEAST	A Bayesian inference program for analysis molecular sequences using MCMC. It is orientated toward rooted, time-measured phylogenies inferred using strict or relaxed molecular clock models. Nucleotide, amino acid sequences, and morphological data can be analyzed	http://beast.bio.ed.ac.uk	Drummond et al. (2012)
RAxML	A model-based rapid and powerful program for maximum likelihood-based phylogenetic inference. It can be used both for nucleotide or amino acid sequences	http://sco.h-its.org/exelixis/web/software/raxml/index.html	Stamatakis (2014)
MEGA	It is an integrated tool for conducting sequence alignment, inferring phylogenetic trees, estimating divergence times, mining online databases, estimating rates of molecular evolution, inferring ancestral sequences, and testing evolutionary hypotheses. It includes distance, maximum parsimony, and maximum likelihood approaches for phylogenetic tree reconstruction	http://www.megasoftware.net/	Tamura et al. (2013)
GARLI	A program that uses genetic algorithms for rapid maximum likelihood inference. It implements nucleotide, amino acid, and codon-based models of sequence evolution and runs on all platforms. The latest version adds support for partitioned models and morphology-like data types	http://code.google.com/p/garli/	Bazinet et al. (2014)
PAUP*	PAUP* 4.0 beta is the latest beta release. It implements distance, maximum parsimony, and maximum likelihood methods of phylogeny reconstruction	http://paup.csit.fsu.edu	Swofford (2003)
PhyML	A model-based rapid program for maximum likelihood-based phylogenetic inference. It can be used both for nucleotide or amino acid sequences. The new version PhyML 3.1 implements aBayes branches support algorithm	http://www.atgc-montpellier.fr/phyml/versions.php	Guindon et al. (2010)
ExaBayes	A Bayesian tree inferences program for whole-genome analysis using MCMC. It is more suitable for large-scale analyses on computer clusters. At present, it only supports Linux and MacOS	http://sco.h-its.org/exelixis/web/software/exabayes/index.html	Aberer et al. (2014)
AWTY	It is a system for graphically exploring convergence of Markov Chain Monte Carlo (MCMC) chains in Bayesian phylogenetic inference	http://king2.scs.fsu.edu/CEBProjects/awty/awty_start.php	Nylander et al. (2008b)
Tracer	It is a program for analyzing the trace files generated by Bayesian MCMC runs	http://tree.bio.ed.ac.uk/software/tracer	Rambaut and Drummond (2007)
Mesquite	Reconstruction of ancestral states (MP, ML); tests of process of character evolution, including correlation. It is a program package and also has many features for managing and processing data, including processing of chromatograms, sequence alignment, editing of morphometric data, and others.	http://mesquiteproject.wikispaces.com/installation	Maddison and Maddison (2008)
SIMMAP	Bayesian analyses of trait evolution, stochastic character mapping, covariation between molecular or non-molecular characters, tests of positive selection, while accounting for model and tree uncertainty	http://www.simmap.com/	Bollback (2006)

(continued)

Table 3.1 (continued)

Program/ Package	Brief description	Source	References
BayesTraits	It performs several analyses related to evaluating evolutionary correlation and ancestral state reconstruction in discrete morphological traits. It uses ML and Bayesian inference	http://www.evolution.rdg.ac.uk/BayesTraits.html	Pagel et al. (2004)
jModelTest	It is a tool to carry out statistical selection of best-fit models of nucleotide substitution	http://code.google.com/p/jmodeltest2	Darriba et al. (2012)
Lagrange	It reconstruct geographic range evolution (ancestral areas, rates of dispersal/range expansion, and local extinction/range contraction) on phylogenetic trees	http://www.reelab.net/lagrange/configurator/index	Ree and Smith (2008)
BayArea	A Bayesian method to infer ancestral species ranges using a molecular phylogeny and presence–absence data. It allows the inclusion of hundreds to thousands of areas per analysis	https://sites.google.com/site/mlandis/main/bayareav10release	Landis et al. (2013)
RASP	It is a tool for inferring ancestral state (ancestral area reconstruction) using S-DIVA (Statistical dispersal-vicariance analysis), Lagrange (DEC), Bayes-Lagrange (S-DEC), BayArea, and BBM (Bayesian Binary MCMC) method	http://mnh.scu.edu.cn/soft/blog/RASP/	Yu et al. (2014)
FigTree	Plotting support values and branch lengths, confidence intervals on node times, creating summary graphics	http://tree.bio.ed.ac.uk/software/figtree	
CIPRES Portal	A web server for phylogenetic analyses	http://www.phylo.org/sub_sections/portal	
Lifeportal	A web server for phylogenetic analyses	https://lifeportal.uio.no/root	
Biology WorkBench	A web server access to a wide variety of analysis and modeling tools	http://workbench.sdsc.edu	
BioHPC	A web interfaces for various computational biology tools	http://cbsuapps.tc.cornell.edu/	
Hamstr	It is a profile hidden Markov model-based tool for a directed ortholog search in whole genome data (EST or protein sequence data). The program takes a predefined core group of orthologous sequences (core orthologs) and a set of sequences from a search taxon as input. HaMStR then combines in a two-step strategy a pHMM-based search and a reverse search via BLAST to extend the core ortholog group with novel sequences from the search taxon	http://www.deep-phylogeny.org/hamstr	Ebersberger et al. (2009)

Note BEAST Bayesian evolutionary analysis sampling trees, *MCMC* Markov Chain Monte Carlo, *MEGA* Molecular evolutionary genetic analysis, *GARLI* Genetic algorithm for rapid likelihood inference, *PAUP** Phylogenetic analysis using parsimony* and other methods, *MP* Maximum parsimony, *ML* Maximum likelihood, *RASP* Reconstruct ancestral state in phylogenies, *EST* Expressed sequence tag

Table 3.2 Strengths and weakness of commonly used phylogenetic reconstruction methods

Methods	Strengths	Weakness
Distance (neighbor joining)	Computationally fast and useful for obtaining preliminary phylogenetic estimates	Underestimates true number of substitutions because the method do not consider variation of distance estimates
	Substitution models can be selected to fit the data	It is sensitive to divergent sequences and missing data (large gaps in the alignment)
	Distances can be applied for nucleotide or amino acid datasets	
Maximum parsimony	Tree reconstruction is based on discrete characters rather than distances	Do not include substitution models that make impossible to incorporate any information of sequence evolution in tree reconstruction
	Simple and easy application in describing results	It could suffer from long-branch attraction
		Finding the best tree can sometime be an issue
		Underestimates branch lengths when substitution rates are high
Maximum likelihood	Based on explicit substitution models to understand sequence evolution	Sometimes, the interpretation of bootstrap values can be subjective
	Often fast and powerful approach for testing hypothesis	Poor statistical properties if the model is mis-specified
Bayesian inference	Based on explicit substitution model similar to maximum likelihood	Can be computationally intensive; large datasets require a high number of generations for runs to converge
	Simultaneous reconstruction of tree topology and branch support	Overestimates posterior probabilities
	Prior probability allows the inclusion of informations	Post-analysis evaluation necessary to determine the result accuracy, and analysis must possibly be re-run
	Interpretation of posterior probabilities straightforward	Mixing and convergence problem can be difficult to rectify
	Integration of phylogenetic uncertainty by sampling a large number of trees with similar probabilities	

pair of sequences is calculated, and the resulting distance matrix is used for tree reconstruction. Character-based methods include maximumparsimony, maximum likelihood, and Bayesian inference methods. These approaches simultaneously compare all sequences in the alignment, considering one character—a site in the alignment—at a time to calculate a score for each tree. The tree score is the minimum number of changes for maximum parsimony, the log-likelihood value for maximum likelihood, and the posterior probability for Bayesian inference. The tree with the best score should be identified by comparing all possible trees. These approaches often generate a starting tree using a fast algorithm and then perform local rearrangements to attempt to improve the tree score.

3.2.1 Distance Methods

Distance matrix methods for phylogenetic analysis are based on genetic distances calculation between every pair of sequences in the alignment. Pairwise genetic distances are calculated assuming a Markov chain model of nucleotide substitution. The GTR, HKY85, JC69, and K80 are commonly used substitution models for computing evolutionary distances. The Jukes–Cantor model (JC69, Jukes and Cantor 1969) is the simplest model that assumes an equal rate of substitution between any two nucleotide base, while the K80 model (Kimura 1980) assumes different rates for transitions and transversions. The general time-reversible model, called the GTR model (Tavaré 1986), has six mutation rate parameters, and assumption of equal base frequencies is relaxed. In distance calculation for phylogenetic analysis, mutation rate variation is accommodated by assuming a gamma (Γ) distribution of rates for sites, leading to models such as JC69 + Γ, HKY85 + Γ, or GTR + Γ. It is highly advisable to use a realistic substitution model to calculate the pairwise distances because it could be crucial depending on the DNA sequence dataset.

Assuming trees from distance data using algorithmic methods is computationally very fast and efficient and therefore commonly used, obtaining a preliminary estimate of phylogeny. The most widely used algorithm is neighbor joining (Saitou and Nei 1987), and an efficient implementation is found in the program MEGA6 (Tamura et al. 2013, Table 3.1). This is a cluster algorithm and operates by starting with a star tree and successively choosing a pair of taxa to join together—based on the taxon distances—until a fully resolved tree is obtained. The unweighted pair-group method with arithmetic averages (UPGMA) and the optimization-based distance methods are occasionally used. The neighbor joining NJ is useful for analyzing large dataset that have low levels of sequence divergence. However, pairwise distances are known to significantly underestimate the true number of substitutions, which limits their accuracy at deeper timescales. Further, distance methods are sensitive to gaps in the sequence alignment, and poorly perform for very divergent sequences. Therefore, it is recommendable supporting evolutionary hypothesis on trees obtained from either maximum likelihood or maximum parsimony or Bayesian inference (see discussion below).

3.2.2 Maximum Parsimony

The maximum parsimony (MP) is a method of computing the phylogenetic tree that requires the smallest number of changes or evolutionary events to explain the observed sequence data. The character length is the minimum number of changes required for that site, while the tree score is the sum of character lengths over all sites. The maximum parsimony tree is the tree that minimizes the tree score. Some sites are not useful for tree comparison by parsimony as constant sites (the same nucleotide occurs in all species). The parsimony-informative sites are those at which at least two distinct characters are observed, each at least twice. These sites are useful for tree comparison by parsimony.

Number of heuristics search has been developed such as branch swapping, genetic algorithms, and simulated annealing. The most commonly used algorithm is branch swapping, where an initial tree estimate—often generated by a process of stepwise addition of sequences—is bisected into subtrees, which are then grafted on one another, with varying limitations on the possible rearrangements. It is advisable having the fewer the limitations because it provides better search. For longer search times, it is wise employing the least restrictive algorithm, called 'tree bisection and reconnection.' The obtained results of the search can be summarized using 'consensus tree' in a strict consensus, nodes present in all trees; or a majority rule consensus, nodes shared by a majority of trees. PAUP (Swofford 2003), MEGA6, and TNT (Goloboff et al. 2008) are widely used parsimony programs. Parsimony is still commonly used because it is computationally efficient and often produces reasonable results. A drawback of parsimony is that it lacks explicit assumptions making it nearly impossible to include any knowledge of the process of sequence evolution in tree reconstruction. This makes the parsimony suffering from a problem called 'long-branch attraction' (see Swofford et al. 2001). Thus, it is advisable using model-based methods such as maximum likelihood or Bayesian inference (see Sects. 3.2.3 and 3.2.4). It is worth noting that the model-based methods can also suffer from long-branch attraction if the realistic substitution models have not been selected.

3.2.3 Maximum Likelihood

The maximum likelihood estimates of parameters are the parameter values that maximize the likelihood. The method requires a substitution model to assess the probability of particular mutations. For example, a tree that requires more mutations at interior nodes to explain the observed phylogeny will be assessed as having a lower probability. This method is largely similar to the maximum parsimony method; however, maximum likelihood allows additional statistical flexibility by permitting varying rates of evolution across both lineages and sites. The method is commonly used due to the increased computing power and software implementations but also to the development of increasingly realistic models of sequence evolution. There are two optimization steps in maximum likelihood tree search: (i) optimization of branch lengths to calculate the tree score for each candidate tree and (ii) a search in the tree space for the maximum likelihood tree.

Maximum likelihood requires selecting a model of nucleotide or amino acid evolution prior to the analysis. This also applies for Bayes analysis (see Sect. 3.2.4). Most models used in molecular phylogenetics assume independent evolution of sites in the sequence, so that the likelihood is a product of the probabilities for different sites. The model that could be different for each alignment can be selected using various criteria implemented in a number of programs as ModelTest (Posada and Crandall 1998) or MrModeltest (Nylander 2004) and jModelTest (Posada 2008), for nucleotide data. The latter is the commonly used program. For amino acid dataset, the model can be selected using the program, e.g., ProtTest (Darriba et al. 2011). The availability of wide range of sophisticated evolutionary models in the likelihood and Bayesian method is one of its main advantages over maximum parsimony. Inferences of phylogenies using conserved proteins almost exclusively rely on likelihood and Bayesian methods. A drawback of maximum likelihood is that the method has poor statistical properties if the selected model is inappropriate. This is also applied for Bayesian analysis (see Table 3.2).

Support for relationships estimated by maximum likelihood, parsimony (Sect. 3.2.2) and distance (Sect. 3.2.1) methods can be assessed by re-sampling primarily jackknifing (Farris et al. 1996) or by bootstrapping (Felsenstein 1985). The latter is widely used. Each resampled dataset can be reanalyzed using the same methods as the original data, and the trees saved from each analysis. Finally, a majority rule consensus of these trees—the bootstrap or jackknife tree—is constructed. Bootstrap support values equal or greater than 70 % are generally considered 'strong' support (see Hillis and Bull 1993).

The maximum likelihood method is implemented in the programs PHYLIP, MOLPHY, PAUP* 4.0, PhyML, RAxML, GARLI, and MEGA6 (Table 3.1). The recent implementations in RAxML, PhyML, and GARLI are computationally much faster and also are more effective in finding trees with high likelihood scores (see Guindon and Gascuel 2003; Stamatakis 2006, 2014; Zwickl 2006; Stamatakis et al. 2008). These recent inclusions have made the method more accessible to biologists who are not experienced computer users.

3.2.4 Bayesian Methods

Bayesian methods assume a *prior distribution* (probability) of the possible trees that could simply be the probability of any one tree among all the possible trees generated from the data. Before the analysis of the sequence or amino acid data, parameters are assigned a *prior distribution* that is combined with the likelihood to generate the *posterior distribution*. Bayesian inference produces phylogenetic trees in a similar manner to the maximum likelihood methods. While likelihood methods find the tree that maximizes the probability of the data, a Bayesian method recovers a tree, which represents the most likely clades, by sketching on the *posterior distribution*. Bayesian inference uses Markov Chain Monte Carlo algorithms (MCMC algorithms, Larget and Simon 1999), and due to this, the inference gained popularity in the last two decades. The release of the program MrBayes (Huelsenbeck and Ronquist 2001; Ronquist et al. 2012) made the Bayesian inference more popular among the biologists and the implementation in the program BEAST (Drummond et al. 2006), which uses relaxed-clock models to infer the rooted trees allow the biologist to estimate age of the lineages without fossil records. Another reason that made the inference popular is that it yields both a phylogenetic hypothesis and measures of support for a given dataset, which can be completed much more rapidly than a single likelihood tree search (i.e., without bootstrapping) for the same data.

Markov Chain Monte Carlo (MCMC) algorithms estimate the posterior probability for phylogenetic trees starting from: a random tree and parameter values, and a chain—a set of ≥4 incrementally heated chains—begins to wander through the space of all possible trees and parameter values, always accepting changes that lead to higher likelihood values, but also accepting changes that decrease the likelihood with finite probability. After running generations (e.g., 10^6–10^7), the summary of chain states assembled represents a valid estimate of the posterior distribution. It is advisable to evaluate the adequacy of the sampled generations and parameters with the program, e.g., Tracer (Rambaut and Drummond 2007) or AWTY (Nylander et al. 2008b). The analyses of convergence among runs, mixing, and effective sample sizes are critical adjuncts to Bayesian MCMC analyses. It is therefore important to conduct robustness analysis to assess the impact of the prior on the posterior estimates. It is advisable reaching effective sample size equal or above 200. The result of a Bayesian analysis is a collection of trees, which can be summarized, e.g., in consensus using 'sumt' option in MrBayes. Posterior probabilities values equal or above 95 % are generally considered 'strong' support. It should be noted that Bayesian inference overestimates nodal posterior probabilities, and therefore, it is advisable to compare the results with any bootstrap method, e.g., maximum likelihood (Susko 2008). Bayesian methods are generally held to be superior to parsimony-based methods; they can be more prone to long-branch attraction than maximum likelihood techniques (see Kolaczkowski and Thornton 2009).

3.2.4.1 Divergence Time Estimates

The low number of published studies on the timing of diversification events in lichen-forming fungi is mainly due to the poor fossil record for fungi in general, including lichen-forming groups, and uncertainties in the interpretation of the few known fossil records. The advent of DNA sequencing technologies and advances in molecular phylogenetic methods have made it possible to estimate divergence dates from molecular genetic data with increasing levels of accuracy in lichens (see e.g., Lücking et al. 2009; Berbee and Taylor 2010; Amo de Paz et al. 2011; Sérusiaux et al. 2011; Divakar et al. 2012; Leavitt et al. 2012; Del-Prado et al. 2013; Prieto and Wedin 2013; Beimforde et al. 2014).

The molecular dating analysis can be done using Bayesian relaxed molecular clock model implemented in the program, e.g., BEAST (Drummond et al. 2006, 2012). Under the clock assumption, the distance between sequences increases linearly with the time of divergence, and if a particular divergence can be assigned an absolute geological age based on the fossil record, the substitution rate can be calculated, and all divergences on the tree can be dated. Similar ideas can be used to estimate divergence times of lichen-forming fungi that have poor fossil records. In the last decade, advancements have been made using the Bayesian framework to increase the accuracy level on divergence time estimates from molecular genetic data. As models of evolutionary rate drift over time have been developed to relax the molecular clock (Rannala and Yang 2007), soft age bounds and flexible probability distributions have been implemented to accommodate uncertainties in fossil calibrations (Inoue et al. 2010). The fossil record has also been statistically analyzed to generate calibration densities for molecular dating analysis (Tavaré et al. 2002). For the divergence time estimates, it is advisable to a user-specified chronogram as the starting tree, rather than a randomly generated tree. Moreover, the lognormal distribution has been shown to be the most appropriate for modeling paleontological information because lineage origination should not postdate the fossil occurrence (Ho and Phillips 2009; Divakar et al. 2012).

3.3 Ancestral State Reconstruction

Phylogenetic analyses based on molecular data have provided a frame of reference for studying the evolution of characters and for inferring the type of traits in the organisms including lichens. The multiple origins of lichen symbiosis have been identified, and it has been suggested that some groups of fungi, including Eurotiomycetes, are derived from lichenized ancestors (Lutzoni et al. 2001; Schoch et al. 2009a). Furthermore, it has shown that the likelihood of the loss of lichenization is somewhat more frequent in evolution than the acquisition of that character. Molecular phylogenies are increasing in studying trait evolution, with numerous examples from lichenized fungi (see e.g., Divakar et al. 2006, 2013; Crespo et al. 2007; Gueidan et al. 2007; Schmitt et al. 2009; Schoch et al. 2009a; Baloch et al. 2010; Otálora et al. 2010, 2013; Prieto et al. 2013).

Ancestral state reconstructions of morphological characters on phylogenies can be done using the principles of parsimony, maximum likelihood, and Bayesian inference. The programs MacClade and or Mesquite are ideal for maximum parsimony analysis. However, maximum likelihood and Bayesian inference can be done using the programs Mesquite, BayesTraits, and SIMMAP (Pagel 1999; Lewis 2001; Pagel et al. 2004; see Table 3.1). Reconstructions can be performed either on a single tree such as a maximum parsimony or a maximum likelihood tree or on multiple trees, e.g., a Bayesian posterior tree sample. In contrast to single-tree reconstructions, the Bayesian approach has the advantage of taking uncertainty in the tree topology and branch lengths. However, ancestral state reconstruction on each of the trees in a posterior tree sample can be performed under either the maximum parsimony criterion, or the maximum likelihood criterion, or using a Bayesian approach (see Pagel 1999; Pagel et al. 2004; Ekman et al. 2008; Divakar et al. 2013). Note that only fully Bayesian inferences take phylogenetic uncertainty and uncertainty in the state reconstruction into account. There are two slightly different Bayesian approaches available, by Huelsenbeck and Bollback (2001), by Pagel et al. (2004), and by Pagel and Meade (2006) (see Ekman et al. 2008; Divakar et al. 2013 for more discussion).

While ancestral character state reconstruction methods provide powerful tools for assessing trait evolution, the obtained results should be critically evaluated. Studies have shown that different methods may provide diverging results (Ekman et al. 2008; Divakar et al. 2013). Moreover, coding of characters, either binary or multistate, may affect the outcome of ancestral state reconstructions (Hibbett 2004; Divakar et al. 2013). Binary character coding method provides better results than multistate (see Divakar et al. 2013). For example, Fig. 3.1 shows the evolution of growth forms on a phylogeny of parmelioid lichen-forming fungi using maximum parsimony, maximum likelihood, and Bayesian approach (see Divakar et al. 2013 for detail discussion). It is advisable using binary character coding method and more than one approach to base the evolutionary hypothesis on character state.

3.4 Ancestral Area Reconstruction

Inference of biogeographic history in lichens is challenging because they have generally larger distribution ranges than most vascular plants.

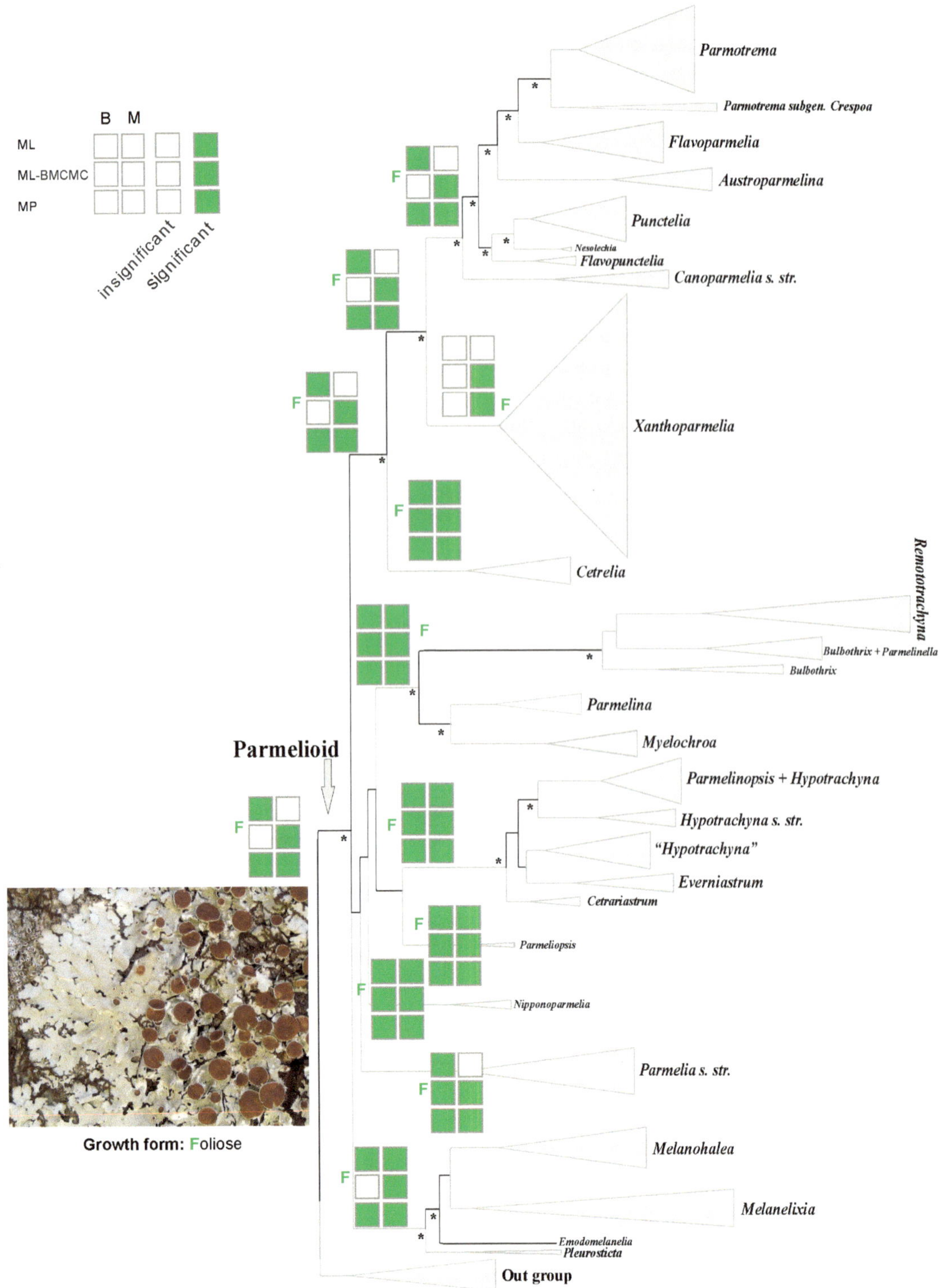

Fig. 3.1 Example of ancestral state reconstruction of growth forms on phylogeny of parmelioid lichen-forming fungi, taken from Divakar et al. (2013). Binary and multistate coding datasets analyzed with ML, ML-BMCMC, and MP approaches. *B* Binary coding, *M* Multistate coding, *F* Foliose

Further, various factors may influence species range such as geological, climatic, and ecological events. However, both the diversity of factors influencing the geographic range of a species and the uncertainty regarding their relative importance motivate pursuit of biogeographic inference within a statistical framework. Lichen biogeography is recently becoming a dynamic field of study in which distribution patterns are analyzed in a phylogenetic framework, using statistical methods to address ancestral range evolution of clades (Lücking et al. 2008; Divakar et al. 2010; Geml et al. 2010; Amo de Paz et al. 2012; Del-Prado et al. 2013).

Ancestral area reconstructions on phylogenies can be done using the principles of parsimony, maximum likelihood, and Bayesian inference. Dispersal-vicariance analysis is implemented in the program DIVA (Ronquist 1997). This is a parsimony 'event-based' biogeographical method that optimizes ancestral areas onto the internal nodes of a phylogeny by minimizing the number of duplication and extinction events required to explain the terminal distributions (Ronquist 2003). Model-based approach as maximum likelihood and Bayesian inference is implemented in the programs LaGrange (Ree et al. 2005; Ree and Smith 2008), Bayes-DIVA (Nylander et al. 2008a), S-DIVA (Yu et al. 2010), and BayArea (Landis et al. 2013). These approaches account for phylogenetic uncertainty and allow a more accurate analysis of the biogeographic history of lineages. The method proposed by Ree et al. (2005) and Ree and Smith (2008) based on stochastic models (LaGrange) is one of the most widely used methods of inferring biogeographic histories in lichen-forming fungi. However, DIVA has been widely used in vascular plants. LaGrange is limited to about ten to twenty areas. A recently developed program BayArea (Landis et al. 2013) claims for analyzing hundreds or thousands of areas. Given that wide distribution range of lichen-forming fungal species, the BayArea seems one of the most appropriate approaches for inferring biogeographic history.

3.5 Phylogenomics, a Future of Lichen Systematics

While phylogenetics compares and analyzes the sequences of single genes, or a small number of genes, the phylogenomics illustrates information by comparing entire genomes, or at least large portions of genomes. Traditional single-gene studies are more effective in establishing phylogenetic relations among closely related taxa/ organisms. However, using phylogenomic approach, it could theoretically be possible to create fully resolved phylogenetic trees.

Molecular phylogenies based on fragment of genes so-called *traditional single-gene* have revolutionized our understanding of the evolution of lichen-forming fungi. There is no doubt that these studies have made enormous advance in lichen systematics and evolution. Nuclear ribosomal rDNA gene fragments as ITS, nuLSU, nuSSU, mitochondrial rDNA mtSSU, mtLSU, and protein coding genes RPB1, RPB2, EF alpha 1, Beta-tubulin, GPD, Mcm7, and Tsr1 are the most widely used markers to resolve the relationships in lichen-forming fungi. While phylogenetic relationships in lichen-forming fungi are widely resolved, the deep level node relationships lacked strong statistical support (see e.g., Lutzoni et al. 2004; James et al. 2006; Spatafora et al. 2006; Schoch et al. 2009b; Crespo et al. 2010; Ertz and Tehler 2011; Rivas Plata et al. 2012; Miadlikowska et al. 2014). More recently, phylogenomic approaches have shown to be outperformed resolving deep level relationship in other organism including fungi (see e.g., Soltis et al. 2011; Ebersberger et al. 2012; Shen et al. 2013; Ampio et al. 2014). This could be a best approach to resolving deep level relationships among the major groups in lichen-forming fungi.

3.5.1 Next Generation Sequencing

Genomic data from lichenized fungi can be obtained using next generation sequencing (NGS) techniques using 454 (http://454.com) or Illumina (http://technology.illumina.com/technology), Life technologies. Detailed methods for obtaining whole genome sequencing can be found in Chap. 4 (Werth et al. 2015).

3.5.2 Phylogenomic Inference

The phylogenomic inference is of two kinds: (i) primary sequence-based methods that are similar to the classical tree reconstruction and (ii) methods based on whole genome features. Both approaches are very similar except the characters used.

3.5.2.1 Methods Based on Whole Genome Features

This approach is sensitive to hidden paralogy, horizontal gene transfer, or tree reconstruction artefacts, and thus, it is important first to search the orthologous genes. The detailed discussion can be found in Philippe et al. (2005). The orthology and copy number in whole genome can be assessed using HaMStR (see Ebersberger et al. 2009). This is an efficient and reliable approach for automated ortholog predictions in genomic data. HaMStR combines a profile Hidden Markov Model search and a subsequent BLAST search to extend existing ortholog cluster with sequences from further taxa. The approach has been shown to be appropriate for fungal phylogeny (see Ebersberger et al. 2012).

3.5.2.2 Primary Sequence-based Approaches

This is also known as *supermatrix and supertree methods* and has been supported for the phylogenetic analysis of hundreds or thousands of genes or proteins, especially when some loci are missing in some species. In supermatrix method, sequences for multiple genes are concatenated to generate a data supermatrix, in which missing data are replaced by question marks, and then, the matrix can be used for tree reconstruction (see de Queiroz and Gatesy 2007). Most supermatrix analyses ignore differences in evolutionary dynamics among the genes. The supermatrix analysis of all genes can be combined using likelihood to accommodate the among gene heterogeneity in the evolutionary process, as described in Ren et al. (2009). The dataset can be partitioned into group genes or sites with similar evolutionary characteristics into the same partition using model tests (e.g., ProtTest or jModelTest). Supertree method separately analyzes each gene and then uses heuristic algorithms to assemble the subtrees for individual genes into a supertree for all species (Bininda-Emonds 2004). The separate analysis could be useful for studying the horizontal gene transfer.

3.6 Concluding Remarks

In lichenized fungi, the advent of DNA sequence technologies and advances in molecular phylogenetic approaches have revolutionized our understanding on systematics and evolution. Based on molecular phylogenies, a number of taxonomic changes have been and are being made. It is advisable basing the evolutionary hypothesis more than one phylogenetic approach. Maximum likelihood (RAxML) and Bayesian inference have shown to be fast and also provide more reliable results. For ancestral area reconstruction, contrasting the results with more than one approach is highly recommended. Moreover, binary coding methods provide better results than multistate coding. Phylogenomic methods could be a most appropriate approach to produce a fully resolved phylogenetic tree of lichen-forming fungi.

Acknowledgments This work was supported by the Spanish Ministerio de Ciencia e Innovación (projects CGL2010-21646/BOS and CGL2013-42498-P), the

Universidad Complutense-Banco Santander (GR 35/10A), and Comunidad Autónoma de Madrid (REMEDINAL S-2009/AMB-1783).

References

Aberer AJ, Kobert K, Stamatakis A (2014) ExaBayes: massively parallel Bayesian tree inference for the whole-genome era. Mol Biol Evol (Advance Access published August 18)

Amo de Paz G, Cubas P, Divakar PK, Lumbsch HT, Crespo A (2011) Origin and diversification of major clades in parmelioid lichens (Parmeliaceae, Ascomycota) during the Paleogene inferred by Bayesian analysis. PLoS ONE 6:e28161

Amo de Paz G, Cubas P, Crespo A, Elix JA, Lumbsch HT (2012) Transoceanic dispersal and subsequent diversification on separate continents shaped diversity of the *Xanthoparmelia pulla* group (Ascomycota). PLoS ONE, e39683

Ampio E, Meusemann K, Szucsich NU et al (2014) Decisive data sets in phylogenomics: lessons from studies on the phylogenetic relationships of primarily wingless insects. Mol Biol Evol 31:239–249

Baloch E, Lücking R, Lumbsch HT, Wedin M (2010) Major clades and phylogenetic relationships between lichenized and non-lichenized lineages in Ostropales (Ascomycota: Lecanoromycetes). Taxon 59:1483–1494

Bazinet AL, Zwickl DJ, Cummings MP (2014) A gateway for phylogenetic analysis powered by grid computing featuring GARLI 2.0. Syst Biol 63(5):812–818

Beimforde C, Feldberg K, Nylinder S et al (2014) Estimating the Phanerozoic history of the Ascomycota lineages: combining fossil and molecular data. Mol Phylogenet Evol 78:386–398

Berbee ML, Taylor JW (2010) Dating the molecular clock in fungi—how close are we? Fungal Biol Rev 24:1–16

Bininda-Emonds ORP (2004) Phylogenetic supertrees: combining information to reveal the tree of life. Kluwer Academic, The Netherlands

Bollback JP (2006) SIMMAP: stochastic character mapping of discrete traits on phylogenies. BMC Bioinform 7:88

Brady A, Salzberg S (2011) Phymm BL expanded: confidence scores, custom databases, parallelization and more. Nat Methods 8:367

Crespo A, Lumbsch HT, Mattsson JE, Blanco O, Divakar PK, Articus K, Wiklund E, Bawingan PA, Wedin M (2007) Testing morphology-based hypotheses of phylogenetic relationships in Parmeliaceae (Ascomycota) using three ribosomal markers and the nuclear RPB1 gene. Mol Phylogenet Evol 44:812–824

Crespo A, Kauff F, Divakar PK et al (2010) Phylogenetic classification of parmelioid lichens (Parmeliaceae, Ascomycota) based on molecular, morphological and chemical evidence. Taxon 59:1735–1753

Crespo A, Divakar PK, Hawksworth DL (2011) Generic concepts in parmelioid lichens, and the phylogenetic value of characters used in their circumscription. Lichenologist 43:511–535

Darriba D, Taboada GL, Doallo R, Posada D (2011) ProtTest 3: fast selection of best-fit models of protein evolution. Bioinformatics 27:1164–1165

Darriba D, Taboada GL, Doallo R, Posada D (2012) jModelTest 2: more models, new heuristics and parallel computing. Nat Methods 9(8):772

de Queiroz A, Gatesy J (2007) The supermatrix approach to systematics. Trends Ecol Evol 22:34–41

Del-Prado R, Blanco O, Lumbsch HT, Divakar PK, Elix JA, Molina MC, Crespo A (2013) Molecular phylogeny and historical biogeography of the lichen-forming fungal genus Flavoparmelia (Ascomycota: Parmeliaceae). Taxon 62:928–939

DePriest P (2004) Early molecular investigations of lichen-forming symbionts: 1986–2001. Annu Rev Microbiol 58:273–301

Divakar PK, Crespo A, Blanco O, Lumbsch HT (2006) Phylogenetic significance of morphological characters in the tropical Hypotrachyna clade of parmelioid lichens (Parmeliaceae, Ascomycota). Mol Phylogenet Evol 40:448–458

Divakar PK, Lumbsch HT, Ferencova Z, Del Prado R, Crespo A (2010) *Remototrachyna*, a newly recognized tropical lineage of lichens in the Hypotrachyna clade (Parmeliaceae, Ascomycota), originated in the Indian subcontinent. Am J Bot 97:579–590

Divakar PK, Del-Prado R, Lumbsch HT, Wedin M, Esslinger TL, Leavitt SD et al (2012) Diversification of the newly recognized lichen-forming fungal lineage Montanelia (Parmeliaceae, Ascomycota) and its relation to key geological and climatic events. Am J Bot 99:2014–2026

Divakar PK, Kauff F, Crespo A, Leavitt SD, Lumbsch HT (2013) Understanding phenotypical character evolution in parmelioid lichenized fungi (Parmeliaceae, Ascomycota). PLoS ONE 8(11):e83115

Drummond AJ, Ho SYW, Phillips MJ, Rambaut A (2006) Relaxed phylogenetics and dating with confidence. PLoS Biol 4:e88

Drummond AJ, Suchard MA, Xie D, Rambaut A (2012) Bayesian phylogenetics with BEAUti and the BEAST 1.7. Mol Biol Evol 29:1969–1973

Ebersberger I, Strauss S, von Haeseler A (2009) HaMStR: profile hidden markov model based search for orthologs in ESTs. BMC Evol Biol 9:e157

Ebersberger I, de Matos Simoes R, Kupczok A, Gube M, Voigt K, Kothe E, von Haeseler A (2012) A consistent phylogenetic backbone for the fungi. Mol Biol Evol 29:1319–1334

Ekman S, Andersen HL, Wedin M (2008) The limitations of ancestral state reconstruction and the evolution of the ascus in the Lecanorales (lichenized Ascomycota). Syst Biol 57:141–156

Ertz D, Tehler A (2011) The phylogeny of Arthoniales (Pezizomycotina) inferred from nucLSU and RPB2 sequences. Fungal Divers 49(1):47–71

Farris JS, Albert VA, Kallersjo M, Lipscomb D, Kluge AG (1996) Parsimony jackknifing outperforms neighbour-joining. Cladistics 12:99–124

Felsenstein J (1985) Confidence-limits on phylogenies—an approach using the bootstrap. Evolution 39:783–791

Gargas A, DePriest PT, Grube M, Tehler A (1995) Multiple origin of lichen symbioses in fungi suggested by SSU rDNA phylogeny. Science 268:1492–1495

Geml J, Kauff F, Brochmann C, Taylor DL (2010) Surviving climate changes: high genetic diversity and transoceanic gene flow in two arctic–alpine lichens, *Flavocetraria cucullata* and *F. nivalis* (Parmeliaceae, Ascomycota). J Biogeogr 37:1529–1542

Goloboff PA, Farris JS, Nixon KC (2008) TNT, a free program for phylogenetic analysis. Cladistics 24: 774–786

Gueidan C, Roux C, Lutzoni F (2007) Using a multigene phylogenetic analysis to assess generic delineation and character evolution in Verrucariaceae (Verrucariales, Ascomycota). Mycol Res 111:1145–1168

Guindon S, Gascuel O (2003) A simple, fast, and accurate algorithm to estimate large phylogenies by maximum likelihood. Syst Biol 52:696–704

Guindon S, Dufayard JF, Lefort V, Anisimova M, Hordijk W, Gascuel O (2010) New algorithms and methods to estimate maximum-likelihood phylogenies: assessing the performance of PhyML 3.0. Syst Biol 59(3):307–321

Hibbett DS (2004) Trends in morphological evolution in homobasidiomycetes inferred using maximum likelihood: a comparison of binary and multistate approaches. Syst Biol 53:889–903

Hibbett DS, Binder M, Bischoff JF et al (2007) A higher-level phylogenetic classification of the fungi. Mycol Res 111:509–547

Hillis DM, Bull JJ (1993) An emperical test of bootstrapping as a method for assessing confidence in phylogenetic analysis. Syst Biol 42:182–192

Ho SYW, Phillips MJ (2009) Accounting for calibration uncertainty in phylogenetic estimates of evolutionary divergence times. Syst Biol 58:367–380

Huelsenbeck JP, Bollback JP (2001) Empirical and hierarchical Bayesian estimation of ancestral states. Syst Biol 50:351–366

Huelsenbeck JP, Ronquist F (2001) MrBayes: Bayesian inference of phylogenetic trees. Bioinformatics 17: 754–755

Inoue J, Donoghue PCH, Yang Z (2010) The impact of the representation of fossil calibrations on Bayesian estimation of species divergence times. Syst Biol 59:74–89

James TY, Kauff F, Schoch C et al (2006) Reconstructing the early evolution of the fungi using a six-gene phylogeny. Nature 443:818–822

Jukes TH, Cantor CR (1969) In mammalian protein metabolism. In: Munro HN (ed) Academic Press, New York, pp 21–123

Kimura M (1980) A simple method for estimating evolutionary rate of base substitution through comparative studies of nucleotide sequences. J Mol Evol 16:111–120

Kolaczkowski B, Thornton JW (2009) Long-branch attraction bias and inconsistency in Bayesian Phylogenetics. PLoS ONE 4(12):e7891

Landis MJ, Matzke NJ, Moore BR, Huelsenbeck JP (2013) Bayesian analysis of biogeography when the number of areas is large. Syst Biol 62:789–804

Larget B, Simon DL (1999) Markov chain Monte Carlo algorithms for the Bayesian analysis of phylogenetic trees. Mol Biol Evol 16:750–759

Leavitt SD, Esslinger TL, Divakar PK, Lumbsch HT (2012) Miocene and Pliocene dominated diversification of the lichen-forming fungal genus Melanohalea (Parmeliaceae, Ascomycota) and Pleistocene population expansions. BMC Evol Biol 12:176

Lewis PO (2001) A likelihood approach to estimating phylogeny from discrete morphological character data. Syst Biol 50:913–925

Lücking R, Papong K, Thammathaworn A, Boonpragob K (2008) Historical biogeography and phenotype-phylogeny of Chroodiscus (lichenized Ascomycota: Ostropales: Graphidaceae). J Biogeogr 35:2311–2327

Lücking R, Huhndorf S, Pfister DH, Plata ER, Lumbsch HT (2009) Fungi evolved right on track. Mycologia 101:810–822

Lumbsch HT (2007) Recent trends in phylogeny and classification of lichen-forming ascomycetes. In: Ganguli BN, Deshmukh SK (eds) Fungi: multifaceted microbes. Anamaya, Delhi, pp 153–168

Lumbsch HT, Huhndorf SM (2010) Myconet volume 14. Part one. Outline of Ascomycota—2009. Part two. Notes on ascomycete systematics. Nos. 4751–5113. Fieldiana Life Earth Sci. 1:1–64

Lutzoni F, Pagel M, Reeb V (2001) Major fungal lineages are derived from lichen symbiotic ancestors. Nature 411:937–940

Lutzoni F, Kauff F, Cox C et al (2004) Assembling the fungal tree of life: progress, classification, and evolution of subcellular traits. Am J Bot 91:1446–1480

Maddison WP, Maddison DR (2008) Mesquite: a modular system for evolutionary analysis. Version 2.5. http://mesquiteproject.org

Miadlikowska J, Kauff F, Högnabba F et al (2014) Multigene phylogenetic synthesis for 1307 fungi representing 1139 infrageneric taxa, 312 genera and 66 families of the class Lecanoromycetes (Ascomycota). Mol Phylogenet Evol 79:132–168

Nylander JAA (2004) MrModeltest v2, Program distributed by the author. Evolutionary Biology Centre, Uppsala University, Sweden

Nylander JAA, Olsson U, Alström P, Sanmartín I (2008a) Accounting for phylogenetic uncertainty in biogeography: a Bayesian approach to dispersal-vicariance Analysis of the thrushes (Aves: Turdus). Syst Biol 57:257–268

Nylander JAA, Wilgenbusch JC, Warren DL, Swofford DL (2008b) AWTY (are we there yet?): a system for graphical exploration of MCMC convergence in Bayesian phylogenetics. Bioinformatics 24:581–583

Otálora GA, Molina MC, Martínez I, Lutzoni F (2010) Disentangling the Collema-Leptogium complex through a molecular phylogenetic study of the Collemataceae (Peltigerales, lichen-forming Ascomycota). Mycologia 102(2):279–290

Otálora MA, Aragón G, Martínez I, Wedin M (2013) Cardinal characters on a slippery slope—a re-evaluation of phylogeny, character evolution, and evolutionary rates in the jelly lichens (*Collemataceae* s. str). Mol Phylogenet Evol 68:185–198

Pagel M (1999) The maximum likelihood approach to reconstructing ancestral character states of discrete characters on phylogenies. Syst Biol 48:612–622

Pagel M, Meade A (2006) Bayesian analysis of correlated evolution of discrete characters by reversible-jump Markov chain Monte Carlo. Am Nat 167:808–825

Pagel M, Meade A, Barker D (2004) Bayesian estimation of ancestral character states on phylogenies. Syst Biol 53:673–684

Paten B et al (2008) Genome-wide nucleotide-level mammalian ancestor reconstruction. Genome Res 18:1829–1843

Philippe H, Delsuc F, Brinkmann H, Lartillot N (2005) Phylogenomics. Annu Rev Ecol Evol Syst 36:541–562

Posada D (2008) jModelTest: phylogenetic model averaging. Mol Biol Evol 25:1253–1256

Posada D, Crandall KA (1998) Modeltest: testing the model of DNA substitution. Bioinformatics 14:817–818

Prieto M, Wedin M (2013) Dating the diversification of the major lineages of Ascomycota (fungi). PLoS ONE 8:e65576

Prieto M, Baloch E, Tehler A, Wedin M (2013) Mazaedium evolution in the Ascomycota (fungi) and the classification of mazaediate groups of formerly unclear relationship. Cladistics 29:296–308

Printzen C (2010) Lichen systematics: the role of morphological and molecular data to reconstruct phylogenetic relationships. In: Lüttge U, Beysch W, Büdel B, Francis D (eds) Progress in botany, vol 71, pp 233–279

Rambaut A, Drummond AJ (2007) Tracer v1.5. Available from http://beast.bio.ed.ac.uk/Tracer

Rannala B, Yang Z (2007) Inferring speciation times under an episodic molecular clock. Syst Biol 56:453–466

Ree RH, Smith SA (2008) Maximum-likelihood inference of geographic range evolution by dispersal, local extinction, and cladogenesis. Syst Biol 57:4–14

Ree RH, Moore BR, Webb CO, Donoghue MJ (2005) A likelihood framework for inferring the evolution of geographic range on phylogenetic trees. Evolution 59:2299–2311

Ren F, Tanaka H, Yang Z (2009) A likelihood look at the supermatrix–supertree controversy. Gene 441:119–125

Rivas Plata E, Lücking R, Lumbsch HT (2012) A new classification for the family Graphidaceae (Ascomycota: Lecanoromycetes: Ostropales). Fung. Divers 52:107–121

Ronquist F (1997) Dispersal-vicariance analysis: a new approach to the quantification of historical biogeography. Syst Biol 46:195–203

Ronquist F (2003) Parsimony analysis of coevolving species associations. In: Page RDM (ed) Tangled trees: phylogeny, cospeciation, and coevolution. University of Chicago Press, Chicago, pp 22–64

Ronquist F, Teslenko M, van der Mark P, Ayres DL, Darling A, Höhna S, Larget B, Liu L, Suchard MA, Huelsenbeck JP (2012) MrBayes 3.2: efficient Bayesian phylogenetic inference and model choice across a large model space. Syst Biol 61:539–542

Saitou N, Nei M (1987) The neighbor-joining method: a new method for reconstructing phylogenetic trees. Mol Biol Evol 4:406–425

Schmitt I, del Prado R, Grube M, Lumbsch HT (2009) Repeated evolution of closed fruiting bodies is linked to ascoma development in the largest group of lichenized fungi (Lecanoromycetes, Ascomycota). Mol Phylogenet Evol 52:34–44

Schoch C, Sung GH, López-Giráldez F et al (2009a) The Ascomycota tree of life: a phylum-wide phylogeny clarifies the origin and evolution of fundamental reproductive and ecological traits. Syst Biol 58:224–239

Schoch CL, Crous PW, Groenewald JZ et al (2009b) A class-wide phylogenetic assessment of Dothideomycetes. Stud Mycol 64:1–15

Sérusiaux E, Villarreal JC, Wheeler T, Goffinet B (2011) Recent origin, active speciation and dispersal for the lichen genus Nephroma (Peltigerales) in Macaronesia. J Biogeogr 38:1138–1151

Shen X, Liang D, Feng J, Chen M, Zhang P (2013) A versatile and highly efficient toolkit including 102 nuclear markers for Vertebrate phylogenomics, tested by resolving the higher level relationships of the Caudata. Mol Biol Evol 30:2235–2248

Soltis D, Smith S, Cellinese N (2011) Angiosperm phylogeny: 17 genes, 640 Taxa. Am J Bot 98:704–730

Spatafora JW, Sung GH, Johnson D et al (2006) A five-gene phylogeny of Pezizomycotina. Mycologia 98 (6):1018–1028

Stamatakis A (2006) RAxML-VI-HPC: maximum likelihood-based phylogenetic analyses with thousands of taxa and mixed models. Bioinformatics 22:2688–2690

Stamatakis A (2014) RAxML version 8: a tool for phylogenetic analysis and post-analysis of large phylogenies. Bioinformatics (Advance Access published January 21)

Stamatakis A, Hoover P, Rougemont J (2008) A rapid bootstrap algorithm for the RAxML web-servers. Syst Biol 75:758–771

Stenroos S, Hyvönen J, Myllys L, Thell A, Ahti T (2002) Phylogeny of the genus *Cladonia* s. lat. (Cladoniaceae, Ascomycetes) inferred from molecular, morphological, and chemical data. Cladistics 18:237–278

Susko E (2008) On the distributions of bootstrap support and posterior distributions for a star tree. Syst Biol 57:602–612

Swofford DL (2003) PAUP*: phylogenetic analysis using parsimony (*and other methods) 4.0 Beta. Sinauer Associates, Sunderland, Massachusetts, USA

Swofford DL et al (2001) Bias in phylogenetic estimation and its relevance to the choice between parsimony and likelihood methods. Syst Biol 50:525–539

Tamura K, Stecher G, Peterson D, Filipski A, Kumar S (2013) MEGA6: molecular evolutionary genetics analysis version 6.0. Mol Biol Evol 30:2725–2729

Tavaré S (1986) Some probabilistic and statistical problems on the analysis of DNA sequences. Lect Math Life Sci 17:57–86

Tavaré S, Marshall CR, Will O, Soligos C, Martin RD (2002) Using the fossil record to estimate the age of the last common ancestor of extant primates. Nature 416:726–729

Thell A, Crespo A, Divakar PK, Kärnefelt I, Leavitt SD, Lumbsch HT, Seaward MRD (2012) A review of the lichen family Parmeliaceae—history, phylogeny and current taxonomy. Nord J Bot 30:1–24

Werth S, Miao VPW, Jónsson ZO, Andrésson OS (2015) Highthroughput sequencing in studies of lichen population biology. In: Upreti DK, Divakar PK, Shukla V, Bajpai R (eds) Recent advances in lichenology: modern methods and approaches in lichen systematics and culture techniques, vol 2. Springer, p 421

Yu Y, Harris AJ, He X (2010) S-DIVA (statistical dispersal-vicariance analysis): a tool for inferring biogeographic histories. Mol Phylogenet Evol 56 (2):848–850

Yu Y, Harris AJ, He X-J (2014) RASP (reconstruct ancestral state in phylogenies) 3.0. Available at http://mnh.scu.edu.cn/soft/blog/RASP

Zwickl D (2006) Genetic algorithm approaches for the phylogenetic analysis of large biological sequence datasets under the maximum likelihood criterion. Thesis. University of Texas, Austin

High-Throughput Sequencing in Studies of Lichen Population Biology

4

Silke Werth, Vivian P.W. Miao, Zophonías O. Jónsson and Ólafur S. Andrésson

Contents

Abstract

The population genetics of lichen fungi and their photobionts have been studied for several decades. In this review, we first focus on basic questions in lichen population genetics and on recent and current experimental approaches. Then, we discuss the utility of single nucleotide polymorphism (SNP) markers and how various high-throughput sequencing (HTS) techniques that allow SNP genotyping can contribute to population studies in lichens, highlighting their respective advantages and limitations for specific types of research questions. We review the emergence of *Lobaria pulmonaria* as a model system in lichen population biology, enabled by the use of microsatellite markers. Finally, we discuss open questions in the field, the steps that could be taken to increase our understanding of population genetics and genomics of lichens, and the merits of HTS for determining intrathalline diversity of lichen symbionts and associated organisms. The era of HTS may turn out to be an exciting time for research in lichenology similar to the period following the invention of the microscope.

S. Werth (✉) · Z.O. Jónsson · Ó.S. Andrésson
Faculty of Life and Environmental Sciences
University of Iceland, Sturlugata 7, 101 Reykjavik, Iceland
e-mail: silke.werth@uni-graz.at; silke.werth@wsl.ch; silke@hi.is

Z.O. Jónsson
e-mail: zjons@hi.is

Ó.S. Andrésson
e-mail: osa@hi.is

S. Werth
Swiss Federal Research Institute WSL, Zürcherstrasse 111, 8903 Birmensdorf, Switzerland

V.P.W. Miao
Department of Microbiology and Immunology, University of British Columbia, Vancouver, Canada
e-mail: vmiao@telus.net

S. Werth
Institute of Plant Sciences, University of Graz, Holteigasse 6, 8010 Graz, Austria

D.K. Upreti et al. (eds.), *Recent Advances in Lichenology*,
DOI 10.1007/978-81-322-2235-4_4

Keywords
Population genetics · Genetic diversity · Phylogeography · Transmission mode · Symbiosis · Horizontal transmission · Clonality · Recombination · High-throughput sequencing · Lichen · Model system · Genetic markers · Microsatellites · SNPs · RAD tag sequencing · Whole genome sequencing · Exon capture · RNA sequencing · Transcriptome · Amplicon sequencing · Photobiont · Mycobiont

4.1 Introduction

Lichens are a fascinating example of symbiosis and complex and intriguing subjects for population genetic studies. Lichens are formed through intimate mutualistic interactions between at least two symbionts, one a fungus (mycobiont) and the other a green algal or cyanobacterial photosynthetic partner (photobiont). Transmission of the symbiosis is considered "vertical" when it occurs via vegetative propagules or thallus fragments that comprise—and thus simultaneously disperse—both mycobiont and photobiont; it is considered "horizontal" when symbionts are individually dispersed, in the case of the lichen-forming fungus, e.g. via pycnidiospores or ascospores, necessitating re-establishment of symbiosis. Given the obligate nature of the symbiosis, population genetic studies of lichens must not only seek to discover the features and parameters characteristic of each symbiont species, but also to understand the roles and relative contributions of vertical versus horizontal transmission. In turn, the study of associations of symbionts across populations is crucial for elucidating factors shaping lichen symbioses.

The next decade holds great promise for extending our understanding of lichen population genetics, as new genetic tools become accessible to more researchers. A large effort devoted to taxonomy and phylogeny (reviewed in DePriest 2004) initiated in the 1990s resulted in a much better understanding of the systematic placement of most groups of lichen-forming fungi as well as their evolutionary history. Currently, molecular biology and genome sequencing are making inroads to advance various aspects of the genetics of lichen symbionts (reviewed in Grube et al. 2013). These methodological advances are also opening ways to address questions about populations of lichens and their interaction with their environments. A number of recent papers have summarized population genetics of mycobionts (Werth 2010b), phylogeography of lichens (Printzen 2010; Werth 2011), and population biology of lichens in the context of conservation (Scheidegger and Werth 2009). Here, we review both fundamental and new research questions concerning lichen population genetics and discuss the merit of recent technological developments for addressing them.

4.2 Population Genetics of Lichens

4.2.1 Why Are Studies at the Population Level Important?

A solid understanding of the factors and processes influencing populations—including, but not limited to the spatial scale and magnitude of gene flow—is of general interest for our understanding of how nature works. For example, population genetic data can allow us to retrace history and recognize founder events or bottlenecks, such as the survival of lineages in refugia, and to understand the microevolutionary changes in populations brought about by mutation, random genetic drift, selection, and gene flow. Knowledge of these processes can be helpful for efficient management of populations for various purposes. If the spatial scale of gene flow is known, for example, biodiversity reserves can be designed and interspersed so that they are functionally connected for an organism of conservation concern. We can also identify the populations and lineages that require immediate conservation action because they are rare and have diverged from others and adapted to a specific environment. Population genetic data help determine whether endangered species are threatened by inbreeding and help substantially with conservation prioritization, especially if they are combined with traditional observational data (population size, demography). We can distinguish local from non-local stock which is important for conservation work, e.g. involving translocations of individuals. With this knowledge, resources can be devoted to the lineages and populations that most urgently require action.

The genetic variability of natural populations is an essential and yet neglected aspect of biodiversity. Low diversity populations are more prone to extinction in a changing climate, as they are less likely to adapt to changed environmental conditions and survive. Genetic diversity is important because the loss of genetic variability is associated with decline of ecosystem resilience in general and may have negative impacts on some ecosystem services in particular (Millennium Ecosystem Assessment 2005). Studies of the genetic variability of natural populations thus address an important aspect of biodiversity with potential implications for human welfare.

4.2.2 Basic Questions

The discipline of population genetics investigates the processes influencing gene frequencies over time and their outcome—population subdivision or the lack of it—in a mathematical framework. Studies typically deal with the distribution of genetic variation in natural populations and the genetic relationships between individuals and populations at various spatial and temporal scales. At the smallest spatio-temporal scale (within a population at one site), common questions are: How closely are individuals related? What is the mating system? How much do specific individuals contribute to parentage? What is the relative importance of outcrossing versus selfing? Is dispersal by means of sexual or vegetative propagules? Which factors influence the genetic diversity of offspring?

If multiple populations of one species are studied in a larger region, questions pertaining to the genetic differences between populations, as well as the spatial scale or frequency of gene flow can be addressed. The term "population structure" refers to the presence and distribution of variation within a study group for one criterion or several, e.g. age, presence of a particular allele, and populations or groups of populations can be described and compared on this basis. How is the genetic variance distributed among hierarchical levels, regions, populations, subpopulations, and individuals? Are differences among groups of populations in distinct biogeographic regions as expected if there are topographic barriers to dispersal? What biotic and abiotic factors influence the genetic diversity of populations? Does habitat management e.g. forestry or grazing lead to lower genetic diversity of populations?

Phylogeography, the genetic relationships between lineages of a species across very large geographic landscapes (Avise et al. 1987) addresses questions at the largest spatio-temporal scale. How did climatic fluctuations, complex topography, and migration shape patterns of contemporary genetic diversity? Do population structures correlate geographically with areas that were potential refugia during periods of adverse climate and have deglaciated areas been colonized from one or several refugia? Are populations in the center of a large range more diverse than populations at the edges? Is a high level of genetic diversity due to long-term persistence or to merging of lineages from different refugial areas, or both? How much time has passed since specific lineages or populations diverged?

All these questions apply to lichens, and population studies of lichens have addressed questions at all three spatio-temporal scales. Below, we describe the experimental approaches that can be used to answer these questions using lichens as a study system.

4.2.3 Experimental Approaches

A number of models and methods have been used to explore population genetics of lichens. Often studies were focused on only one symbiont (typically the mycobiont) (reviewed by Werth 2010b), but concomitant studies of both symbionts are increasingly feasible and recognized as being more informative. The earliest studies were not based on DNA markers, but rather used lichen chemistry or proteins (isozymes) (Table 4.1) and established that lichens exhibit genetic variability, within and between populations. Fahselt (1986) investigated enzymatic differences between morphotypes of the lichen *Cladonia cristatella* and found no evidence to support genetic divergence among morphotypes. Other studies demonstrated gene flow between certain subspecies (i.e., chemotypes) of *Cladonia*, but reproductive isolation between others (Culberson et al. 1988, 1993). Not all lichens exhibit chemical variability, limiting its universal applicability for studies of lichen population biology. Also isozyme studies were somewhat limited as a large amount of material was needed for analyses, such that the technique was mainly applicable to large, foliose or fruticose species.

Later studies utilized DNA-based markers. DNA-based studies have the advantage of working for any species including those that are small in size, and while it is advantageous to use freshly collected material, lichen thalli can usually be analyzed successfully after several months of storage at room temperature or after years of freezer storage. Until recently, there were only a few established DNA markers. The first (Table 4.1) were variable regions associated with well-characterized loci, e.g. the ITS region of ribosomal RNA genes for mycobionts and photobionts, or the *rbc*LX region specific to cyanobacterial photobionts (and cephalodia). The utility of these approaches is constrained, as they possess low–resolving power, evolve neutrally (unlikely to be under selection, but, suitable for assessing population structure), and are not representative of the genome.

Anonymous markers (such as randomly amplified polymorphic DNA or inter–simple sequence repeats) are more representative of the genomes, but may be of an uncertain taxonomic origin when thallus material is used, and lack information on associated gene function. Although in some cases markers were obtained by using parts of a thallus containing only one symbiont (Heibel et al. 1999; Printzen et al. 1999) or axenic cultures of symbionts (e.g. Dyer et al. 2001; Itten and Honegger 2010; Nyati et al. 2013), these options are laborious and not fruitful for all species.

With microsatellites, a new generation of markers became available. This type of marker is based on 2–6 base pair (bp), tandem repeated DNA sequences that are abundant in the genomes of most organisms (Jarne and Lagoda 1996). Alleles differentiated by the number of repeats are numerous and genotypes can be distinguished by monitoring the size of PCR amplicons. Microsatellites are most useful for comparing closely related individuals and populations, although they have also in some cases been used to study populations situated on different continents, on a phylogeographic scale. The first microsatellite markers reported for lichens were developed for the mycobiont of *Lobaria pulmonaria* (Walser et al. 2003); more recently, this method was also adopted for investigation of its green algal photobiont, *Dictyochloropsis reticulata* (Dal Grande et al. 2009, 2012; Widmer et al. 2012). They offered a significant advantage over previous markers in providing a means by which a large number of individuals could be genotyped at a moderate cost per individual and have been employed successfully in studying genetic differentiation between populations and small-scale geographic variability within populations. Lately, studies of fungal and algal clones have also been undertaken, in order to study the fundamental processes shaping the genetic variability of lichen populations (Dal Grande et al. 2012). Microsatellites have helped to establish a body of population genetic data for *Lobaria pulmonaria* and

Table 4.1 Overview of various population level studies of lichen-forming fungi and their photobionts

Lichen species	Symbiont	Scale	Question type	Method	References
Anaptychia ciliaris	Photobiont	Region	Photobiont diversity	Eight microsatellites; DNA sequences of nrITS	Dal Grande et al. (2014a)
Biatora helvola	Mycobiont	Region	Genetic distance reflects glacial refugia of phorophyte	RAPD	Printzen et al. (1999)
Cavernularia hultenii	Mycobiont	Region (continents)	Phylogeography	DNA sequences of nrITS and nrIGS	Printzen et al. (2003), Printzen and Ekman (2002)
Cetraria spp.	Mycobiont	Region (continents)	Genetic variability within populations and genetic differences among species	Protein banding patterns	Mattsson (1991)
Cetraria aculeata	Mycobiont and photobiont	Region (continents)	Phylogeography	DNA sequences. Mycobiont: nrITS, mtLSU, GPD; photobiont: nrITS, actin, mtCOX2	Fernández-Mendoza et al. (2011)
Cetraria arenaria	Lichen	Landscape	No differentiation between two collecting sites	Protein banding patterns	Fahselt and Hagemann (1983)
Cladonia spp.	Mycobiont	Region (continents)	Mating system is heterothallic	RAPD, amplified fragment-length polymorphisms, DNA sequences of mating-type locus	Seymour et al. (2005a)
Cladonia arbuscula	Mycobiont and photobiont	Landscape	Population subdivision in a landscape	Mycobiont: presence/absence of introns in nrSSU. Photobiont: PCR-RFLP of nrITS region	Kotelko et al. (2008)
Cladonia arbuscula	Mycobiont and photobiont	Landscape	Population subdivision and symbiont transmission mode in ten transects within a 2-km area	Mycobiont: presence/absence of introns in nrSSU. Photobiont: PCR-RFLP of nrITS region	Robertson and Piercey-Normore (2007)
Cladonia arbuscula	Mycobiont	Landscape	Genetic diversity in four sampling locations	Presence/absence of introns in nrSSU. Photobiont: PCR-RFLP of nrITS region	Piercey-Normore (2006b)
Cladonia chlorophaea	Mycobiont	Site	Gene flow among some chemotypes of lichens and reproductive isolation of others	Analysis of secondary products in offspring of different chemotypes	Culberson et al. (1988)

(continued)

Table 4.1 (continued)

Lichen species	Symbiont	Scale	Question type	Method	References
Cladonia chlorophaea	Mycobiont	Site	Genetic diversity in one sampling location	Intron presence/ absence and PCR-RFLP of nrSSU	DePriest (1993)
Cladonia chlorophaea	Mycobiont	Site	Genetic diversity in one sampling location	PCR-RFLP of nrSSU and southern hybridization	DePriest (1994)
Cladonia cristatella	Lichen	Site	Genetic diversity of morphotypes	Protein banding patterns	Fahselt (1986)
Cladonia gracilis	Mycobiont and photobiont	Region	Genetic diversity of sampling locations and population subdivision	Mycobiont: presence/absence of introns in nrSSU. Photobiont: PCR-RFLP of nr ITS region	Piercey-Normore (2004)
Cladonia multiformis	Mycobiont and photobiont	Region	Genetic diversity and population subdivision	Mycobiont: presence/absence of introns in nrSSU. Photobiont: PCR-RFLP of nr ITS region	Piercey-Normore (2004)
Cladonia rangiferina	Mycobiont and photobiont	Region	Genetic diversity and population subdivision	Mycobiont: presence/absence of introns in nrSSU. Photobiont: PCR-RFLP of nr ITS region	Piercey-Normore (2004)
Cladonia rangiferina	Lichen	Landscape	Seasonal variability in enzyme polymorphisms in two sites	Protein banding patterns	Fahselt and Trembley (1999)
Cladonia subcervicornis	Mycobiont	Landscape	Population subdivision	DNA sequences (mt COX1)	Printzen and Ekman (2003)
Cladonia subtenuis	Mycobiont	Region	Genetic diversity of sampling locations	PCR-RFLP and presence of introns in nrSSU	Beard and DePriest (1996)
Cliostomum corrugatum	Mycobiont	Landscape	High gene flow within a landscape	DNA sequences of nrSSU intron	Lättman et al. (2009)
Degelia atlantica, D. plumbea	Mycobiont and photobiont	Region	Genetic diversity and transmission mode	Mycobiont: nrITS, RPB1, RPB2. Photobiont: *rbc*LX	Otálora et al. (2013)
Dictyonema glabratum	Mycobiont	Region	One tropical lichen consists of hundreds of unrecognized species	DNA sequencing of nrITS (Sanger sequencing and 454 pyrosequencing)	Lücking et al. (2014)
Evernia mesomorpha	Mycobiont and photobiont	Landscape	Genetic diversity and population subdivision	PCR-RFLP of nrITS	Piercey-Normore (2006a)
Flavocetraria nivalis	Mycobiont and photobiont	Region	Genetic diversity of sampling locations	DNA sequences. Mycobiont: nrITS, nrIGS; nrLSU, unidentified locus. Photobiont: nrITS	Opanowicz and Grube (2004)

(continued)

Table 4.1 (continued)

Lichen species	Symbiont	Scale	Question type	Method	References
Hypogymnia physodes	Mycobiont	Region	Genetic diversity of sampling locations	DNA sequences (nrITS)	Mattsson et al. (2009)
Hypogymnia tubulosa	Mycobiont	Region	Genetic diversity of sampling locations and differences between substrate types	DNA sequences and intron presence (nrITS)	Mattsson et al. (2009)
Lasallia pustulata	Mycobiont	Region	Marker development for studies of genetic diversity and symbiont selectivity	DNA sequences. Mycobiont: mtSSU, mtLSU, MCM7, TSR1; photobiont: nrITS, cp COX2, cp *psb*J-L, cp *rbc*L	Sadowska-Deś et al. (2013)
Lasallia pustulata	Photobiont	Region	Range-wide analysis of photobiont diversity; photobiont sharing with other lichens	DNA sequences of nrITS, cp *psb*J-L, cp COX2, cp *rbc*L	Sadowska-Deś et al. (2014)
Lecanora rupicola	Photobiont	Region (continents)	Low degree of selectivity for photobionts	nrITS	Blaha et al. (2006)
Leptogium spp.	Mycobiont	Region (continents)	Phylogeography	DNA sequences of nrITS, nrLSU, RPB2	Otálora et al. (2010)
Lepraria spp.	Mycobiont	Region (continents)	Genetic variability and species delimitation	DNA sequences of nrITS	Tretiach et al. (2009)
Lepraria spp.	Mycobiont and photobiont	Region (continents)	Horizontal photobiont transmission in co-dispersed lichen symbionts	DNA sequences; mycobiont: nrITS, mtSSU; photobiont: nrITS, actin 1	Nelsen and Gargas (2008)
Letharia columbiana	Mycobiont and photobiont	Region	Phylogeography, genetic variability of populations and population subdivision	DNA sequences; mycobiont: nrITS and two anonymous loci; photobiont: nr ITS, actin 1	Altermann (2009)
Letharia "lupina," L. "gracilis"	Mycobiont	Region	Test for recombined population structure	DNA sequences of ten anonymous loci, nrDNA and chitin synthetase	Kroken and Taylor (2001)
Letharia spp.	Mycobiont	Region (continents)	Genetic variability and species delimitation; cryptic species	DNA sequences of multiple loci	Altermann et al. (2014)
Letharia vulpina	Mycobiont	Region (continents)	Genetic diversity of sampling locations and post-glacial recolonization of Europe from North America	DNA sequences of eight anonymous loci	Högberg et al. (2002)

(continued)

Table 4.1 (continued)

Lichen species	Symbiont	Scale	Question type	Method	References
Letharia vulpina	Mycobiont	Region (continents)	Genetic diversity of sampling locations and post-glacial recolonization of Europe from Moroccan and Caucasus refugia	DNA sequences of eight anonymous loci	Arnerup et al. (2004)
Lobaria pulmonaria	Mycobiont	Landscape	Genetic diversity and community assembly in old and young forests	Three microsatellites	Gjerde et al. (2012)
Lobaria pulmonaria	Lichen	Landscape	Dendrochronological reconstruction of disturbance history and genetic diversity of populations	Six microsatellites	Bolli et al. (2008)
Lobaria pulmonaria	Lichen	Landscape	Spatial genetic structure of populations	Six microsatellites	Wagner et al. (2005)
Lobaria pulmonaria	Lichen	Landscape	Modeling of genetic diversity under disturbance scenarios	Six microsatellites	Wagner et al. (2006)
Lobaria pulmonaria	Lichen	Landscape	Population subdivision at the landscape level	Six microsatellites	Werth et al. (2007)
Lobaria pulmonaria	Lichen	Landscape	Genetic diversity and spatial genetic structure are affected by historic forest disturbance	Six microsatellites	Werth et al. (2006b)
Lobaria pulmonaria	Mycobiont and photobiont	Landscape	Population subdivision and transmission mode of mycobiont and photobiont	Mycobiont: eight microsatellites; photobiont: ten microsatellites	Werth and Scheidegger (2012)
Lobaria pulmonaria	Mycobiont and photobiont	Landscape	Genetic diversity of populations	Mycobiont: three microsatellites; photobiont: three microsatellites	Werth (2010a)
Lobaria pulmonaria	Mycobiont	Landscape	Genetic diversity of populations	Eight microsatellites	Jüriado et al. (2011)
Lobaria pulmonaria	Mycobiont	Landscape	High diversity and high ongoing dispersal. Low genetic differentiation among forest fragments	Eight microsatellites	Hilmo et al. (2012)
Lobaria pulmonaria	Mycobiont and photobiont	Region	Skewed distribution of mating-type idiomorphs leads to forced clonality in populations	Mycobiont: Mating-type idiomorphs, eight microsatellites; photobiont: seven microsatellites	Singh et al. (2012)

(continued)

Table 4.1 (continued)

Lichen species	Symbiont	Scale	Question type	Method	References
Lobaria pulmonaria	Mycobiont	Region	Genetic diversity and population subdivision	DNA sequences of nrITS and nrLSU	Zoller et al. (1999)
Lobaria pulmonaria	Mycobiont	Region	Genetic diversity of populations and population subdivision	Three microsatellites	Otálora et al. (2011)
Lobaria pulmonaria	Mycobiont and photobiont	Region	Transmission mode and importance of mutations in fungal and algal genotypes	Mycobiont: eight microsatellites; photobiont: seven microsatellites	Dal Grande et al. (2012)
Lobaria pulmonaria	Mycobiont and photobiont	Region	Phylogeography in Europe and hot spots of genetic diversity	Mycobiont: eight microsatellites; photobiont: seven microsatellites	Widmer et al. (2012)
Lobaria immixta, L. macaronesica, L. pulmonaria	Mycobiont	Region (continents)	Species delimitation in Macaronesian species of *Lobaria*	DNA sequences of nrITS, RPB2 and EF-1a	Cornejo and Scheidegger (2010)
Lobaria immixta, L. macaronesica, L. pulmonaria	Mycobiont	Region (continents)	Propagule size and regional population subdivision of island populations	Six microsatellite loci	Werth et al. (2014)
Lobaria retigera group	Mycobiont	Region (continents)	Loss of vegetative propagules over evolutionary time	DNA sequences of nrITS	Cornejo et al. (2009)
Lobariaceae	Photobiont	Region (continents)	Symbiont associations and spatial distribution of algal lineages	DNA sequences of nrITS, nrSSU, and cp *rbc*L; seven microsatellites	Dal Grande et al. (2014b)
Lobathallia radiosa	Lichen	Region	Genetic diversity of populations and delimitation between closely related species	RAPD and inter-simple sequence repeats	Yüzbaşıoğlu et al. (2011)
Nephroma spp.	Mycobiont and photobiont	Region (continents)	Genetic diversity of *Nostoc* photobionts and fungal-cyanobacterial associations	DNA sequences; mycobiont: nrITS; photobiont: tRNA (Leu) (UAA) intron	Fedrowitz et al. (2012a)
Nephroma spp.	Photobiont	Landscape	Genetic diversity of *Nostoc* photobionts	DNA sequences of tRNA(Leu) (UAA) intron	Fedrowitz et al. (2011)
Nephroma spp., *Peltigera* spp.	Mycobiont	Region (continents)	The same fungal species associates with green algae or cyanobacteria (photomorphs)	DNA sequences of nrITS	Goffinet and Bayer (1997)
Nephroma spp.	Mycobiont	Region (continents)	Macaronesian endemics are neoendemics and some have expanded their ranges to the mainland	DNA sequences of nrITS, mtSSU, nrLSU	Sérusiaux et al. (2011)

(continued)

Table 4.1 (continued)

Lichen species	Symbiont	Scale	Question type	Method	References
Parmelia sulcata	Mycobiont	Region	Genetic diversity of recently colonized and low-air pollution sites	Presence of introns in nrDNA	Crespo et al. (1999)
Pannaria spp.	Photobiont	Region (continents)	Symbiont associations and relationship to habitat ecology and taxonomy	DNA sequences of 16S rRNA	Elvebakk et al. (2008)
Parmotrema perforatum, P. hypotrypum, P. hypoleucinum	Lichen	Landscape	Genetic variation within and differences among sampling locations	Protein banding patterns	Fahselt and Jancey (1977)
Parmotrema tinctorum	Mycobiont and photobiont	Landscape	Genetic diversity and population subdivision; fungal selectivity for photobiont lineages	Mycobiont and photobiont: DNA sequences of nrITS	Ohmura et al. (2006)
Parmotrema tinctorum	Mycobiont and photobiont	Landscape	Intrathalline variation; genetic diversity and population subdivision	Mycobiont: four microsatellites; photobiont: five microsatellites	Mansournia et al. (2011)
Peltigera spp.	Photobiont	Region (continents)	Fungal selectivity and photobiont associations	DNA sequences of 16S rRNA, *rbc*LX, and *trn*L	O'Brien et al. (2005)
Peltigera spp.	Mycobiont	Region	Reproductive isolation among morphospecies of *Peltigera*	DNA sequences of nrITS, RPB1, and betatubulin	O'Brien et al. (2009)
Porina spp.	Mycobiont	Region	Population subdivision in five species of tropical epiphyllic *Porina* lichens	DNA sequences of mtSSU	Baloch and Grube (2009)
Porpidia flavicunda	Mycobiont	Region (continents)	Population subdivision among continents and inter-continental gene flow	DNA sequences of nrLSU, RPB2, and beta-tubulin	Buschbom (2007)
Ramalina farinacea	Mycobiont and photobiont	Region (continents)	Population subdivision among Iberian Peninsula and Canary Islands; ecological diversification in lichen algae	DNA sequences. Mycobiont: nrITS and RPB2; photobiont: nrITS	del Campo et al. (2013)
Ramalina farinacea	Photobiont	Region (continents)	Two species of algae in all thalli	DNA sequences of cpLSU and nrITS	Casano et al. (2011)
Ramalina menziesii	Mycobiont	Landscape	Genetic diversity and population subdivision	DNA sequences of three nuclear genes and an anonymous locus	Werth and Sork (2008)

(continued)

Table 4.1 (continued)

Lichen species	Symbiont	Scale	Question type	Method	References
Ramalina menziesii	Mycobiont	Region	Phylogeography. Population subdivision among ecoregions and origins of high genetic diversity	DNA sequences of three nuclear genes and an anonymous locus	Sork and Werth (2014)
Ramalina menziesii	Photobiont	Landscape	Population subdivision between sampling locations and phorophyte species	DNA sequences of nr nrITS, chloroplast *rbc*L, and *psb*J-L	Werth and Sork (2010)
Ramalina menziesii	Photobiont	Region	Population subdivision among ecoregions and origins of high genetic diversity	DNA sequences of nrITS and chloroplast *rbc*L	Werth and Sork (2014)
Ramalina menziesii	Photobiont	Landscape	Photobiont sharing with other lichen fungi	DNA sequences of nrITS	Werth (2012)
Ramalina siliquosa	Mycobiont	Site	Reproductive isolation among most and gene flow among some chemotypes	Analysis of secondary products in offspring of different chemotypes	Culberson et al. (1993)
Rhizoplaca chrysoleuca	Mycobiont	Region	Genetic differences among chemotypes	DNA sequences of nrITS and presence of introns	Zhou et al. (2006)
Roccellina capensis	Mycobiont	Region	"Species pair" of sterile vs. fertile specimens is conspecific	DNA sequences of nrITS and RAPD	Lohtander et al. (1998)
Tephromela atra	Mycobiont and photobiont	Region	Genetic diversity of photobionts in different habitats and fungal–algal associations	DNA sequences; mycobiont: nr ITS, betatubulin, polyketide synthase; photobiont: nrITS	Muggia et al. (2008)
Tephromela atra	Photobiont	Region	Genetic diversity of photobionts	DNA sequences of nrITS, actin 1, and cp *rbc*L	Muggia et al. (2010)
Tephromela atra	Mycobiont	Region (continents)	Delimitation of species within a species complex	DNA sequences of nrITS, betatubulin, and MCM7	Muggia et al. (2014)
Thamnolia subuliformis	Mycobiont	Landscape	Population subdivision within a landscape	ISSR, intron presence/absence in nrDNA	Cassie and Piercey-Normore (2008)
Trapeliopsis glaucolepidea	Mycobiont	Region (continents)	Population differentiation between continents	DNA sequences of nrITS	Palice and Printzen (2004)
Umbilicaria spp.	Lichen	Landscape	Higher genetic variability in populations of sexual species, relative to vegetative species	Protein banding patterns	Fahselt (1989)

(continued)

Table 4.1 (continued)

Lichen species	Symbiont	Scale	Question type	Method	References
Umbilicaria cylindrical	Lichen	Landscape	Genetic variability in populations and enzymatic activity in subfossil lichens that had been ice-covered for ~1,350 years	Protein banding patterns	Fahselt et al. (1995)
Umbilicaria mammulata, U. vellea	Lichen	Site	Intrathalline variation	Protein banding patterns	Larson and Carey (1986)
Umbilicaria mammulata	Lichen	Landscape	Geographic patterns of genetic variability	Protein banding patterns	Hageman and Fahselt (1992)
Umbilicaria mammulata	Lichen	Landscape	No seasonal variability in enzyme polymorphisms	Protein banding patterns	Fahselt and Trembley (1999)
Umbilicaria vellea	Lichen	Region	Genetic variability of populations	Protein banding patterns	Hageman and Fahselt (1990)
Usnea filipendula	Mycobiont	Region	Recolonization of a formerly polluted region	RAPD	Heibel et al. (1999)
Xanthomendoza borealis	Photobiont	Region (continents)	Photobiont lineage on different continents	DNA sequences of nrITS	Lindblom and Søchting (2013)
Xanthoria spp.	Mycobiont	Region (continents)	Mating system of *X. parietina* homothallic, five species heterothallic	RAPD analysis of single-spore isolates; phenotypic variation	Honegger et al. (2004a)
Xanthoria elegans	Mycobiont	Region (continents)	Population differentiation between continents	RAPD and nrITS sequences; growth rate	Murtagh et al. (2002)
Xanthoria parietina	Photobiont	Region	Photobiont diversity	Eight microsatellites and DNA sequences of nrITS	Dal Grande et al. (2014a)
Xanthoria parietina	Mycobiont	Region (continents)	Genetic diversity of worldwide locations	RAPD	Honegger et al. (2004b)
Xanthoria parietina	Mycobiont	Landscape	Genetic diversity and population subdivision	DNA sequences of nrITS and IGS	Lindblom and Ekman (2007), (2006)
Xanthoria parietina	Mycobiont	Landscape	Genetic diversity and population subdivision	RAPD	Itten and Honegger (2010)
Xanthoria parietina	Photobiont	Landscape	Genetic diversity and population subdivision	RAPD and nrITS sequences	Nyati et al. (2013)

cp Chloroplast, *nr* nuclear ribosomal DNA, *mt* mitochondrial, *ITS* internal transcribed spacer region of rDNA, *IGS* intergenic spacer region of rDNA, *LSU* large subunit of rDNA, *SSU* small subunit of rDNA

Parmotrema tinctorum with their photobionts (Table 4.1) and for *Trebouxia decolorans* (Dal Grande et al. 2014a). More microsatellite markers have recently been published for additional lichen fungi, i.e., *Bryoria* Section *Implexae*, *Buellia frigida*, *Lobaria pindarensis*, *Nephroma laevigatum* and *N. parile*, *Peltigera dolichorhiza*, *Protoparmeliopsis muralis*, and *Usnea subfloridana* (Devkota et al. 2014; Guzow-Krzeminska and Stocker-Wörgötter 2013; Jones et al. 2012; Magain et al. 2010; Nadyeina et al. 2014a; Tõrra et al. 2014, Belinchón et al. 2014). Microsatellites have played an important role in the emergence of lichen model systems, with the best-studied case being *L. pulmonaria*. They will likely contribute to the development of additional model systems in lichens in the near future.

4.3 Single Nucleotide Polymorphisms and High-Throughput Methods

Markers based on the most fundamental level of variation, single nucleotide polymorphisms (SNPs). High-throughput sequencing (HTS) platforms for both marker development and data acquisition are increasingly being used in population genetics. SNPs are abundant and have a broad utility and can supply high-resolution data (Garvin et al. 2010; Kim and Misra 2007; Kwok 2001). In general, little a priori knowledge of the genome is required to use SNPs, although care must be taken to avoid genotyping loci that are linked or selecting loci based on too few individuals ("ascertainment bias", Nielsen 2000). Most earlier SNP-based studies were limited to a relatively small number, e.g. dozens of loci, by resource constraints, (Helyar et al. 2011) but several economical SNP genotyping techniques are now available. For analysis of dozens to a few hundred loci, a Fluidigm dynamic array platform, Nanogen Centaurus assay, or similar, may be very cost-effective, and other options are developing rapidly. The drastic drops in the cost of sequencing (Hudson 2008; Liu et al. 2012; Metzker 2010; Shendure and Ji 2008) that have made HTS much more accessible have also revolutionized the scale and scope of SNP genotyping. Partial or whole genome sequencing of lichen thalli and symbionts has already yielded, as primary or collateral benefits, data for identification of new microsatellite markers (e.g. Dal Grande et al. 2013; Devkota et al. 2014; Werth et al. 2013) as well as thousands of SNPs distributed across the entire genome for population genetic studies. When analyzed with HTS technology, SNPs should yield high-resolution data that can provide scope and depth for phylogenies (phylogenomics, Chaps. 2 and 3 in this volume) and for studies of individuals and populations (population genomics). One possibility of efficiently sequencing many individuals is to sequence pools of individuals, e.g. containing an entire population (Druley et al. 2009; Futschik and Schlotterer 2010; Sham et al. 2002). Results from pool sequencing provide accurate allele frequency estimates at a much reduced cost, making pool sequencing an efficient method if the aim is looking at differences between populations (Futschik and Schlotterer 2010).

A number of HTS technologies are currently available and more are under development. All these techniques simultaneously sequence a large collection of DNA fragments that are derived in different ways: directly from DNA, from selective amplicons (targeted, exon capture, etc.) or via cDNA from RNA (total RNA, rRNA depleted RNA, mRNA, micro RNA, etc.). Several of these techniques utilize DNA amplification at some stage (e.g. Illumina, 454, SOLiD and Ion Torrent) rather than reading of individual molecules (PacBio, Oxford Nanopore). All of the methods can be used for the tasks listed in Table 4.2, but some are more suited to particular tasks. The PacBio technique may be especially useful for de novo WGS, as it allows sequencing of template DNA without prior amplification and generates long reads of >10,000 bases; these advantages are, however, offset by relatively high error rates. Nanopore sequencing holds a similar promise although not yet commercially available. The other commonly used methods provide from $\sim$100 base to $\sim$600 base reads from each DNA molecule. The reads generated by SOLiD are the shortest, which consequently limits its use in

Table 4.2 Overview of high-throughput sequencing strategies[a]

Method	Molecule	Applications	Data required for symbiont identification	Drawbacks	Advantages
RNA-seq	RNA	Genotyping and SNP discovery, gene expression	None (transcripts can be assigned to fungal/green algal/ cyanobacterial proteins by blastx)	Requires RNA; unequal representation of transcripts between samples (differential expression)	Straightforward method to reduce the genome to its transcribed parts
Amplicon sequencing	DNA	Genotyping of specific loci	None (blast to GenBank)	Selection of loci non-random; usually a smaller number of loci are sequenced than using other approaches	Useful to determine composition of organisms from environmental samples
Genome reduction approaches (RAD-seq, GBS)	DNA	Genotyping, SNP discovery	Genome reduction sequencing profiles from axenic cultures of symbionts (or genome data for the symbiont species)	For symbiotic taxa, individual genomes of the species or a close relative are required to be able to bioinformatically assign sequences to the respective organism	Useful to genotype many individuals at a large number of loci
Exon capture	DNA	Genotyping, SNP discovery	Mycobiont and photobiont transcriptome data, in chlorolichens best from axenic cultures	Requires substantial knowledge of the genome of target organism	Useful way to reduce the genome and cover all exons. Especially valuable for species with large genomes
Whole genome sequencing	DNA	Genotyping, SNP discovery	Best to sequence axenic cultures. Metagenomic sequencing possible, but bioinformatically complex	Prohibitively expensive, considerable bioinformatics effort	Useful for comparative studies looking at presence/absence of genes in certain lineages

[a] The technical aspects of template preparation, sequencing, and assembly have been reviewed by Metzker (2010) and applications of HTS for molecular studies of the ecology of non-model organisms have been reviewed by Ekblom and Galindo (2011)

WGS and in generating haplotypes (series of linked markers). Currently, Illumina provides the best combination of cost, read-length, and accuracy and is most commonly used for both DNA and RNA sequencing. Most sequencing providers also offer complementary technology for adapting to various uses, e.g. indexing to enable multiplexing of samples. Table 4.2 compares the common uses of HTS methods and Table 4.3 assists in choice of molecular method; below, we discuss how HTS can be applied to population studies of lichens.

4.3.1 Whole Genome Sequencing

Whole genome sequencing (WGS) has been used to compare genomes within and among species (Altshuler et al. 2010; Ellison et al. 2011; Jaillon et al. 2004, 2007; Lawniczak et al. 2010; Neafsey et al. 2010), e.g. to study the distribution of genes across multiple lineages (McDonald et al. 2013). Although examples remain few, both axenically cultured lichen symbionts and whole lichen thalli have been used (reviewed in Grube et al. 2013, Table 4.4). Pure cultures are preferred as there is

Table 4.3 Choice of molecular method for different question types

Question type	Number of loci	Number of individuals	Relatedness between samples	Method of choice
Gene expression	100–1,000	10	Same species	RNA-seq
Gene expression	10–100	100	Same species	qPCR
Individual delimitation, clonality	100–1,000	100	Same species	RAD-seq, (RNA-seq, Exon capture, WGS)
Gene flow and population subdivision	10–100	100	Same species	SNP genotyping assays, amplicon sequencing, RAD-seq, microsatellites
Local adaptation and loci under selection	1,000	100	Same species	RAD-seq, RNA-seq
Landscape genetics	10–100	1,000	Same species	SNP genotyping assays, amplicon sequencing, RAD-seq, microsatellites
Phylogeography	10–100	100	Multiple lineages	Amplicon sequencing, RAD-seq
Divergence time and coalescent analysis	100–1,000	100	Multiple lineages	RAD-seq, (exon capture, RNA-seq, WGS)
Genes involved in ecological speciation	All	10	Closely related species	WGS
Community genetics	10–100	1,000	Various taxonomic groups	Amplicon sequencing

no ambiguity with respect to the source of DNA sequences, but in most cases, this is not practical. It may be possible to generate genomic sequences that represent essentially only one symbiont (e.g. using rhizines or other tissue devoid of photobionts (Santhini Basil and Andrésson, unpublished data) or by micro-dissection of thalli), but usually whole thalli will be used. It is essential to consider that thallus DNA comprises a metagenome, i.e., genomic DNA from mixtures of organisms, including not only the mycobiont and photobiont(s), but also organisms found on or in the lichen. In addition to potential challenges in assigning the taxonomic source of certain sequences, commensurately more sequencing is needed than for pure cultures. WGS of a thallus sample to a coverage of at least 10X may require sequencing 2–4 billion bases; an optimal run on an Illumina MiSeq can accommodate 2–4 such samples. It would provide data on the >15,000 genes of each eukaryotic partner or the >6,000 genes of a cyanobacterial photobiont, but the bioinformatic processing would be formidable. There may also be a considerable contribution from endolichenic and other associated eukaryotes and bacteria (Grube et al. 2013); these organisms comprised ~30 % of the metagenomic data from sequencing of thallus material in *Peltigera membranacea* (Kampa et al. 2013) and elsewhere, have been suggested to be important in the lichen symbiosis (Grube et al. 2014) are HTS studies at the population level could help to clarify the role that bacteria are playing in lichen metabolism. Software developed for assembly of single genomes may not perform well with metagenomes, e.g. leading to greater frequencies of chimeric contigs, but improvements are being made to overcome the issues associated with metagenomic libraries (e.g. Boisvert et al. 2012). It should be noted that while de novo genome assembly is a large undertaking, processing re-sequencing data (from additional specimens) is much easier once a reference genome has been created.

Table 4.4 Studies of lichens using HTS

Topic	Species	HTS data	References
Horizontal transmission mode of ammonium transporter genes from prokaryotes and high rates of retention in lichen fungi	Multiple lichen-forming fungi	Whole genome sequencing	McDonald et al. (2013)
Discovery of genes involved in important biological pathways	*Cladonia rangiferina* and *Asterochloris* sp. photobiont	RNA sequencing	Junttila and Rudd (2012)
Taxonomic placement of species	*Lepraria* spp.	Amplicon sequencing of ITS and LSU	Hodkinson and Lendemer (2013)
Discovery of genes, focussing on those involved in drought tolerance	*Endocarpon pusillum*	Whole genome sequencing	Wang et al. (2014)
Comparison of mitochondrial genome sequences among species and with other ascomycete fungi	*Peltigera membranacea, P. malacea*	Metagenome sequencing	Xavier et al. (2012)
High variability in the *lec-2* gene, a gene thought to code for a galectin involved in interaction with the Nostoc photobiont	*Peltigera membranacea*	Metagenome sequencing	Manoharan et al. (2012)
Differential expression of lectin gene *lec-1* in different thallus parts	*Peltigera membranacea*	RNA sequencing	Miao et al. (2012)
Discovery of genes involved in secondary metabolite synthesis	*Peltigera membranacea* and *Nostoc* sp. photobiont	Metagenome sequencing	Kampa et al. (2013)

4.3.2 Sequencing of Partial Genomes

WGS still represents a major effort and provides information on many more loci than generally needed for population genetic studies. A number of methods are available to reduce the complexity of the data by specifically sequencing only part of the genome(s) and in some cases, constrain data acquisition to parts of the genome that are more informative for the organisms used or questions studied. Approaches in which DNA is processed before HTS offer greater flexibility and control over data quality, but generally are more demanding in planning and execution. In addition, individual features of each method may require special consideration for lichens, especially if thalli are used, as the various genomes may behave differently with particular methods.

The most general of the genotyping-by-sequencing approaches is restriction-site-associated DNA sequencing (RAD-seq) (Miller et al. 2007). In this method, the genome is digested with one (Elshire et al. 2011) or two (Peterson et al. 2012) restriction enzymes and the resulting fragments are ligated to sample bar-coded adapters before multiplex HTS. The fraction of the genome that will be sequenced can be controlled by size selection of fragments (either before or after adapter ligation) and the choice of restriction enzymes. Conservation of restriction sites (e.g. 6 bases) typically leads to recovery of the same genomic fragments (demarcated by the chosen enzymes) from each sample for analysis, while sequencing reads (30–300 bases inwards from the restriction sites) will frequently contain one or more SNPs that can be used to characterize genetic variability and divergence in natural populations. The number of loci studied can be adjusted, but typically ~1,000 loci are used and a large number of samples can therefore be practically accommodated, e.g. 400 samples at 20X mean coverage in one MiSeq run (assuming an output of 1.2 Gb in a run with 2X 150 cycles). The proportion of the genome represented is usually low, often <1 %. For species with large and complex genomes, using restriction enzymes that do not cut methylated DNA can help to avoid generating sequencing data from repetitive regions (which are generally methylated) and to increase the sequence coverage of other parts of the genome (Elshire et al. 2011). In lichens, this may not confer a significant advantage with respect to mycobionts, as fungi typically have little methylated DNA (Armaleo and Miao 1999; Zemach et al. 2010, Santhini Basil and Andrésson, unpublished data) but it may be a useful consideration for algal photobionts, as model green algae are known to be heavily methylated and in unusual ways (Feng et al. 2010; Zemach et al. 2010).

In general, neither WGS nor genome reduction approaches such as RAD-seq, discriminate between coding and non-coding DNA, and it may be problematic, if thallus DNA is used as the source, to confidently assign a majority of the non-coding sequences to the mycobiont or photobionts by bioinformatic comparisons to characterized genomes. Contributions arising from associated organisms are less of a problem, as they will be very diverse, and their sequences will have very low coverage. In addition, non-coding DNA tends to evolve rapidly. As it becomes less likely that restriction-site-associated loci will be shared as species diverge, it is worthwhile to contemplate genome reduction methods that are based on enrichment for informative regions, e.g. exon capture or RNA sequencing (RNA-seq) approaches (see below) which may be better suited for studies with a phylogenetic scope that include different species.

Exon capture, the targeted sequencing of coding DNA (Albert et al. 2007; Hodges et al. 2007; Metzker 2010; Ng et al. 2010, 2009; Turner et al. 2009), is a strategy that reduces sequencing effort and targets it toward protein-coding sequences. A vast part of the genomes of many eukaryotic organisms is composed of intergenic or repetitive regions (Wegrzyn et al. 2014), which should evolve more or less neutrally. Exon capture depends on hybridization (capture) of sheared and PCR linker-ligated sample DNA to arrayed probes that represent coding regions, followed by recovery and processing of the bound portion of the sample DNA for HTS. Because the method relies on hybridization, the capture array can be developed from genome information of the target species or sequences from a related organism (Bi et al. 2012; Cosart et al. 2011). In contrast to RNA-seq where sequencing depth depends heavily on gene expression levels, exon capture should deliver equal depth for all loci in the array, which is an advantage for determining the amount of sequencing required for a certain coverage of the target loci. Whether this method is efficient for lichens is unclear. Fungal genomes are smaller (e.g. 40 Mb) than those of higher eukaryotes and may comprise only 1–5 % repetitive DNA (Wöstemeyer and Kreibich 2002). Lichen symbionts may also have rather small genomes that mostly comprise coding sequences (64 % in the green alga *Coccomyxa subelliptica*; 53 % in the lichenized fungus *Cladonia grayi* according to www.jgi.doe.gov, accessed June 2014).

"Transcriptome" sequencing is another way to focus on coding components of the genome. Total RNA is extracted from a sample and selectively enriched for mRNA. rRNA depletion kits have been developed for various organisms to alleviate problems associated with the high level of rRNA. For lichen thalli, however, the traditional method of capturing poly-A containing mRNA on oligo-dT-coated surfaces will continue to be the method of choice, as it removes most of the mitochondrial, bacterial, and archaeal RNAs in addition to tRNA and rRNAs which make up the bulk of the RNA and which otherwise would obscure signals from nuclear genes with low expression levels. In lichens with a green algal photobiont, both mycobiont and green algal mRNAs will be recovered, but in species with a primary cyanobacterial photobiont, only mycobiont mRNA will be obtained; mRNA from eukaryotic endophytes or epibionts might be present but is not expected to pose a serious problem owing to low individual abundance relative to mycobiont and photobiont cells.

As a tool to detect sequence variation, RNA sequencing has three advantages over exon capture: Preparation is much simpler, no prior genome sequence is required, and it provides rich information about gene expression levels (reflected in sequence coverage or reads per base position). The last point is the main reason for most RNA-seq studies, with SNP information being a fortuitous by-product. This combination may allow investigators to study genetic variability, differential gene expression (under various natural or experimental conditions), and environmental heterogeneity as part of the same experiment (Scott et al. 2009). The requirement of a large amount of high-quality RNA for transcriptome sequencing may limit its application where deficiencies in RNA quality or quantity exist, e.g. depauperate specimens, thalli that are very small or tightly embedded in a substrate, and lichens for which RNA extraction methods are not optimized. As the transcriptome reflects only the fraction of the genome that is actively transcribed under a particular set of conditions, additional SNPs may be revealed under other conditions. Strategies for analyzing RNA-seq data for both expression levels and SNP marker development have been reviewed by De Wit et al. (2012).

In amplicon sequencing, one to many genes may be sequenced in parallel for a large number of pooled individuals on an HTS platform. A few loci that contain species-specific information (e.g. ribosomal genes) can be used, for example, to reveal the diversity of intrathalline organisms (Bates et al. 2011; Hodkinson et al. 2012; Sigurbjörnsdóttir et al. 2014), but for phylogeography or population genetics, dozens or hundreds of loci can be used. After or before amplification, samples are ligated to linkers containing bar code indices to distinguish individuals. Indexed products from multiple individuals are then pooled and often subjected to additional amplification before HTS. After sequencing, individuals are identified by their bar codes, and loci are separated based on their sequences, i.e., by mapping reads to reference sequences for each gene. Although this approach can be applied at various taxonomic levels and is very efficient for working with several dozen loci from many individuals, it has been used less frequently than the other HTS approaches in population genomic studies.

While HTS has been used to develop new microsatellite markers for lichens, it is also becoming increasingly attractive and economically feasible to use HTS data directly to obtain genomic sequences and perform population level analyses for many individuals on thousands of genes simultaneously (Ellison et al. 2011; Hohenlohe et al. 2011; Neafsey et al. 2010; Wagner et al. 2013; Yoder et al. 2014). The results will be qualitatively similar to those from Sanger-type DNA sequences or microsatellites, but the accuracy and the level of confidence will increase substantially with the number of loci sampled, making inferences less prone to the influence of "outlier" loci. HTS methodology is especially promising where there is currently a lack of genetic resolution, e.g. for defining individuals and lineages at high resolution (fungal, algal and lichen clones), for estimating divergence time between populations or lineages, and for resolving phylogenetic relationships.

4.3.3 Genomic Resources for Lichen Symbionts

In order to take full advantage of the wealth of data generated by HTS, genomic resources such as annotated genomes are a prerequisite. If the genome of the organism of interest has not been sequenced, it is often possible to map sequence reads to the genome of a related organism. The genomic resources publically available for lichen-forming fungi and their photobionts are summarized in Table 4.5, and lichen genomic studies are listed in Table 4.4. Genome sequences, together with mRNA-based annotation, are publically available for several species of lichen-forming fungi (*Cladonia grayi*, *C. macilenta*, *C. metacorallifera*, *Endocarpon pusillum*, *Gyalolechia flavorubescens*, *Trypethelium eluteriae*, and *Xanthoria parietina*) and for two green algal photobiont species (*Asterochloris* sp., *Dictyochloropsis reticulata*). Moreover, transcriptome data have been published for *Cladonia rangiferina* and its *Asterochloris* photobiont (Junttila and Rudd 2012), and a c-DNA library has been described for *Endocarpon pusillum* (Wang et al. 2011).

Table 4.5 Genomic resources for lichen-forming fungi and their photobionts

Species	Taxonomic group	Status	Resources	Link
Acarospora strigata	Ascomycota, lecanoromycetes, acarosporales	In progress	Genome	http://www.genomesonline.org/projects?id=102217
Arthonia rubrocincta	Ascomycota, arthoniomycetes, arthoniales	In progress	Genome	http://www.genomesonline.org/projects?id=102211
Cladonia grayi	Ascomycota, lecanoromycetes, lecanorales	Finished	Genome, transcriptome, EST	http://genome.jgi-psf.org/Clagr2/Clagr2.home.html
Cladonia macilenta	Ascomycota, lecanoromycetes, lecanorales	Finished	Genome	http://www.genomesonline.org/projects?id=42523
Cladonia metacorallifera	Ascomycota, lecanoromycetes, lecanorales	In progress	Genome	http://www.genomesonline.org/projects?id=49795
Cladonia rangiferina	Ascomycota, lecanoromycetes, lecanorales	Published	Transcriptome	http://www.biomedcentral.com/1471-2164/13/575
Endocarpon pallidulum	Ascomycota, eurotiomycetes, verrucariales	In progress	Genome	http://www.genomesonline.org/projects?id=102116
Endocarpon pusillum	Ascomycota, eurotiomycetes, verrucariales	In progress	Genome	http://www.genomesonline.org/projects?id=90027
Graphis scripta	Ascomycota, lecanoromycetes, ostropales	In progress	Genome	http://www.genomesonline.org/projects?id=102159
Lobaria pulmonaria	Ascomycota, lecanoromycetes, peltigerales	In progress	Genome, transcriptome	http://genome.jgi-psf.org/Lobpulcupartners/Lobpulcupartners.info.html
Trypethelium eluteriae	Ascomycota, dothideomycetes, trypethiales	Finished	Genome, EST	http://genome.jgi.doe.gov/Tryvi1/Tryvi1.home.html
Umbilicaria muehlenbergii	Ascomycota, lecanoromycetes, umbilicariales	In progress	Genome	http://www.genomesonline.org/projects?id=90033
Xanthoria parietina	Ascomycota, lecanoromycetes, teloschistales	Finished	Genome, transcriptome, EST	http://genome.jgi-psf.org/Xanpa1/Xanpa1.home.html
Asterochloris sp. (photobiont of *C. grayi*)	Chlorophyta, trebouxiophyceae, microthamniales	Finished	Genome, transcriptome, EST	http://genome.jgi-psf.org/Astpho1/Astpho1.home.html
Asterochloris sp. (photobiont of *C. rangiferina*)	Chlorophyta, trebouxiophyceae, microthamniales	Published	Transcriptome	http://www.biomedcentral.com/1471-2164/13/575
Dictyochloropsis reticulata (photobiont of *L. pulmonaria*)	Chlorophyta, trebouxiophyceae, microthamniales	Finished	Transcriptome	http://genome.jgi.doe.gov/pages/home.jsf?core=genome&query=1016107&searchType=JGI%20Project%20ID

4.4 Model Systems and New Questions

Different model systems are needed for different groups of lichens (Table 4.1), but a few have served as prototypes to guide development of others. By far, the largest body of work comes from studies of *L. pulmonaria* based on microsatellite markers (see Sect. 4.2.3). This epiphytic macrolichen is a tripartite symbiosis: in addition to the mycobiont and the green algal photobiont, cyanobacterial symbionts (*Nostoc* sp.) are also found in cephalodia inside the thalli. It occurs in temperate and boreal environments throughout the Northern hemisphere (Yoshimura 1971); in continental Europe, it is predominantly found in old forests (Rose 1976, 1988), although in certain circumstances, it can also occur in younger forests (Kalwij et al. 2005; Werth et al. 2006b). Deforestation and habitat fragmentation in its natural range has generated metapopulations (Fedrowitz et al. 2012b; Gu et al. 2001; Snäll et al. 2005), and *L. pulmonaria* is currently a flagship species for conservation (Scheidegger and Werth 2009).

L. pulmonaria is a good model for examining both vertical and horizontal transmissions of symbionts as well as their role in maintaining population structures or changing it in response to environmental perturbations and selection, as reproduction and establishment of vegetative propagules are affected by macro- and microclimatic factors (Martínez et al. 2012; Scheidegger 1995; Werth et al. 2006a). The mycobiont is capable of local and long-distance dispersal, although local dispersal predominates (Wagner et al. 2006; Walser 1999, 2004; Walser et al. 2001; Werth et al. 2006a). If the mycobiont reproduces with vegetative propagules, the green algal photobiont is vertically transmitted and co-dispersed, but not the *Nostoc* photobiont which is thought to be taken up from the environment. The mycobiont is heterothallic, i.e., for sexual reproduction, a compatible mating partner with opposite mating type is needed (Honegger et al. 2004a; Seymour et al. 2005b; Singh et al. 2012). In addition to novel establishment of symbiosis by ascospore germlings, horizontal transmission may also occur if the photobiont is exchanged during establishment from vegetative propagules (Werth and Scheidegger 2012; Wornik and Grube 2010).

Population genetic studies of *L. pulmonaria* have been conducted at all spatio-temporal scales, in many cases taking advantage of microsatellite markers (Werth 2010b). These investigations demonstrated that *L. pulmonaria* has a mixed reproductive strategy, with clonality predominating (Dal Grande et al. 2012; Scheidegger and Werth 2009; Werth and Scheidegger 2012). Observations of both high rates of gene flow among forest patches separated by a centuries old meadow and concurrent autocorrelation in genotype diversity at a local scale were hypothesized to reflect a predominately local dispersal of propagules, leading to genetic similarity at short distances (up to ~200 m), coupled with less frequent long-distance dispersal equalizing allele frequencies at larger spatial scales (Werth et al. 2006b, 2007).

Human activity influences the local abundance of *L. pulmonaria* and genetic diversity is greatly reduced in managed forests, compared to natural or primeval forests (Jüriado et al. 2011; Kalwij et al. 2005; Otálora et al. 2011; Scheidegger et al. 2012; Werth et al. 2006b). Although fragmented forest stands may retain genetic diversity to a certain degree (Hilmo et al. 2012; Otálora et al. 2011; Werth and Scheidegger 2012; Werth et al. 2006b), populations affected by large-scale cutting and forest fires exhibit very low genetic diversity even after a long period (Werth et al. 2006b). In such instances, reproduction of the mycobiont may be essentially clonal (Singh et al. 2012) because of the absence of compatible mating types within a short distance (Singh et al. in press).

On a larger spatial scale (several km^2), multiple, intermingled fungal and algal gene pools have been observed (Nadyeina et al. 2014b; Werth and Scheidegger 2012) and gene pool associations with elevation have been interpreted as evidence of climate-driven local adaptations in the mycobionts (Scheidegger et al. 2012). On a landscape scale, climatic associations of fungal and algal gene pools have also been found for

L. pulmonaria in the largest continuous primeval forest of Europe (Nadyeina et al. 2014b). At the largest spatial scale, genetic differentiation of populations of *L. pulmonaria* mycobionts and photobionts has been shown within Europe and between continents (Scheidegger et al. 2012; Walser et al. 2005; Widmer et al. 2012).

All these studies show that *Lobaria pulmonaria* has served as a workhorse to address new questions based on microsatellite markers. However, in recent years, HTS technology has been developed which allows researchers to address old questions with a higher level of confidence, but also to explore an entirely different set of questions (see below).

4.4.1 Wide Distributions in Lichens

We can expect substantial future work using HTS methods on the geographic distribution of and divergence among lineages of lichen symbionts (phylogeography). The analytical power to answer questions related to divergence time and historic gene flow increases dramatically through the use of thousands of loci based on HTS, thus it should be possible to make substantial progress in this field.

One interesting question with respect to lichen biogeography that we do not have a satisfactory answer to is why some lichens have extraordinarily wide distributions, spanning multiple climatic zones on continents, or across multiple continents. For example, in the genus *Melanelia*, several species are widespread on different continents in the same habitat type throughout the Holarctic, while one species of *Melanelia* is endemic to Europe (Otte et al. 2005). Species of *Physconia* occurring in Europe show a diversity of distribution types, but include several species with wide, intercontinental distributions (Otte et al. 2002). Some lichens show profound gaps in their distributions (disjunctions). For example, the epiphytic lichen *Cavernularia hultenii* occurs disjunctly in spruce (*Picea* spp.) forests of the Pacific Northwest, Newfoundland, and Scandinavia (Printzen et al. 2003). The terricolous lichen *Cetraria aculeata* is disjunctly distributed on different continents in Arctic, maritime Antarctic, and Mediterranean environments characterized by low competition from vascular plants (Fernández-Mendoza et al. 2011). Examples of species with a continuous range across multiple climate zones within a continent are *Niebla cephalota* and *Ramalina menziesii*. Both of these lichens occur across multiple climatic zones and eco-geographic regions within a continent (Brodo et al. 2001).

The wide distributions in certain lichen fungi could be the result of repeated and ongoing long-distance dispersal, including inter-continental dispersal, which would remove any tendencies for populations to be differentiated. It has traditionally been assumed that microorganisms (including eukaryotic microorganisms such as fungi and algae) are not dispersal-limited: "Everything is everywhere—the environment selects" (Baas Becking 1934). Also some modern authors argue that microorganisms should not exhibit biogeographic patterns due to their great abundances, very short generation times, and high dispersal rates (e.g. Finlay 2002). In line with this hypothesis, evidence points toward ongoing and repeated intercontinental dispersal in certain species of bryophytes (Shaw et al. 2014) and Arctic fungi, including lichen-forming taxa (Buschbom 2007; Geml 2011; Geml et al. 2012a, b).

However, most studies of lichen populations find some genetic differentiation among populations and between continents (reviewed in Werth 2010b, 2011), which makes repeated and ongoing long-distance or even intercontinental dispersal a less likely explanation for these taxa, unless the observed biogeographic structure is due to selection by different environments (Geml 2011). For the lichen-forming fungus *R. menziesii,* substantial migration was inferred from the northern part of its large range toward southern and inland sites, but no migration was inferred in the southernmost part of the range. Geml (2011) found evidence for frequent long-distance dispersal in Arctic-Alpine, but not in boreal-temperate fungi including lichens, which was attributed to Arctic taxa being selected for high movement during the various glacial periods.

Some studies show that the wide distributions of lichens are composed of sets of locally adapted symbiont lineages. Sequence analyses of a few loci suggested that while the spatial distribution of photobiont lineages associating with C. aculeata and R. menziesii was influenced by local climatic and ecological conditions, e.g. phorophyte or habitat type, that of the mycobionts showed less ecological specialization but exhibited substantial geographic structure across the range (Fernández-Mendoza et al. 2011; Werth and Sork 2014). These results indicate that associations with locally adapted photobionts may give lichens a broad ecological amplitude and allow them to occupy wide ranges. It remains to be seen if the habitat ecology and climate-related structures found in the photobionts of *C. aculeata* and *R. menziesii* represent a general trend applying to other species. HTS-based studies of the ecological specialization, geographic distribution, and associations among lichen mycobiont and photobiont lineages in other widely distributed lichen species using far more loci would help to reveal how general the observed pattern is.

Many studies of lichens have focused on the description of patterns occurring in nature, providing very valuable information from a natural history perspective, and generating hypotheses on the processes shaping the observed patterns. Field or laboratory experiments can be set up to test specific hypotheses on the performance of symbiont lineages under different experimental conditions. Also, equipped with the new toolset of HTS, we can finally set out to address some of the questions that the descriptive studies have brought up, such as, what are the genes underlying observed morphological or chemical differences among individuals? The discipline of lichenology will benefit greatly from hypotheses-oriented research complementing earlier descriptive studies.

4.4.2 Life History and Population Structure

We hope that a large body of work will be performed to improve our understanding of population structure and genetic diversity in populations of lichen symbionts differing in important traits. For example, only one study has compared population structure across multiple species differing in the size distribution of vegetative propagules (Werth et al. 2014). Another study contrasted the genetic diversity of lichen fungi with clonal and sexual modes of propagation (Otálora et al. 2013). It remains unknown how the heterothallic versus homothallic mating systems of lichen fungi influence population structure or the diversity of populations. In sexual species, we would predict that homothallic, i.e., self-fertile lichen fungi show more population subdivision as selfing would lead to spatial accumulation of the same fungal genotypes, as long as most spores are dispersed locally. The logical next step is to compare species differing in various traits directly affecting reproduction.

We still lack a basic understanding of parentage and of relatedness in populations of lichen fungi. Which individuals contribute as fathers to offspring, and how far can a paternal thallus be away from a maternal one to allow fertilization? Do juvenile cohorts have similar genetic structuring and diversity as the adult populations? Do environmental factors such as forest stand structure influence the genetic diversity of progeny of lichen mycobionts? Applying HTS to the study of the genetic structure and diversity of populations should help to obtain firmer understanding of many important processes influencing lichen populations, and new insights may be gained by using a large number of loci.

4.4.3 Local Adaptations to the Environment and Phenotypic Plasticity

When faced with a changing global climate, organisms can tolerate the new conditions in situ via phenotypic plasticity, migrate to favorable sites, or adapt to the changed conditions (Aitken et al. 2008; Davis et al. 2005; Savolainen et al. 2007). The relative roles of phenotypic plasticity, adaptation, and migration for the persistence of populations in a changing climate are still poorly understood. This represents one of the great challenges to population genetic research.

Green algal photobionts show preferences for specific ecological conditions such as phorophyte species, habitats, and altitudinal zones (Nadyeina et al. 2014b; Peksa and Škaloud 2011; Werth and Sork 2010, 2014; Yahr et al. 2006). As elevation is closely correlated with several macroclimatic parameters (temperature, precipitation and wind), altitudinal distributions of genotypes likely result from local adaptation of the symbionts' genotypes to specific local climatic conditions. The finding of ecological specialization of photobionts provides a basis for future research which can to a large degree be based on controlled experiments, e.g. of the survival and relative growth rate of lichen thalli carrying putatively adaptive photobiont genotypes, transplanted to different environments.

How can HTS be used to investigate phenotypic plasticity and local adaptations to the environment in lichen symbionts? Stapley et al. (2010) point out important aspects of the genomics of adaptations to the environment. Total or partial sequencing of the symbionts' genomes (e.g. via RNA-seq or exon capture approaches, see Sect. 4.3.2) can be used to identify loci under selection. Prior to HTS, finding genes under selection was a matter of sheer luck ("finding the needle in a haystack"). With HTS, it has finally become feasible to identify loci under selection by using methods that result in sequence data of a high number of loci (e.g. RAD-seq, RNA-seq). We think this is one of the most interesting aspects of using HTS. In lichens, discovering whether adaptive changes in one symbiont are paralleled by changes in the other symbiont, opens up exciting avenues for research. One can make use of the principle that loci under divergent selection show a higher than average degree of genetic differentiation between populations as compared to neutrally evolving loci. If the sampling of individuals is based on populations, this property can be used to test whether loci are under divergent selection using "outlier analyses" (Beaumont and Balding 2004; Beaumont and Nichols 1996; Manel et al. 2009). One problematic aspect with using multiple samples per population to determine variants under selection is that individuals are not independent. Both genetic correlations between individuals and large-scale population structure may introduce spurious significances in statistical tests of loci under selection. Thus, it may be better to sample single individuals from many localities for HTS and test for associations of loci with environmental variables after accounting for population structure (Sork et al. 2013). Specialized software has been developed for this purpose (Coop et al. 2010; Eckert et al. 2010; Fumagalli et al. 2011; Joost et al. 2007; Kang et al. 2010). Thus utilized, HTS provides a powerful tool to identify loci under selection.

However, identifying such loci with HTS is not enough. Most studies of local adaptation of genotypes to the environment fail to demonstrate that genotypes have relevance for adaptation, such as increased fitness in specific environments. Testing whether specific genotypes have a fitness advantage in specific environments calls for experiments, e.g. (i) studying relative growth rates (as a measure of fitness) of thalli possessing these variants in different environments, or (ii) comparing the growth of different variants in a common garden, or under experimental conditions. Although growth rates can be determined in short-term (14 day) laboratory experiments (Bidussi et al. 2013), the most promising estimate of fitness in lichens is growth during an extended time period (e.g. relative annual growth rate). Studies of lichen growth based on transplanted lichens are numerous, with firmly established methods (Antoine and McCune 2004; Gauslaa and Goward 2012; Gauslaa et al. 2009;

Hallingbäck 1990; Larsson et al. 2012; McCune et al. 1996; Walser and Scheidegger 2002). A crucial factor for understanding lichen growth (and fitness) is being able to separate the effects of acclimation of thalli to specific local environmental conditions from the effect of the genetic background. Transplanted lichens can acclimate to their new environments within one year (Caldiz 2004). To eliminate effects of local acclimation, individuals carrying specific genetic variants should be grown in the same environment for at least one year before assessment of growth rate (Y. Gauslaa, pers. comm.). Based on genetically different specimens acclimated to the same environment, the relative growth rate can be measured as an estimate of fitness, as it reflects the long-term positive carbon balance of both symbionts.

Another crucial factor determining whether individuals and populations are successful is the survival and establishment of propagules under various conditions (Hilmo et al. 2011; Larsson and Gauslaa 2011; Scheidegger 1995; Sillett et al. 2000; Werth et al. 2006a; Zoller et al. 2000). In-depth studies of survival and growth of diaspores carrying different genetic variants could provide valuable information on whether they are viable in a given environment. This information is essential for understanding the role of genetic variation for the local adaptation of lichens.

Ultimately, the persistence of individuals carrying a specific "adaptive" genetic variant in an environment depends on their fitness in these environmental circumstances. The growth and survival rates of a transplanted thallus fragment will reflect the combined fitness of the symbiosis partners. Either of them could be locally adapted to its environment, but we can only measure the survival and growth rate of the symbiotic association as a whole.

One extension from studies of local adaptations to the environment is the study of recent, or ongoing, speciation processes. In other fields, considerable attention has been devoted to the study of ecological speciation in sympatry, a situation where reproductive barriers arise between species despite ongoing gene flow (reviewed in Rice et al. 2011). Whole genome resequencing is the method of choice to find genomic regions involved in speciation. By studying SNPs along entire linkage groups, one can find the regions in a genome that are involved in preventing gene flow and thus in establishing new species. By genomic comparisons of taxa differing in phenotype, it is now possible to determine the genomic regions that are responsible for complex phenotypes (Mardis 2008), which allows researchers to better understand phenotypic trait variation. HTS holds great potential to reveal new exciting insights about the underlying genomic foundations for complex phenotypic traits, phenotypic plasticity, and ecological speciation.

4.4.4 Diversity of Symbiont Associations and Associated Organisms

One intriguing but simultaneously complicating aspect of using lichens as a study system is intrathalline genetic variability. Each lichen thallus contains many photobiont cells (Honegger 1998; Schwendener 1868). These can be derived by multiple cell divisions and be genetically identical. For example, thalli of *L. pulmonaria* are genetically homogeneous (Walser 2003), and thalli of *P. membranacea* are composed of a single mycobiont and a single photobiont genotype (Andrésson and Werth, unpublished). However, based on microscopy, researchers have concluded that thalli of *Hypogymnia physodes*, *Parmelia sulcata,* and *Physcia tenella* may develop from the fusion of several vegetative propagules (Ott 1987; Schuster and Jahns 1985) or from the incorporation of algal cells by rhizinomorphs (Sanders and Rico 1992). Also, initial developmental stages can contain different photobionts than the adult thallus (Friedl 1987). This can result in multiple photobiont or fungal genotypes within a lichen thallus; such intrathalline variability has indeed been reported by several genetic studies (Casano et al. 2011; Dal Grande et al. 2014a; del Campo et al. 2013; Larson and Carey 1986; Mansournia et al. 2011; Piercey-Normore 2006a). HTS of lichen samples is likely to provide information on intrathalline diversity of photobionts and mycobionts in a variety of lichen species. Prior to HTS technology, the detection of intrathalline variability and the discovery of occurrences of additional symbiont genotypes in a thallus would only have been possible through a painstaking effort, involving PCR, cloning of PCR products, and Sanger sequencing. In the era of HTS, comparable analyses have become straightforward and affordable.

We expect that HTS-based population studies will allow researchers to obtain a greater in-depth understanding of the associations among lichen symbionts, their transmission modes (Dal Grande et al. 2012; Werth and Scheidegger 2012), and the population biology and ecology of the individual symbionts. Using genome reduction approaches, it has become feasible to sequence a fraction of a genome for many bar-coded individuals at a time, yielding thousands of loci per individual that can be utilized to define mycobiont, photobiont, and lichen clones at a high level of statistical confidence. HTS data may reveal that lichens are far more complex study systems than anticipated.

4.5 Conclusions

The development of technologies, allowing sequencing at a fraction of the cost of traditional Sanger sequencing now facilitates sequencing of whole genomes and has the potential to revolutionize the way in which we perceive the molecular and population biology of lichens. HTS data allow identifying the few loci in a genome that are under selection, which has not been feasible with data sets containing a moderate amount of loci. Moreover, using WGS, it is now possible to find the parts of a genome that are changing during population divergence and speciation processes. Thus, HTS will provide essential information for understanding the nature of lichen symbioses and of lichen population biology. The most exciting times for lichenology so far followed the invention of the microscope in the late nineteenth century which lead to the discovery that lichens represent a symbiosis (Schwendener 1868), and later, the application of DNA sequencing (Sanger et al. 1965) resulting in a detailed understanding of evolutionary and taxonomic relationships in lichens. We have great hopes for HTS to lead the way to yet another exciting era for lichenology, with many intriguing avenues of research in population biology.

Acknowledgments SW was supported through a Marie Curie Fellowship ("LICHENOMICS," #302589) within the framework of FP7 of the European Commission. SW, ÓSA, and ZOJ received funding through project grants from the Icelandic Research Fund (#120247023 and #141102-051).

References

Aitken SN, Yeaman S, Holliday JA, Wang TL, Curtis-McLane S (2008) Adaptation, migration or extirpation: climate change outcomes for tree populations. Evol Appl 1:95–111. doi:10.1111/j.1752-4571.2007.00013.x

Albert TJ, Molla MN, Muzny DM et al (2007) Direct selection of human genomic loci by microarray hybridization. Nat Methods 4:903–905. doi:10.1038/nmeth1111

Altermann S (2009) Geographic structure in a symbiotic mutualism. PhD thesis, University of California at Santa Cruz

Altermann S, Leavitt SD, Goward T, Nelsen MP, Lumbsch HT (2014) How do you solve a problem like *Letharia*? A new look at cryptic species in lichen-forming fungi using Bayesian clustering and SNPs from multilocus sequence data. PLoS ONE 9. doi:10.1371/journal.pone.0097556

Altshuler D et al (2010) A map of human genome variation from population-scale sequencing. Nature 467:1061–1073. doi:10.1038/nature09534

Antoine ME, McCune B (2004) Contrasting fundamental and realized ecological niches with epiphytic lichen transplants in an old-growth *Pseudotsuga* forest. Bryologist 107:163–172

Armaleo D, Miao V (1999) Symbiosis and DNA methylation in the *Cladonia* lichen fungus. Symbiosis 26:143–163

Arnerup J, Högberg N, Thor G (2004) Phylogenetic analysis of multiple loci reveal the population structure within *Letharia* in the Caucasus and Morocco. Mycol Res 108:311–316. doi:10.1017/S0953756204009360

Avise JC, Arnold J, Ball RM et al (1987) Intraspecific phylogeography—the mitochondrial-DNA bridge between population genetics and systematics. Annu Rev Ecol Syst 18:489–522

Baas Becking LGMB (1934) Geobioligie; of inleiding tot de milieukunde. Van Stockum and Zoon, Den Haag

Baloch E, Grube M (2009) Pronounced genetic diversity in tropical epiphyllous lichen fungi. Mol Ecol 18:2185–2197. doi:10.1111/j.1365-294X.2009.04183.x

Bates ST, Cropsey GWG, Caporaso JG, Knight R, Fierer N (2011) Bacterial communities associated with the lichen symbiosis. Appl Environ Microbiol 77:1309–1314. doi:10.1128/aem.02257-10

Beard KH, DePriest PT (1996) Genetic variation within and among mats of the reindeer lichen *Cladina subtenuis*. Lichenologist 28:171–181

Belinchón R, Ellis CJ, Yahr R (2014) Microsatellite loci in two epiphytic lichens with contrasting dispersal modes: Nephroma laevigatum and N. parile (Nephromataceae) App Plant Sci 2, doi: 10.3732/apps.1400080

Beaumont MA, Balding DJ (2004) Identifying adaptive genetic divergence among populations from genome scans. Mol Ecol 13:969–980. doi:10.1111/j.1365-294X.2004.02125.x

Beaumont MA, Nichols RA (1996) Evaluating loci for use in the genetic analysis of population structure. Proc R Soc B 263:1619–1626

Bi K, Vanderpool D, Singhal S, Linderoth T, Moritz C, Good JM (2012) Transcriptome-based exon capture enables highly cost-effective comparative genomic data collection at moderate evolutionary scales. BMC Genomics 13. doi:10.1186/1471-2164-13-403

Bidussi M, Gauslaa Y, Solhaug K (2013) Prolonging the hydration and active metabolism from light periods into nights substantially enhances lichen growth. Planta 237:1359–1366. doi:10.1007/s00425-013-1851-y

Blaha J, Baloch E, Grube M (2006) High photobiont diversity associated with the euryoecious lichen-forming ascomycete *Lecanora rupicola* (Lecanoraceae, Ascomycota). Biol J Linn Soc 88:283–293

Boisvert S, Raymond F, Godzaridis E, Laviolette F, Corbeil J (2012) Ray Meta: scalable de novo metagenome assembly and profiling. Genome Biol 13:R122. doi:10.1186/gb-2012-13-12-r122

Bolli JC, Wagner HH, Kalwij JM, Werth S, Cherubini P, Scheidegger C, Rigling A (2008) Growth dynamics after historic disturbance in a montane forest and its implications for an endangered epiphytic lichen. Bot Helv 118:111–127. doi:10.1007/s00035-008-0834-2

Brodo IM, Sharnoff SD, Sharnoff S (2001) Lichens of North America. Yale University Press, New Haven

Buschbom J (2007) Migration between continents: geographical structure and long-distance gene flow in *Porpidia flavicunda* (lichen-forming Ascomycota). Mol Ecol 16:1835–1846. doi:10.1111/j.1365-294X.2007.03258.x

Caldiz MS (2004) Seasonal growth pattern in the lichen *Pseudocyphellaria berberina* in north-western Patagonia. Lichenologist 36:435–444. doi:10.1017/s0024282904014422

Casano LM, del Campo EM, Garcia-Breijo FJ et al (2011) Two *Trebouxia* algae with different physiological performances are ever-present in lichen thalli of *Ramalina farinacea*. Coexistence versus competition? Environ Microbiol 13:806–818. doi:10.1111/j.1462-2920.2010.02386.x

Cassie DM, Piercey-Normore MD (2008) Dispersal in a sterile lichen-forming fungus, *Thamnolia subuliformis* (Ascomycotina: Icmadophilaceae). Botany-Botanique 86:751–762. doi:10.1139/b08-032

Coop G, Witonsky D, Di Rienzo A, Pritchard JK (2010) Using environmental correlations to identify loci underlying local adaptation. Genetics 185:1411–1423

Cornejo C, Chabanenko S, Scheidegger C (2009) Phylogenetic analysis indicates transitions from vegetative to sexual reproduction in the *Lobaria retigera* group (Lecanoromycetidae, Ascomycota). Lichenologist 41:1–10

Cornejo C, Scheidegger C (2010) *Lobaria macaronesica* sp. nov., and the phylogeny of *Lobaria* sect. *Lobaria* (Lobariaceae) in Macaronesia. Bryologist 113:590–604. doi:10.1639/0007-2745-113.3.590

Cosart T, Beja-Pereira A, Chen SY, Ng SB, Shendure J, Luikart G (2011) Exome-wide DNA capture and next generation sequencing in domestic and wild species. BMC Genomics 12. doi:10.1186/1471-2164-12-347

Crespo A, Bridge PD, Hawksworth DL, Grube M, Cubero OF (1999) Comparison of rRNA genotype frequencies of *Parmelia sulcata* from long established and recolonizing sites following sulphur dioxide amelioration. Plant Syst Evol 217:177–183

Culberson CF, Culberson WL, Johnson A (1988) Gene flow in lichens. Am J Bot 75:1135–1139

Culberson WL, Culberson CF, Johnson A (1993) Speciation in lichens of the *Ramalina siliquosa* complex (Ascomycotina, Ramalinaceae): gene flow and reproductive isolation. Am J Bot 80:1472–1481

Dal Grande F, Alors D, Divakar PK, Bálint M, Crespo A, Schmitt I (2014a) Insights into intrathalline genetic diversity of the cosmopolitan lichen symbiotic green alga *Trebouxia decolorans* Ahmadjian using microsatellite markers. Mol Phylogenet Evol 72:54–60. doi:10.1016/j.ympev.2013.12.010

Dal Grande F, Beck A, Cornejo C, Singh G, Cheenacharoen S, Nelsen MP, Scheidegger C (2014b) Molecular phylogeny and symbiotic selectivity of the green algal genus *Dictyochloropsis* s.l. (Trebouxiophyceae): a polyphyletic and widespread group forming photobiont-mediated guilds in the lichen family Lobariaceae. New Phytol 202:455–470. doi:10.1111/nph.12678

Dal Grande F, Beck A, Singh G, Schmitt I (2013) Microsatellite primers in the lichen symbiotic alga *Trebouxia decolorans* (Trebouxiophyceae). Appl Plant Sci 1:1200400. doi:10.3732/apps.1200400

Dal Grande F, Widmer I, Beck A, Scheidegger C (2009) Microsatellite markers for *Dictyochloropsis reticulata* (Trebouxiophyceae), the symbiotic alga of the lichen *Lobaria pulmonaria* (L.). Conserv Genet 11:1147–1149. doi:10.1007/s10592-009-9904-2

Dal Grande F, Widmer I, Wagner HH, Scheidegger C (2012) Vertical and horizontal photobiont transmission within populations of a lichen symbiosis. Mol Ecol 21:3159–3172. doi:10.1111/j.1365-294X.2012.05482.x

Davis MB, Shaw RG, Etterson JR (2005) Evolutionary responses to changing climate. Ecology 86:1704–1714

De Wit P, Pespeni MH, Ladner JT et al (2012) The simple fool's guide to population genomics via RNA-Seq: an introduction to high-throughput sequencing data analysis. Mol Ecol Resour 12:1058–1067. doi:10.1111/1755-0998.12003

del Campo EM, Catala S, Gimeno J et al (2013) The genetic structure of the cosmopolitan three-partner lichen *Ramalina farinacea* evidences the concerted diversification of symbionts. FEMS Microbiol Ecol 83:310–323. doi:10.1111/j.1574-6941.2012.01474.x

DePriest PT (1993) Small subunit rDNA variation in a population of lichen fungi due to optional group-I introns. Gene 134:67–74

DePriest PT (1994) Variation in the *Cladonia chlorophaea* complex. II: Ribosomal DNA variation in a southern Appalachian population. Bryologist 97:117–126. doi:10.2307/3243749

DePriest PT (2004) Early molecular investigations of lichen-forming symbionts: 1986–2001. Annu Rev Microbiol 58:273–301

Devkota S, Cornejo C, Werth S, Chaudhary RP, Scheidegger C (2014) Characterization of microsatellite loci in the Himalayan lichen fungus *Lobaria pindarensis* (Lobariaceae). Appl Plant Sci 2:1300101. doi:10.3732/apps.1300101

Druley TE, Vallania FLM, Wegner DJ et al (2009) Quantification of rare allelic variants from pooled genomic DNA. Nat Methods 6:263–265. doi:10.1038/nmeth.1307

Dyer PS, Murtagh GJ, Crittenden PD (2001) Use of RAPD-PCR DNA fingerprinting and vegetative incompatibility tests to investigate genetic variation within lichen-forming fungi. Symbiosis 31:213–229

Eckert AJ, Bower AD, González-Martínez SC, Wegrzyn JL, Coop G, Neale DB (2010) Back to nature: ecological genomics of loblolly pine (*Pinus taeda*, Pinaceae). Mol Ecol 19:3789–3805. doi:10.1111/j.1365-294X.2010.04698.x

Ekblom R, Galindo J (2011) Applications of next generation sequencing in molecular ecology of non-model organisms. Heredity 107:1–15. doi:10.1038/hdy.2010.152

Ellison CE, Hall C, Kowbel D, Welch J, Brem RB, Glass NL, Taylor JW (2011) Population genomics and local adaptation in wild isolates of a model microbial eukaryote. Proc Natl Acad Sci USA 108:2831–2836. doi:10.1073/pnas.1014971108

Elshire RJ, Glaubitz JC, Sun Q, Poland JA, Kawamoto K, Buckler ES, Mitchell SE (2011) A robust, simple genotyping-by-sequencing (GBS) approach for high diversity species. PLoS ONE 6:e19379. doi:10.1371/journal.pone.0019379

Elvebakk A, Papaefthimiou D, Robertsen EH, Liaimer A (2008) Phylogenetic patterns among *Nostoc* cyanobionts within bi- and tripartite lichens of the genus *Pannaria*. J Phycol 44:1049–1059. doi:10.1111/j.1529-8817.2008.00556.x

Fahselt D (1986) Multiple enzyme forms of morphotypes in a population of *Cladonia cristatella* Tuck. Bryologist 89:139–143

Fahselt D (1989) Enzyme polymorphism in sexual and asexual umbilicate lichens from Sverdrup Pass, Ellesmere Island, Canada. Lichenologist 21:279–285. doi:10.1017/S0024282989000496

Fahselt D, Alstrup V, Tavares S (1995) Enzyme polymorphism in *Umbilicaria cylindrica* in northwestern Greenland. Bryologist 98:118–122

Fahselt D, Hageman C (1983) Isozyme banding patterns in two stands of *Cetraria arenaria* Kärnef. Bryologist 86:129–134

Fahselt D, Jancey RC (1977) Polyacrylamide gel electrophoresis of protein extracts from members of the *Parmelia perforata* complex. Bryologist 80:429–438

Fahselt D, Trembley M (1999) Effect of sampling time on isozyme banding patterns in *Cladonia rangiferina* and *Umbilicaria mammulata*. Lichenologist 31:397–402

Fedrowitz K, Kaasalainen U, Rikkinen J (2011) Genotype variability of *Nostoc* symbionts associated with three epiphytic *Nephroma* species in a boreal forest landscape. Bryologist 114:220–230. doi:10.1639/0007-2745-114.1.220

Fedrowitz K, Kaasalainen U, Rikkinen J (2012a) Geographic mosaic of symbiont selectivity in a genus of epiphytic cyanolichens. Ecol Evol 2:2291–2303. doi:10.1002/ece3.343

Fedrowitz K, Kuusinen M, Snall T (2012b) Metapopulation dynamics and future persistence of epiphytic cyanolichens in a European boreal forest ecosystem. J Appl Ecol 49:493–502. doi:10.1111/j.1365-2664.2012.02113.x

Feng SH, Cokus SJ, Zhang XY et al (2010) Conservation and divergence of methylation patterning in plants and animals. Proc Natl Acad Sci USA 107:8689–8694. doi:10.1073/pnas.1002720107

Fernández-Mendoza F, Domaschke S, García MA, Jordan P, Martín MP, Printzen C (2011) Population structure of mycobionts and photobionts of the widespread lichen *Cetraria aculeata*. Mol Ecol 20:1208–1232. doi:10.1111/j.1365-294X.2010.04993.x

Finlay BJ (2002) Global dispersal of free-living microbial eukaryote species. Science 296:1061–1063. doi:10.1126/science.1070710

Friedl T (1987) Thallus development and phycobionts of the parasitic lichen *Diploschistes muscorum*. Lichenologist 19:183–191

Fumagalli M, Sironi M, Pozzoli U, Ferrer-Admettla A, Pattini L, Nielsen R (2011) Signatures of environmental genetic adaptation pinpoint pathogens as the main selective pressure through human evolution. PLoS Genet 7:e1002355. doi:10.1371/journal.pgen.1002355

Futschik A, Schlotterer C (2010) The next generation of molecular markers from massively parallel sequencing of pooled DNA samples. Genetics 186:207–218. doi:10.1534/genetics.110.114397

Garvin MR, Saitoh K, Gharrett AJ (2010) Application of single nucleotide polymorphisms to non-model species: a technical review. Mol Ecol Resour 10:915–934. doi:10.1111/j.1755-0998.2010.02891.x

Gauslaa Y, Goward T (2012) Relative growth rates of two epiphytic lichens *Lobaria pulmonaria* and *Hypogymnia occidentalis*, transplanted within and outside of Populus dripzones. Botany 90:954–965. doi:10.1139/b2012-062

Gauslaa Y, Palmqvist K, Solhaug KA, Hilmo O, Holien H, Nybakken L, Ohlson M (2009) Size-dependent growth of two old-growth associated macrolichen species. New Phytol 181:683–692. doi:10.1111/j.1469-8137.2008.02690.x

Geml J (2011) Coalescent analyses reveal contrasting patterns of intercontinental gene flow in arctic-alpine and boreal-temperate fungi. In: Fontaneto D (ed) Biogeography of micro-organisms. Is everything small everywhere? Cambridge University Press, New York, pp 177–190

Geml J, Kauff F, Brochmann C, Lutzoni F, Laursen GA, Redhead SA, Taylor DL (2012a) Frequent circumarctic and rare transequatorial dispersals in the lichenised agaric genus *Lichenomphalia* (*Hygrophoraceae*, *Basidiomycota*). Fungal Biol 116:388–400. doi:10.1016/j.funbio.2011.12.009

Geml J, Timling I, Robinson CH et al (2012b) An arctic community of symbiotic fungi assembled by long-distance dispersers: phylogenetic diversity of ectomycorrhizal basidiomycetes in Svalbard based on soil and sporocarp DNA. J Biogeogr 39:74–88. doi:10.1111/j.1365-2699.2011.02588.x

Gjerde I, Blom HH, Lindblom L, Saetersdal M, Schei FH (2012) Community assembly in epiphytic lichens in early stages of colonization. Ecology 93:749–759

Goffinet B, Bayer RJ (1997) Characterization of mycobionts of photomorph pairs in the Peltigerineae (lichenized ascomycetes) based on internal transcribed spacer sequences of the nuclear ribosomal DNA. Fungal Genet Biol 21:228–237

Grube M, Berg G, Andrésson ÓS, Vilhelmsson O, Dyer PS, Miao VPW (2013) Lichen genomics: prospects and progress. In: Martin F (ed) The ecological genomics of fungi. Wiley, Hoboken, pp 191–212. doi:10.1002/9781118735893.ch9

Grube M, Cernava T, Soh J et al (2014) Exploring functional contexts of symbiotic sustain within lichen-associated bacteria by comparative omics. ISME J. doi:10.1038/ismej.2014.138

Gu WD, Kuusinen M, Konttinen T, Hanski I (2001) Spatial pattern in the occurrence of the lichen *Lobaria pulmonaria* in managed and virgin boreal forests. Ecography 24:139–150

Guzow-Krzeminska B, Stocker-Wörgötter E (2013) Development of microsatellite markers in *Protoparmeliopsis muralis* (lichenized Ascomycete)—a common lichen species. Lichenologist 45:791–798. doi:10.1017/S002428291300042X

Hageman C, Fahselt D (1990) Multiple enzyme forms as indicators of functional sexuality in the lichen *Umbilicaria vellea*. Bryologist 93:389–394

Hageman C, Fahselt D (1992) Geographical distribution of enzyme polymorphisms in the lichen *Umbilicaria mammulata*. Bryologist 95:316–323

Hallingbäck T (1990) Transplanting Lobaria pulmonaria to new localities and a review on the transplanting of lichens. Windahlia 18:57–64

Heibel E, Lumbsch HT, Schmitt I (1999) Genetic variation of *Usnea filipendula* (Parmeliaceae) populations in western Germany investigated by RAPDs suggests reinvasion from various sources. Am J Bot 86:753–757

Helyar SJ, Hemmer-Hansen J, Bekkevold D et al (2011) Application of SNPs for population genetics of nonmodel organisms: new opportunities and challenges. Mol Ecol Resour 11:123–136. doi:10.1111/j.1755-0998.2010.02943.x

Hilmo O, Lundemo S, Holien H, Stengrundet K, Stenøien HK (2012) Genetic structure in a fragmented Northern Hemisphere rainforest: large effective sizes and high connectivity among populations of the epiphytic lichen *Lobaria pulmonaria*. Mol Ecol 21:3250–3265. doi:10.1111/j.1365-294X.2012.05605.x

Hilmo O, Rocha L, Holien H, Gauslaa Y (2011) Establishment success of lichen diaspores in young and old boreal rainforests: a comparison between *Lobaria pulmonaria* and *L. scrobiculata*. Lichenologist 43:241–255. doi:10.1017/S0024282910000794

Hodges E, Xuan Z, Balija V et al (2007) Genome-wide in situ exon capture for selective resequencing. Nat Genet 39:1522–1527. doi:10.1038/ng.2007.42

Hodkinson BP, Gottel NR, Schadt CW, Lutzoni F (2012) Photoautotrophic symbiont and geography are major factors affecting highly structured and diverse bacterial communities in the lichen microbiome. Environ Microbiol 14:147–161

Hodkinson BP, Lendemer JC (2013) Next-generation sequencing reveals sterile crustose lichen phylogeny. Mycosphere 4:1028–1039

Högberg N, Kroken S, Thor G, Taylor JW (2002) Reproductive mode and genetic variation suggest a North American origin of European *Letharia vulpina*. Mol Ecol 11:1191–1196. doi:10.1046/j.1365-294X.2002.01513.x

Hohenlohe PA, Amish SJ, Catchen JM, Allendorf FW, Luikart G (2011) Next-generation RAD sequencing identifies thousands of SNPs for assessing hybridization between rainbow and westslope cutthroat trout. Mol Ecol Resour 11:117–122. doi:10.1111/j.1755-0998.2010.02967.x

Honegger R (1998) The lichen symbiosis—what is so spectacular about it? Lichenologist 30:193–212. doi:10.1017/S002428299200015X

Honegger R, Zippler U, Gansner H, Scherrer S (2004a) Mating systems in the genus *Xanthoria* (lichen-forming ascomycetes). Mycol Res 108:480–488

Honegger R, Zippler U, Scherrer S, Dyer PS (2004b) Genetic diversity in *Xanthoria parietina* (L.) Th. Fr. (lichen-forming ascomycete) from worldwide locations. Lichenologist 36:381–390

Hudson ME (2008) Sequencing breakthroughs for genomic ecology and evolutionary biology. Mol Ecol Resour 8:3–17. doi:10.1111/j.1471-8286.2007.02019.x

Itten B, Honegger R (2010) Population genetics in the homothallic lichen-forming ascomycete *Xanthoria parietina*. Lichenologist 42:751–761. doi:10.1017/s0024282910000411

Jaillon O, Aury JM, Brunet F et al (2004) Genome duplication in the teleost fish Tetraodon nigroviridis reveals the early vertebrate proto-karyotype. Nature 431:946–957. doi:10.1038/nature03025

Jaillon O, Aury JM, Noel B et al (2007) The grapevine genome sequence suggests ancestral hexaploidization in major angiosperm phyla. Nature 449:463–465. doi:10.1038/nature06148

Jarne P, Lagoda PJL (1996) Microsatellites, from molecules to populations and back. Trends Ecol Evol 11:424–429

Jones TC, Green TGA, Hogg ID, Wilkins RJ (2012) Isolation and characterization of microsatellites in the lichen *Buellia frigida* (Physciaceae), an Antarctic endemic. Am J Bot 99:131–133. doi:10.3732/ajb.1100440

Joost S, Bonin A, Bruford MW, Despres L, Conord C, Erhardt G, Taberlet P (2007) A spatial analysis method (SAM) to detect candidate loci for selection: towards a landscape genomics approach to adaptation. Mol Ecol 16:3955–3969. doi:10.1111/j.1365-294X.2007.03442.x

Junttila S, Rudd S (2012) Characterization of a transcriptome from a non-model organism *Cladonia rangiferina*, the grey reindeer lichen, using high-throughput next generation sequencing and EST sequence data. BMC Genomics 13:575. doi:10.1186/1471-2164-13-575

Jüriado I, Liira J, Csencsics D, Widmer I, Adolf C, Kohv K, Scheidegger C (2011) Dispersal ecology of the endangered woodland lichen *Lobaria pulmonaria* in managed hemiboreal forest landscape. Biodivers Conserv 20:1803–1819. doi:10.1007/s10531-011-0062-8

Kalwij JM, Wagner HH, Scheidegger C (2005) Effects of stand-level disturbances on the spatial distribution of a lichen indicator. Ecol Appl 15:2015–2024. doi:10.1890/04-1912

Kampa A, Gagunashvili AN, Gulder TAM et al (2013) Metagenomic natural product discovery in lichen provides evidence for a family of biosynthetic pathways in diverse symbioses. Proc Natl Acad Sci USA 110:E3129–3137. doi: 10.1073/pnas.1305867110

Kang HM, Sul JH, Service SK et al (2010) Variance component model to account for sample structure in genome-wide association studies. Nat Genet 42:348–354. doi:10.1038/ng.548

Kim S, Misra A (2007) SNP genotyping: technologies and biomedical applications. In: Annual review of biomedical engineering, vol 9, pp 289–320. doi:10.1146/annurev.bioeng.9.060906.152037

Kotelko R, Doering M, Piercey-Normore MD (2008) Species diversity and genetic variation of terrestrial lichens and bryophytes in a boreal jack pine forest of central Canada. Bryologist 111:594–606. doi:10.1639/0007-2745-111.4.594

Kroken S, Taylor JW (2001) Outcrossing and recombination in the lichenized fungus *Letharia*. Fungal Genet Biol 34:83–92. doi:10.1006/fgbi.2001.1291

Kwok PY (2001) Methods for genotyping single nucleotide polymorphisms. Annu Rev Genomics Hum Genet 2:235–258. doi:10.1146/annurev.genom.2.1.235

Larson DW, Carey CK (1986) Phenotypic variation within "individual" lichen thalli. Am J Bot 73:214–223. doi:10.2307/2444174

Larsson P, Gauslaa Y (2011) Rapid juvenile development in old forest lichens. Botany-Botanique 89:65–72. doi:10.1139/b10-086

Larsson P, Solhaug KA, Gauslaa Y (2012) Seasonal partitioning of growth into biomass and area expansion in a cephalolichen and a cyanolichen of the old forest genus *Lobaria*. New Phytol 194:991–1000. doi:10.1111/j.1469-8137.2012.04130.x

Lättman H, Lindblom L, Mattsson JE, Milberg P, Skage M, Ekman S (2009) Estimating the dispersal capacity of the rare lichen *Cliostomum corrugatum*. Biol Conserv 142:1870–1878. doi:10.1016/j.biocon.2009.03.026

Lawniczak MKN, Emrich SJ, Holloway AK et al (2010) Widespread divergence between incipient *Anopheles gambiae* species revealed by whole genome sequences. Science 330:512–514. doi:10.1126/science.1195755

Lindblom L, Ekman S (2006) Genetic variation and population differentiation in the lichen-forming ascomycete *Xanthoria parietina* on the island Storfosna, central Norway. Mol Ecol 15:1545–1559. doi:10.1111/j.1365-294X.2006.02880.x

Lindblom L, Ekman S (2007) New evidence corroborates population differentiation in *Xanthoria parietina*. Lichenologist 39:259–271. doi:10.1017/S0024282907006780

Lindblom L, Søchting U (2013) Genetic diversity of the photobiont of the bipolar lichen-forming ascomycete *Xanthomendoza borealis*. Herzogia 26:307–322

Liu L, Li YH, Li SL et al (2012) Comparison of next-generation sequencing systems. J Biomed Biotech 251364. doi:10.1155/2012/251364

Lohtander K, Kallersjo M, Tehler A (1998) Dispersal strategies in *Roccellina capensis* (Arthoniales). Lichenologist 30:341–350

Lücking R, Dal-Forno M, Sikaroodi M et al (2014) A single macrolichen constitutes hundreds of unrecognized species. Proc Natl Acad Sci USA 111:11091–11096. doi: 10.1073/pnas.1403517111

Magain N, Forrest LL, Serusiaux E, Goffinet B (2010) Microsatellite primers in the *Peltigera dolichorhiza* complex (lichenized ascomycete, Peltigerales). Am J Bot 97:e102–e104

Manel S, Conord C, Després L (2009) Genome scan to assess the respective role of host-plant and environmental constraints on the adaptation of a widespread insect. BMC Evol Biol 9:288. doi:10.1186/1471-2148-9-288

Manoharan SS, Miao VPW, Andrésson ÓS (2012) LEC-2, a highly variable lectin in the lichen *Peltigera membranacea*. Symbiosis 58:91–98

Mansournia MR, Wu B, Matsushita N, Hogetsu T (2011) Genotypic analysis of the foliose lichen *Parmotrema tinctorum* using microsatellite markers: association of mycobiont and photobiont, and their reproductive modes. Lichenologist 44:1–22. doi:10.1017/S0024282911000909

Mardis ER (2008) The impact of next-generation sequencing technology on genetics. Trends Genet 24:133–141. doi:10.1016/j.tig.2007.12.007

Martínez I, Flores T, Otálora MAG, Belinchón R, Prieto M, Aragón G, Escudero A (2012) Multiple-scale environmental modulation of lichen reproduction. Fungal Biol 116:1192–1201

Mattsson J-E (1991) Protein banding patterns in some American and European species of *Cetraria*. Bryologist 94:261–269

Mattsson JE, Hansson A-C, Lindblom L (2009) Genetic variation in relation to substratum preferences of *Hypogymnia physodes*. Lichenologist 41:547–555. doi:10.1017/S0024282909990247

McCune B, Derr CC, Muir PS, Shirazi A, Sillett SC, Daly WJ (1996) Lichen pendants for transplant and growth experiments. Lichenologist 28:161–169

McDonald TR, Mueller O, Dietrich FS, Lutzoni F (2013) High-throughput genome sequencing of lichenizing fungi to assess gene loss in the ammonium transporter/ammonia permease gene family. BMC Genom 14:225. doi:10.1186/1471-2164-14-225

Metzker ML (2010) Applications of next-generation sequencing. Sequencing technologies—the next generation. Nat Rev Genet 11:31–46. doi:10.1038/nrg2626

Miao VPW, Manoharan SS, Snaebjarnarson V, Andrésson ÓS (2012) Expression of lec-1, a mycobiont gene encoding a galectin-like protein in the lichen *Peltigera membranacea*. Symbiosis 57:23–31

Millennium Ecosystem Assessment (2005) Ecosystems and human wellbeing: wetlands and water synthesis. World Resources, Washington, DC

Miller MR, Dunham JP, Amores A, Cresko WA, Johnson EA (2007) Rapid and cost-effective polymorphism identification and genotyping using restriction site associated DNA (RAD) markers. Genome Res 17:240–248. doi:10.1101/gr.5681207

Muggia L, Grube M, Tretiach M (2008) Genetic diversity and photobiont associations in selected taxa of the *Tephromela atra* group (Lecanorales, lichenised Ascomycota). Mycol Prog 7:147–160. doi:10.1007/s11557-008-0560-6

Muggia L, Perez-Ortega S, Fryday A, Spribille T, Grube M (2014) Global assessment of genetic variation and phenotypic plasticity in the lichen-forming species *Tephromela atra*. Fungal Divers 64:233–251. doi:10.1007/s13225-013-0271-4

Muggia L, Zellnig G, Rabensteiner J, Grube M (2010) Morphological and phylogenetic study of algal partners associated with the lichen-forming fungus *Tephromela atra* from the Mediterranean region. Symbiosis 51:149–160

Murtagh GJ, Dyer PS, Furneaux PA, Crittenden PD (2002) Molecular and physiological diversity in the bipolar lichen-forming fungus *Xanthoria elegans*. Mycol Res 106:1277–1286

Nadyeina O, Cornejo C, Boluda CG, Myllys L, Rico VJ, Crespo A, Scheidegger C (2014a) Characterization of microsatellite loci in lichen-forming fungi of *Bryoria* Section *Implexae* (Parmeliaceae). Appl Plant Sci 2:1400037. doi:10.3732/apps.1400037

Nadyeina O, Dymytrova L, Naumovych A, Postoyalkin S, Werth S, Cheenacharoen S, Scheidegger C (2014b) Microclimatic differentiation of genepools in the *Lobaria pulmonaria* symbiosis in a primeval forest landscape. Mol Ecol 23(21):5164–5178. doi: 10.1111/mec.12928

Neafsey DE, Barker BM, Sharpton TJ et al (2010) Population genomic sequencing of *Coccidioides* fungi reveals recent hybridization and transposon control. Genome Res 20:938–946. doi:10.1101/gr.103911.109

Nelsen MP, Gargas A (2008) Dissociation and horizontal transmission of codispersing lichen symbionts in the genus *Lepraria* (Lecanorales: Stereocaulaceae). New Phytol 177:264–275. doi:10.1111/j.1469-8137.2007.02241.x

Ng SB, Buckingham KJ, Lee C et al (2010) Exome sequencing identifies the cause of a mendelian disorder. Nat Genet 42:30–41. doi:10.1038/ng.499

Ng SB, Turner EH, Robertson PD et al (2009) Targeted capture and massively parallel sequencing of 12 human exomes. Nature 461:153–272. doi:10.1038/nature08250

Nielsen R (2000) Estimation of population parameters and recombination rates from single nucleotide polymorphisms. Genetics 154:931–942

Nyati S, Werth S, Honegger R (2013) Genetic diversity of sterile cultured *Trebouxia* photobionts associated with the lichen-forming fungus *Xanthoria parietina* visualized with RAPD-PCR fingerprinting techniques. Lichenologist 45:825–840. doi:10.1017/S0024282913000546

O'Brien HE, Miadlikowska J, Lutzoni F (2005) Assessing host specialization in symbiotic cyanobacteria associated with four closely related species of the lichen fungus *Peltigera*. Eur J Phycol 40:363–378. doi:10.1080/09670260500342647

O'Brien HE, Miadlikowska J, Lutzoni F (2009) Assessing reproductive isolation in highly diverse communities of the lichen-forming fungal genus *Peltigera*. Evolution 63:2076–2086. doi:10.1111/j.1558-5646.2009.00685.x

Ohmura Y, Kawachi M, Kasai F, Watanabe MM, Takeshita S (2006) Genetic combinations of symbionts in a vegetatively reproducing lichen, *Parmotrema tinctorum*, based on ITS rDNA sequences. Bryologist 109:43–59. doi:10.1639/0007-2745(2006)109[0043:GCOSIA]2.0.CO;2

Opanowicz M, Grube M (2004) Photobiont genetic variation in *Flavocetraria nivalis* from Poland (Parmeliaceae, lichenized Ascomycota). Lichenologist 36:125–131. doi:10.1017/S0024282904013763

Otálora MAG, Martínez I, Aragón G, Molina MC (2010) Phylogeography and divergence date estimates of a lichen species complex with a disjunct distribution pattern. Am J Bot 97:216–223

Otálora MAG, Salvador C, Martínez I, Aragón G (2013) Does the reproductive strategy affect the transmission and genetic diversity of bionts in cyanolichens? A case study using two closely related species. Microb Ecol 65:517–530. doi:10.1007/s00248-012-0136-5

Otálora MG, Martínez I, Belinchón R, Widmer I, Aragón G, Escudero A, Scheidegger C (2011) Remnants fragments preserve genetic diversity of the old forest lichen *Lobaria pulmonaria* in a fragmented Mediterranean mountain forest. Biodivers Conserv 20:1239–1254. doi:10.1007/s10531-011-0025-0

Ott S (1987) The juvenile development of lichen thalli from vegetative diaspores. Symbiosis 3:57–74

Otte V, Esslinger TL, Litterski B (2002) Biogeographical research on European species of the lichen genus *Physconia*. J Biogeogr 29:1125–1141

Otte V, Esslinger TL, Litterski B (2005) Global distribution of the European species of the lichen genus *Melanelia* Essl. J Biogeogr 32:1221–1241. doi:10.1111/j.1365-2699.2005.01268.x

Palice Z, Printzen C (2004) Genetic variability in tropical and temperate populations of *Trapeliopsis glaucolepidea*: evidence against long-range dispersal in a lichen with disjunct distribution. Mycotaxon 90:43–54

Peksa O, Škaloud P (2011) Do photobionts influence the ecology of lichens? A case study of environmental preferences in symbiotic green alga *Asterochloris* (Trebouxiophyceae). Mol Ecol 20:3936–3948. doi:10.1111/j.1365-294X.2011.05168.x

Peterson BK, Weber JN, Kay EH, Fisher HS, Hoekstra HE (2012) Double digest RADseq: an inexpensive method for de novo SNP discovery and genotyping in model and non-model species. PLoS ONE 7. doi:10.1371/journal.pone.0037135

Piercey-Normore MD (2004) Selection of algal genotypes by three species of lichen fungi in the genus *Cladonia*. Can J Bot 82:947–961. doi:10.1139/B04-084

Piercey-Normore MD (2006a) The lichen-forming ascomycete *Evernia mesomorpha* associates with multiple genotypes of *Trebouxia jamesii*. New Phytol 169:331–344. doi:10.1111/j.1469-8137.2005.01576.x

Piercey-Normore MD (2006b) Lichens from the Hudson Bay lowlands: diversity in the southeastern peatlands of Wapusk National Park, Manitoba. Can J Bot 84:1781–1793. doi:10.1111/j.1469-8137.2005.01576.x

Printzen C (2010) Uncharted terrain: the phylogeography of arctic and boreal lichens. Plant Ecol Div 1:265–271. doi:10.1080/17550870802328702

Printzen C, Ekman S (2002) Genetic variability and its geographical distribution in the widely disjunct *Cavernularia hultenii*. Lichenologist 34:101–111

Printzen C, Ekman S (2003) Local population subdivision in the lichen *Cladonia subcervicornis* as revealed by mitochondrial cytochrome oxidase subunit 1 intron sequences. Mycologia 95:399–406

Printzen C, Ekman S, Tønsberg T (2003) Phylogeography of *Cavernularia hultenii*: evidence of slow genetic drift in a widely disjunct lichen. Mol Ecol 12:1473–1486

Printzen C, Lumbsch HT, Schmitt I, Feige GB (1999) A study on the genetic variability of Biatora helvola using RAPD markers. Lichenologist 31:491–499

Rice AM, Rudh A, Ellegren H, Qvarnstrom A (2011) A guide to the genomics of ecological speciation in natural animal populations. Ecol Lett 14:9–18. doi:10.1111/j.1461-0248.2010.01546.x

Robertson J, Piercey-Normore MD (2007) Gene flow in symbionts of *Cladonia arbuscula*. Lichenologist 39:69–82. doi:10.1017/S0024282906005809

Rose F (1976) Lichenological indicators of age and environmental continuity in woodlands. In: Brown DH, Hawksworth DL, Bailey RH (eds) Lichenology: progress and problems systematics association special, vol 8. Academic Press, London, pp 279–307

Rose F (1988) Phytogeographical and ecological aspects of Lobarion communities in Europe. Bot J Linn Soc 96:69–79

Sadowska-Deś AD, Balint M, Otte J, Schmitt I (2013) Assessing intraspecific diversity in a lichen-forming fungus and its green algal symbiont: evaluation of eight molecular markers. Fungal Ecol 6:141–151. doi:10.1016/j.funeco.2012.12.001

Sadowska-Deś AD, Dal Grande F, Lumbsch HT et al (2014) Integrating coalescent and phylogenetic approaches to delimit species in the lichen photobiont *Trebouxia*. Mol Phylogenet Evol 76:202–210. doi:10.1016/j.ympev.2014.03.020

Sanders WB, Rico VJ (1992) Lichenizing rhizomorphs and thallus development in the squamulose lichen *Aspicilia crespiana* Rico ined. (Lecanorales, Ascomycetes). Bot Acta 105:449–456. doi:10.1111/j.1438-8677.1992.tb00327.x

Sanger F, Brownlee GG, Barrell BG (1965) A two-dimensional fractionation procedure for radioactive nucleotides. J Mol Biol 13:373–398

Savolainen O, Pyhajarvi T, Knurr T (2007) Gene flow and local adaptation in trees. Annu Rev Ecol Evol Syst 38:595–619. doi:10.1146/annurev.ecolsys.38.091206.095646

Scheidegger C (1995) Early development of transplanted isidioid soredia of *Lobaria pulmonaria* in an endangered population. Lichenologist 27:361–374

Scheidegger C, Bilovitz PO, Werth S, Widmer I, Mayrhofer H (2012) Hitchhiking with forests: population genetics of the epiphytic lichen Lobaria pulmonaria in primeval and managed forests in Southeastern Europe. Ecol Evol 2:2223–2240. doi:10.1002/ece3.341

Scheidegger C, Werth S (2009) Conservation strategies for lichens: insights from population biology. Fungal Biol Rev 23:55–66. doi:10.1016/j.fbr.2009.10.003

Schuster G, Jahns H (1985) Artificial cultures of lichens in the natural environment. Lichenologist 17:247–253

Schwendener S (1868) Ueber die Beziehungen zwischen Algen und Flechtengonidien. Bot Ztg 26:289–292

Scott CP, Williams DA, Crawford DL (2009) The effect of genetic and environmental variation on metabolic gene expression. Mol Ecol 18:2832–2843. doi:10.1111/j.1365-294X.2009.04235.x

Sérusiaux E, Villarreal AJC, Wheeler T, Goffinet B (2011) Recent origin, active speciation and dispersal for the lichen genus *Nephroma* (Peltigerales) in Macaronesia. J Biogeogr 38:1138–1151. doi:10.1111/j.1365-2699.2010.02469.x

Seymour FA, Crittenden PD, Dickinson MJ, Paoletti M, Montiel D, Cho L, Dyer PS (2005a) Breeding systems in the lichen-forming fungal genus *Cladonia*. Fungal Genet Biol 42:554–563

Seymour FA, Crittenden PD, Dyer PS (2005b) Sex in the extremes: lichen-forming fungi. Mycologist 19:51–58

Sham P, Bader JS, Craig I, O'Donovan M, Owen M (2002) DNA pooling: a tool for large-scale association studies. Nat Rev Genet 3:862–871. doi:10.1038/nrg930

Shaw AJ, Golinski GK, Clark EG, Shaw B, Stenøien HK, Flatberg KI (2014) Intercontinental genetic structure in the amphi-Pacific peatmoss *Sphagnum miyabeanum* (Bryophyta: Sphagnaceae). Biol J Linn Soc 111:17–37. doi:10.1111/bij.12200

Shendure J, Ji HL (2008) Next-generation DNA sequencing. Nat Biotechnol 26:1135–1145. doi:10.1038/nbt1486

Sigurbjörnsdóttir MA, Heiðmarsson S, Jónsdóttir AR, Vilhelmsson O (2014) Novel bacteria associated with Arctic seashore lichens have potential roles in nutrient scavenging. Can J Microbiol 60:307–317. doi:10.1139/cjm-2013-0888

Sillett SC, McCune B, Peck JE, Rambo TR, Ruchty A (2000) Dispersal limitations of epiphytic lichens result in species dependent on old-growth forests. Ecol Appl 10:789–799

Singh G, Dal Grande F, Cornejo C, Schmitt I, Scheidegger C (2012) Genetic basis of self-incompatibility in the lichen-forming fungus *Lobaria pulmonaria* and skewed frequency distribution of mating-type idiomorphs: implications for conservation. PLoS ONE 7:e51402. doi:10.1371/journal.pone.0051402

Singh G, Dal Grande F, Werth S, Scheidegger C Long term consequences of disturbances on reproductive strategies of the rare epiphytic lichen *Lobaria pulmonaria*: clonality a gift and a curse FEMS Microbiol Ecol. doi: 10.1093/femsec/fiu009 (in press)

Snäll T, Pennanen J, Kivistö L, Hanski I (2005) Modelling epiphyte metapopulation dynamics in a dynamic forest landscape. Oikos 109:209–222

Sork VL, Aitken SN, Dyer RJ, Eckert AJ, Legendre P, Neale DB (2013) Putting the landscape into the genomics of trees: approaches for understanding local adaptation and population responses to changing climate. Tree Genet Genomes 9:901–911. doi:10.1007/s11295-013-0596-x

Sork VL, Werth S (2014) Phylogeography of *Ramalina menziesii*, a widely distributed lichen-forming fungus in western North America. Mol Ecol 23:2326–2339. doi:10.1111/mec.12735

Stapley J, Reger J, Feulner PGD et al (2010) Adaptation genomics: the next generation. Trends Ecol Evol 25:705–712. doi:10.1016/j.tree.2010.09.002

Tõrra T, Cornejo C, Cheenacharoen S, Grande FD, Marmor L, Scheidegger C (2014) Characterization of fungus-specific microsatellite markers in the lichen fungus *Usnea subfloridana* (Parmeliaceae). Appl Plant Sci 2:1400034. doi:10.3732/apps.1400034

Tretiach M, Muggia L, Baruffo L (2009) Species delimitation in the *Lepraria isidiata-L. santosii* group: a population study in the Mediterranean-Macaronesian region. Lichenologist 41:1–15. doi:10.1017/s0024282909008093

Turner EH, Lee CL, Ng SB, Nickerson DA, Shendure J (2009) Massively parallel exon capture and library-free resequencing across 16 genomes. Nat Methods 6:315–316. doi:10.1038/nmeth.f.248

Wagner CE, Keller I, Wittwer S et al (2013) Genome-wide RAD sequence data provide unprecedented resolution of species boundaries and relationships in the Lake Victoria cichlid adaptive radiation. Mol Ecol 22:787–798. doi:10.1111/mec.12023

Wagner HH, Holderegger R, Werth S, Gugerli F, Hoebee SE, Scheidegger C (2005) Variogram analysis of the spatial genetic structure of continuous populations using multilocus microsatellite data. Genetics 169:1739–1752. doi:10.1534/genetics.104.036038

Wagner HH, Werth S, Kalwij JM, Bolli JC, Scheidegger C (2006) Modelling forest recolonization by an epiphytic lichen using a landscape genetic approach. Landsc Ecol 21:849–865. doi:10.1007/s10980-005-5567-7

Walser JC (1999) An assay for detection of *Lobaria pulmonaria* diaspores with a species-specific DNA-marker: implications for dispersal biology of a threatened forest lichen. Diploma thesis, Universität Bern

Walser JC (2003) Population genetic processes and ecological adaptation in a lichenized fungus assessed by microsatellite analysis and transplantation experiments. PhD thesis, University of Berne

Walser JC (2004) Molecular evidence for limited dispersal of vegetative propagules in the epiphytic lichen *Lobaria pulmonaria*. Am J Bot 91:1273–1276

Walser JC, Holderegger R, Gugerli F, Hoebee SE, Scheidegger C (2005) Microsatellites reveal regional population differentiation and isolation in *Lobaria pulmonaria*, an epiphytic lichen. Mol Ecol 14:457–467. doi:10.1111/j.1365-294X.2004.02423.x

Walser JC, Scheidegger C (2002) Transplanting lichen fragments for provenance-clone tests. In: Nimis PL, Scheidegger C, Wolseley PA (eds) Monitoring with lichens—monitoring lichens. Kluwer Academic Publishers, Dordrecht, pp 385–390

Walser JC, Sperisen C, Soliva M, Scheidegger C (2003) Fungus-specific microsatellite primers of lichens: application for the assessment of genetic variation on different spatial scales in *Lobaria pulmonaria*. Fungal Genet Biol 40:72–82. doi:10.1016/S1087-1845(03)00080-X

Walser JC, Zoller S, Büchler U, Scheidegger C (2001) Species-specific detection of *Lobaria pulmonaria* (lichenized ascomycete) diaspores in litter samples trapped in snow cover. Mol Ecol 10:2129–2138

Wang YY, Liu B, Zhang XY et al (2014) Genome characteristics reveal the impact of lichenization on lichen-forming fungus *Endocarpon pusillum* Hedwig (Verrucariales, Ascomycota). BMC Genomics 15. doi:10.1186/1471-2164-15-34

Wang YY, Zhang T, Zhou QM, Wei JC (2011) Construction and characterization of a full-length cDNA library from mycobiont of *Endocarpon pusillum* (lichen-forming Ascomycota). World J Microbiol Biotechnol 27:2873–2884. doi:10.1007/s11274-011-0768-5

Wegrzyn JL et al (2014) Unique features of the loblolly pine (*Pinus taeda* L.) megagenome revealed through sequence annotation. Genetics 196:891–909

Werth S (2010a) Optimal sample sizes and allelic diversity in studies of the genetic variability of mycobiont and photobiont populations. Lichenologist 43:1–9. doi:10.1017/S0024282910000563

Werth S (2010b) Population genetics of lichen-forming fungi—a review. Lichenologist 42:499–519. doi:10.1017/S0024282910000125

Werth S (2011) Biogeography and phylogeography of lichen fungi and their photobionts. In: Fontaneto D (ed) Biogeography of micro-organisms. Is everything small everywhere? Cambridge University Press, New York, pp 191–208

Werth S (2012) Fungal-algal interactions in *Ramalina menziesii* and its associated epiphytic lichen community. Lichenologist 44:543–560. doi:10.1017/S0024282912000138

Werth S, Cheenacharoen S, Scheidegger C (2014) Propagule size is not a good predictor for regional population subdivision or fine-scale spatial structure in lichenized fungi. Fungal Biol 118:126–138. doi:10.1016/j.funbio.2013.10.009

Werth S, Cornejo C, Scheidegger C (2013) Characterization of microsatellite loci in the lichen fungus *Lobaria pulmonaria* (Lobariaceae). Appl Plant Sci 1:1200290. doi:10.3732/apps.1200290

Werth S, Gugerli F, Holderegger R, Wagner HH, Csencsics D, Scheidegger C (2007) Landscape-level gene flow in *Lobaria pulmonaria*, an epiphytic lichen. Mol Ecol 16:2807–2815. doi:10.1111/j.1365-294X.2007.03344.x

Werth S, Scheidegger C (2012) Congruent genetic structure in the lichen-forming fungus *Lobaria pulmonaria* and its green-algal photobiont. Mol Plant-Microbe Interact 25:220–230. doi:10.1094/MPMI-03-11-0081

Werth S, Sork VL (2008) Local genetic structure in a North American epiphytic lichen, *Ramalina menziesii* (Ramalinaceae). Am J Bot 95:568–576. doi:10.3732/ajb.2007024

Werth S, Sork VL (2010) Identity and genetic structure of the photobiont of the epiphytic lichen *Ramalina menziesii* on three oak species in southern California. Am J Bot 97:821–830. doi:10.3732/ajb.0900276

Werth S, Sork VL (2014) Ecological specialization in *Trebouxia* (Trebouxiophyceae) photobionts of *Ramalina menziesii* (Ramalinaceae) across six range-covering ecoregions of western North America. Am J Bot 101:1127–1140. doi:10.3732/ajb.1400025

Werth S, Wagner HH, Gugerli F, Holderegger R, Csencsics D, Kalwij JM, Scheidegger C (2006a) Quantifying dispersal and establishment limitation in a population of an epiphytic lichen. Ecology 87:2037–2046. doi:10.1890/0012-9658(2006)87[2037:QDAELI]

Werth S, Wagner HH, Holderegger R, Kalwij JM, Scheidegger C (2006b) Effect of disturbances on the genetic diversity of an old-forest associated lichen. Mol Ecol 15:911–921. doi:10.1111/j.1365-294X.2006.02838.x

Widmer I, Dal Grande F, Excoffier L, Holderegger R, Keller C, Mikryukov VS, Scheidegger C (2012) European phylogeography of the epiphytic lichen fungus *Lobaria pulmonaria* and its green algal symbiont. Mol Ecol 21:5827–5844. doi:10.1111/mec.12051

Wornik S, Grube M (2010) Joint dispersal does not imply maintenance of partnerships in lichen symbioses. Microb Ecol 59:150–157. doi:10.1007/s00248-009-9584-y

Wöstemeyer J, Kreibich A (2002) Repetitive DNA elements in fungi (Mycota): impact on genomic architecture and evolution. Curr Genet 41:189–198. doi:10.1007/s00294-0020-0306-y

Xavier BB, Miao VPW, Jónsson ZO, Andrésson ÓS (2012) Mitochondrial genomes from the lichenized fungi *Peltigera membranacea* and *Peltigera malacea*: features and phylogeny. Fungal Biol 116:802–814

Yahr R, Vilgalys R, DePriest PT (2006) Geographic variation in algal partners of *Cladonia subtenuis* (Cladoniaceae) highlights the dynamic nature of a lichen symbiosis. New Phytol 171:847–860. doi:10.1111/j.1469-8137.2006.01792.x

Yoder JB, Stanton-Geddes J, Zhou P, Briskine R, Young ND, Tiffin P (2014) Genomic signature of adaptation to climate in *Medicago truncatula*. Genetics 196:1263–1275

Yoshimura I (1971) The genus Lobaria of Eastern Asia. J Hattori Bot Lab 34:231–364

Yüzbaşıoğlu E, Halıcı MG, Karabacak M, Aksoy A (2011) RAPD and ISSR markers indicate high genetic variation within Lobathallia radiosa in Turkey. Mycol Prog 10:219–228. doi:10.1007/s11557-010-0691-4

Zemach A, McDaniel IE, Silva P, Zilberman D (2010) Genome-wide evolutionary analysis of eukaryotic DNA methylation. Science 328:916–919. doi:10.1126/science.1186366

Zhou QM, Guo SY, Huang MR, Wei JC (2006) A study of the genetic variability of *Rhizoplaca chrysoleuca* using DNA sequences and secondary metabolic substances. Mycologia 98:57–67

Zoller S, Frey B, Scheidegger C (2000) Juvenile development and diaspore survival in the threatened epiphytic lichen species *Sticta fuliginosa*, *Leptogium saturninum* and *Menegazzia terebrata*: conclusions for in situ conservation. Plant Biol 2:496–504

Zoller S, Lutzoni F, Scheidegger C (1999) Genetic variation within and among populations of the threatened lichen *Lobaria pulmonaria* in Switzerland and implications for its conservation. Mol Ecol 8:2049–2059

5 Type I NR-PKS Gene Characterization of the Cultured Lichen Mycobiont *Xanthoparmelia Substrigosa* (Ascomycota)

Christina Hametner and Elfie Stocker-Wörgötter

Contents

C. Hametner (✉) · E. Stocker-Wörgötter
Department of Organismic Biology,
University of Salzburg, Hellbrunnerstraße 34,
5020 Salzburg, Austria
e-mail: Christina.Hametner2@sbg.ac.at

E. Stocker-Wörgötter
e-mail: Elfriede.Stocker@sbg.ac.at

D.K. Upreti et al. (eds.), *Recent Advances in Lichenology*,
DOI 10.1007/978-81-322-2235-4_5

Abstract

Secondary metabolites play an important role in drug discovery. Lichens, in the first instance the fungal symbiotic partner (referred to as mycobiont), produce a multitude of metabolites with interesting biological functions. A large share of these organic compounds represents polyketides (e.g., anthraquinones, depsides, and depsidones) biosynthesized by typical polyketide synthase (PKS) enzymes. This study addresses the detection of PKS genes that control the formation of secondary metabolites (polyketide-type compounds), specifically in the lichen species *Xanthoparmelia substrigosa* (Parmeliaceae). As a first step, the isolated mycobiont of *X. substrigosa* was cultured in large scale. For the molecular detection and identification of the PKS gene from the cultured mycobiont, the highly conserved β-ketoacyl synthase (KS) domain was used as starting point. Then, the SMARTer RACE cDNA amplification was performed using extracted RNA of the fungal cultures by means of specific primers generated from the obtained sequence of the KS domain. Since the PCR products were over several 1,000 base pairs (bps) long, cloning of the products was the next step to attain sufficient material for the sequencing procedure. The located type I PKS gene of the mycobiont (Xsm PKSI) comprises 2,273 amino acids and is composed of one KS, one acyltransferase (AT), two acyl carrier proteins (ACPs), and one thioesterase (TE) domain. Similar domain formation of the PKS gene has been found in the original lichen thallus of *X. substrigosa* (XsTPKSI), but differs from the cultured mycobiont sequence with five inserted introns arranged along the first 2,120 bps. Additionally, HPLC analyses of lichen thallus and fungal culture confirmed usnic acid, norstictic acid, and salazinic acid as secondary metabolites, which could be biosynthesized by the detected PKS gene.

Keywords

Fungal culture • HPLC analyses • KS domain • Mycobiont • PKS type I • RNA isolation • Secondary metabolites • *Xanthoparmelia*

5.1 Introduction

Secondary metabolites are a substantial group of natural products in pharmaceutical and industrial sectors. One important group is specified as "small" molecules according to their low molecular weight (Pichersky and Gang 2000) produced by one or a combination of various large enzyme complexes found in plants and microorganisms (Croteau et al. 2000; Pietra 1997; Scharf et al. 2014). The secondary compounds in plants are classified into three major groups based on their biosynthetic origins (mevalonic acid pathway, 2-C-methyl-D-erythritol 4-phosphate/1-deoxy-D-xylulose 5-phosphate pathway, berberine synthesis pathway, morphine biosynthesis, phenylpropanoid pathway, phenylpropanoid-acetate pathway, etc.), i.e., terpenoids, alkaloids, and phenylpropanoids, which comprise the majority of phenolic metabolites as well as allied phenolic substances (Croteau et al. 2000). For the pharmaceutical industry, secondary plant compounds play an important role in the production of diverse antibiotics, antitussives, sedatives, etc.; however, the industrial implementation of such substances includes the creation of new combinations of products and chemical hybrid compounds that can be introduced as flavors, perfumes, color, antifeedants against herbivory, and for further applications (Croteau et al. 2000; Pichersky and Gang 2000). In particular, secondary metabolites produced by diverse microorganisms are of high interest. They find applications as antibiotic agents against diseases caused by bacteria and fungi, antitumor agents (e.g., breast and ovarian cancer), immunosuppressive drugs, anti-inflammatory agents, other enzyme inhibitors to lower cholesterol, antiparasitic agents for animals and humans, food additives for improved cell development, etc. (Demain 1999; Pietra 1997). They have also significant roles in biological interactions in form of signal molecules leading the communication between human immune system and fungal pathogen, plant andendophytic bacteria, and the capability to influence performance, growth, and stress tolerance of the host (Brader et al. 2014; Scharf et al. 2014).

One of these groups of organisms which have the ability to produce a various range of secondary compounds is represented by filamentous fungi. Fungi can be found in diverse habitats and substrata acting as saprophyte, parasite, pathogen, and symbionts (Fox and Howlett 2008). Lichens in particular are symbiosis of filamentous fungi (ascomycetes), which associate with one or even more taxa of green algae or cyanobacteria. Nowadays, it is well known that lichens produce and harbor a great variety of secondary substances that are unique to lichens (Huneck and Yoshimura 1996). A considerable number of secondary metabolites are used as chemical markers in taxonomy and systematics (Culberson 1969; Elix 1993; Fehrer et al. 2008; Nylander 1866; Piercey-Normore 2007; Schmitt and Lumbsch 2004; Schmull et al. 2011). One of the extensively investigated group of lichen substances constitutes polyketides, which are thought to originate from the biosynthetic activity of the fungal symbiotic partner (mycobiont) in lichens (Elix 1996; Huneck 1999) and are deposited as extracellular microcrystals on the outer surface of the hyphae in the lichen cortex or located inside the medulla, the internal fungal layer of the lichen thallus. Most secondary metabolites are derived from the acetyl-polymalonyl pathway, although the majority of substances from this pathway constitutes polyketides (Molnár and Farkas 2010; Stocker-Wörgötter et al. 2004). The expression of lichen polyketides is regulated by PKS genes which control the formation of polyketide-type lichen metabolites by polyketide synthases (Grube and Blaha 2003). In the case of fungi, polyketide synthases (PKSs) belong to type I and type III PKSs (Katsuyama and Ohnishi 2012; Yu et al. 2012a). The type I enzymes were further categorized into non-reducing (NR-PKS), partially reducing (PR-PKS), and highly reducing PKSs (HR-PKS), as exemplified by the reports of Bingle et al. (1999) and Nicholson et al. (2001).

Studies of Armaleo et al. (2011), Brunauer et al. (2009), Chooi et al. (2008), Gagunashvili et al. (2009), Miao et al. (2001), Timsina et al. (2012), Valarmathi et al. (2009), Wang et al. (2012), and Yu et al. (2012b) have already investigated a few polyketide synthesized from various lichen species in more detail. These include the investigations of *Xanthoparmelia cumberlandia* (Miao et al. 2001) dealing with the β-ketoacyl synthase (KS) domain and of the fully encoded PKS gene from *X. semiviridis* (Chooi et al. 2008). We had the opportunity to collect *Xanthoparmelia* species in Australia, and for this reason, we selected the lichen *X. substrigosa* for the screening of PKS genes. The lichen *X. substrigosa* is a very common species in Australia, growing on rock and soil. The thallus is characterized by a yellow-green and shiny color, broad and strap-shaped lobes, and dense furcated rhizines and is lacking soredia and isidia. The secondary metabolites norstictic acid (major), connorstictic acid, salazinic acid (±), consalazinic acid (±), and usnic acid are usually found in *X. substrigosa* (Elix 1994; Hale 1990). Interestingly, under less favorable conditions in the environment, for still unknown reasons, the complete set of substances is not expressed and satellite or even major compounds can be absent (Stocker-Wörgötter unpublished results).

The aim of this study was to isolate and culture the mycobiont of *X. substrigosa* to avoid contamination originating from other organisms like the photobiont, other fungi, and bacteria living in or on the lichen thallus, which are known to produce type I and type III PKSs as well. Degenerate primers were used targeting KS domains of *X. substrigosa* using the SMART-rapid amplification of cDNA ends (RACE) as applied in Brunauer et al. (2009) to achieve full length of the PKS genes. High-performance liquid chromatography (HPLC) analyses of the mycobiont culture and lichen thallus were accomplished for comparison and discussion.

5.2 Preparation of the Voucher Specimen *Xanthoparmelia Substrigosa*

5.2.1 Collection Information of the Lichen *X. Substrigosa*

The voucher specimen *X. substrigosa* (Fig. 5.1) was collected by Elfie Stocker-Wörgötter on December 3, 2011, at Mount Ainslie, Canberra, Australia (35°16′10.92″S, 149°9′27.85″E). The specimen is maintained in the herbarium of Elfie Stocker-Wörgötter (HNr. 9182), which is publicly accessible through the Department of Organismic Biology at the University of Salzburg.

Fig. 5.1 Voucher specimen *Xanthoparmelia substrigosa*

5.2.2 Isolation of the Mycobiont Under Axenic Conditions

5.2.2.1 Lichen Tissue Culture Method

The isolation of the mycobiont *X. substrigosa* was performed according to the method of Yamamoto (1990) with some modifications suggested by Stocker-Wörgötter (2002). In this technique, lichen fragments (2–3 mm in size) were washed in double-distilled water for 15 min. Then, one drop Tween 80 (detergent) was added and the fragments were washed again for 10 min to remove dust and dirt particles from the thallus surface. These cleaned lichen fragments were transferred into fresh double-distilled water, washed for further 20 min and subsequently softly grinded with some drops of fresh sterile double-distilled water using an autoclaved mortar and pestle. The homogenized suspension was filtered primarily through a sieve with 500 μm mesh size and afterward through a sieve with 150 μm mesh size. Twenty-five test tubes with slanted LBM medium (Lillly and Barnett 1951) prepared with Bacto-agar (Difco Laboratories, USA; LBMDifco) were inoculated with one lichen piece picked up from the 150-μm sieve by sterile bamboo sticks under a Leica MS5 dissecting microscope (Leica Microsysteme Handelsges.m.b. H, Austria). All steps except the washing procedure were performed under a laminar flow bench to attain axenic conditions for the mycobiont isolation.

5.2.2.2 Incubation of the Test Tubes

The test tubes, loaded with lichen pieces, were incubated for several months in a plant growth chamber APT.line™KBW (Binder GmbH, Germany) under regular light–dark and temperature regimes of 14 h at 21 °C and 10 h at 10 °C with a radiant flux density of 375 Lux/(W/m^2).

5.2.3 Subcultivation of the Isolated Mycelia

As soon as the fungal hyphae, growing in the test tubes, reached a size of 3–4 mm in diameter, the fungal isolates were subcultured. Under a laminar flow bench, some hyphae were removed by an inoculation needle and softly grinded in some drops of sterile double-distilled water. The obtained fungal suspension was absorbed by a Pasteur pipette (the glass tip was cut off to broaden the opening for the grinded hyphae) and then dropped onto solid LBMDifco medium, prepared in petri dishes. The petri dishes remained under the laminar flow bench until the residual water from the dropped fungal suspensions, which were placed before on the agar medium, was evaporated. Then, the petri dishes were closed using some parafilm stripes and incubated in the plant growth chamber under the same conditions as mentioned above.

5.2.4 Identification of the Fungal Isolations

DNA analyses were performed to confirm the identities of the fungal isolates. DNA was extracted from the hyphal subcultures, which were first frozen in liquid nitrogen and pulverized using the grinding mill MM301 (Retsch GmbH, Germany). The grinded material was processed with a modified CTAB-method (Doyle and Doyle 1987). The ITS region was amplified by PCR with the primer pair ITS1F (Gardes and Bruns 1993) and ITS4 (White et al. 1990). The PCR products were cleaned with the Wizard® SV Gel and PCR Clean-Up System (Promega, Austria) and sequenced by Eurofins MWG Operon (Germany). The obtained sequences were compared with sequences of the NCBI GenBank.

5.3 Molecular Characterization of Fungal PKS Genes

5.3.1 Amplification and Primer Design from the KS Domain Using Mycobiont Culture (Ainslie Mycobiont)

5.3.1.1 PCR Settings for KS Domain Amplification

The first step of the molecular characterization was the amplification of the KS domain from the Ainslie mycobiont isolated from *X. substrigosa*. The same DNA extraction of the Ainslie culture which was prepared for mycobiont identification was used for the PCR. The PCR contained 6 µl 5x GoTaq-Reaction Buffer (Promega, Austria), 0.6 µl dNTPs (10 mM each; Fermentas, Austria), 0.12 µl LC1-Im (100 pmol/µl; Schmitt et al. 2005), 0.12 µl LC2c-Im (100 pmol/µl; Schmitt et al. 2005), 0.1 µl GoTaq DNA polymerase (5u/µl; Promega, Austria), 2 µl DNA template, and 21.06 µl sterile water in a total volume of 30 µl. The cycles of the touchdown PCR consisted of an initial denaturing step at 94 °C for 3 min, followed by 6 cycles of denaturation at 94 °C for 30 s, annealing at variable temperatures for 30 s which comprised temperature settings at 61 °C for the first 2 cycles and a continuous decrease by 1 °C for each 2 subsequent cycles, and an extension at 72 °C for 50 s. This was followed by 12 cycles with equal settings as used before but with a temperature decrease of 1 °C for each 3 subsequent cycle (i.e., temperature varied from 58 to 55 °C). The next 3 cycles had an annealing temperature of 55.5 °C, followed by 25 cycles of denaturation at 94 °C for 30 s, annealing temperature at 54 °C, an extension at 72 °C for 50 s, and ended with an extension at 72 °C for 7 min.

5.3.1.2 Cloning and Sequencing of the KS Sequence

The PCR product was cloned with pGEM®-T Easy Vector System (Promega, Austria), and the plasmids were cleaned by Wizard® *Plus* SV Minipreps DNA Purification System (Promega, Austria) according to the manufacturer′s protocols. The sequencing was performed by Eurofins MWG Operon (Germany).

5.3.1.3 Specific Primers from the KS Domain of *Xanthoparmelia Substrigosa*

The identity of the obtained sequences was checked by standard Nucleotide BLAST search in NCBI GenBank. Moreover, the identified sequences from the KS domain of the mycobiont

X. substrigosa were used to design the following specific primers for the SMARTer™ RACE cDNA amplification using the program Geneious version R6 (Biomatters): XsubKS5 (3′-ACA TCG TGT GGG TCT TTT CCA GCA TCG TCC-5′) and XsubKS3 (5′-GTC TCC TCG CGA AGC AAC ACA AAC AGA TCC-3′), additionally for nested PCR XsubKS5_nes (3′-GAG TCG GTG CGT GAG GAT GTG TGA TGG-5′) and XsubKS3_nes (5′-AGC CCT AGA AAT GGC CGG CTA CGT ACC-3′).

5.3.2 RNA Isolation from the Cultured Ainslie Mycobiont

Fungal RNA was isolated from a subculture (not older than 6 months) of *X. substrigosa* (Fig. 5.2). First of all, 100 mg of hyphal material was transferred with a sterile inoculation needle to an Eppendorf tube under a laminar flow bench. Subsequently, the loaded Eppendorf tube was frozen with liquid nitrogen and 667 μl of TRIzol® Reagent (Life Technologies, Austria) was pipetted into it. Then, the content of the tube was grinded by a micropistill (Eppendorf, Switzerland) on ice. The RNA isolation procedure was performed according to the manufacturer´s protocol and some modifications described in Savchuk (2006) to attain sufficient fungal RNA. The concentration and condition of the RNA isolation were controlled by gel electrophoresis and a Nanodrop 2000c Spectrophotometer (Peqlab Biotechnologie GmbH, Germany).

Fig. 5.2 Isolated and cultured mycobiont of *Xanthoparmelia substrigosa* on solid LBMDifco, 14 months old. The grayish zones (*arrows*) indicate that the cultured mycobionts have started to differentiate lobelike and more compact layered mycelial structures

5.3.3 SMARTer CDNA Synthesis and RACE-PCR

The RNA isolated from the subcultured Ainslie mycobiont was used as template for the cDNA synthesis using SMARTer™ RACE cDNA Amplification Kit (Clontech Takara Bio Europe, France) according to the manufacturer's protocol. In this technique, two first-strand reaction products diluted in 20 μl Tricine-EDTA Buffer were produced, the 5′-RACE ready cDNA and the 3′-RACE ready cDNA. Thereafter, the RACE-PCR was prepared with specific primers designed from the KS domain (see Sect. 5.3.1) using Advantage® 2PCR Enzyme System (Clontech Takara Bio Europe, France). For the first PCR, the primer combination UPM (Universal Primer A Mix; SMARTer™ RACE cDNA Amplification Kit) and XsubKS5 or XsubKS3 was used for the 5′-RACE or 3′-RACE. To increase the content of the PCR products, nested variations with primers NUP (Nested Universal Primer A; SMARTer™ RACE cDNA Amplification Kit) and XsubKS5_nes or XsubKS3_nes were accomplished by using 1 μl PCR product. The cycles of all RACE-PCRs consisted of an initial denaturing step at 95 °C for 2 min, followed by 30 cycles of denaturation at 94 °C for 30 s, annealing temperature at 69 °C for 30 s, extension at 72 °C for 5 min, and a final extension at 72 °C for 7 min.

5.3.4 Cloning Procedure and Plasmid Preparation

5.3.4.1 Cloning of Long PCR Products

The PCR products of the RACE amplification were cleaned with the NucleoTraP®CR PCR cleanup and NucleoTraP® Gel Extraction Kit supplied by the SMARTer™ RACE cDNA

Amplification Kit. Since the 5′-RACE product was over 2,000 bp and the 3′-RACE product over 6,000 bp long, the TOPO® XL PCR Cloning Kit (Invitrogen by Life Technologies, Austria) was used for cloning. The cloning procedure per se was performed according to the manufacturer's protocol. To control the effective cloning of the inserts, colony PCR was accomplished using GoTaq DNA polymerase (Promega, Austria) or KAPAHIFI PCR Kit (Peqlab Biotechnologie GmbH, Germany) with the primer combination M13 Forward (–20) and M13 Reverse supplied by the TOPO® XL PCR Cloning Kit.

5.3.4.2 DNA Purification of Long Inserts

The clones with complete target insert, which were preserved by rescue plates, were grown in overnight liquid cultures and afterward cleaned using Wizard® *Plus* SV Minipreps DNA Purification System (Promega, Austria) to attain the plasmid DNA.

5.3.5 Sequencing and Specification of the Fungal PKS Gene

5.3.5.1 Sequencing Approach to Obtain Full Length Ainslie PKS Gene

Sequencing of the cloned plasmid DNA from the 5′-RACE and 3′-RACE was performed by Eurofins MWG Operon (Germany) by initiating the procedure at the standard priming sites within the pCR-XL-TOPO® vector. Consecutively, the sequencing was finished by primer walking on the basis of primers specified in Table 5.1.

5.3.5.2 Procedure of the Fungal PKS Gene Specification

Three individual clones of each RACE-PCR were used to eliminate any incorrect base exchanges that were arisen during the cDNA syntheses, RACE-PCR, cloning and sequencing procedure. The individual sequences were edited and assembled using the KS domain as overlap area for the consolidation of the 5′-RACE and 3′-RACE products with the program Geneious version R6 (Biomatters). The specification of the domains from the assembled PKS sequence (6,817 bp long) was implemented by the search program Domain Enhanced Lookup Time Accelerated BLAST (DELTA-BLAST) in NCBI GenBank and ScanProsite tool in Expasy.

5.3.6 Molecular Approach to Identify the PKS Architecture of the *X. Substrigosa* Thallus

5.3.6.1 Amplification of the Thallus PKS Gene

Genomic DNA was extracted from 100 mg frozen lichen thallus with GeneJET Plant Genomic DNA

Table 5.1 List of primers which were used for primer walking to obtain the PKS gene of the cultured mycobiont *Xanthoparmelia substrigosa*

PCR product	Primer name	Primer sequence	Tm (°C)
5′-RACE	Xsub5rev_1	3′-TCA TAG TGA GTC TCC ACG TTG-5′	57.9
	Xsub5rev_2	3′-TCC ACC AAC ATA TTG AAT ATA CTG C-5′	58.1
3′-RACE	Xsub3for_1	5′-CGG CAT CAA GAA GAC GAT-3′	53.7
	Xsub3for_2	5′-CTG TCG CCG TTA ATT GTC C-3′	56.7
	Xsub3for_3	5′-TGT CCA TCG ACT GGA ACG AGT-3′	59.8
	Xsub3rev_1	3′-ATC TGC CGG CAC GTA ATC-5′	56.0
	Xsub3rev_2	3′-TTG GGA GAG CCT TGA AGG-5′	56.0
	Xsub3rev_3	3′-CAG GAC ATG AGT AGT GAC TGA-5′	57.9
	Xsub3_refin1	5′-TGC ACG GCA CTG GGA CA-3′	57.6
	Xsub3_refin2	5′-TAC CTG CGA TCG AAG CGG-3′	58.2
	Xsub3_refin3	5′-GTA CGA TGC AGT GCC CAT 3′	56.0

The primers named with the abbreviation "refin" were used for improvements of the PKS sequence

Table 5.2 List of primer combinations and cycle annealing temperatures, which were used for the amplification of the PKS gene from the thallus DNA extraction of *Xanthoparmelia substrigosa*

Primer combination	Primer name	Primer sequence	Cycle Tm (°C)
1	XsubT_A	5′-ATG GCT GAG ACG CGG ATC TTC CTC TTT GGG GAT-3′	68
	XsubKS5	see Sect. 5.3.1	
2	XsubKS3_nes	see Sect. 5.3.1	59
	Xsub3for_3rev	3′-ACT CGT TCC AGT CGA TGG ACA-5′	
3	Xsub3for_2	see Table 5.1	56
	XsubT_stop	3′-AAC CCT GTC AAG CGA TCC ACA TAC GGC GCC T-5′	

Purification Kit (Thermo Scientific, Austria) according to the manufacturer's protocol. Primers were designed on the basis of the PKS sequence from the cultured mycobiont *X. substrigosa* for the amplification of the thallus PKS gene. The PCR consisted of 25 μl KAPA3G Plant PCR Buffer (2x) with dNTPs (PeqlabBiotechnologie GmbH, Germany), 1.5 μl forward primer (10 pmol/μl), 1.5 μl reverse primer (10 pmol/μl), 0.4 μl KAPA3G Plant DNA polymerase (2.5 u/μl; PeqlabBiotechnologie GmbH, Germany), 2 μl DNA template, and 19.6 μl sterile water. The primer combinations which were used for the amplification are listed in Table 5.2.

The cycles for all PCRs consisted of an initial denaturing step at 95 °C for 3 min, followed by 35 cycles of denaturation at 95 °C for 20 s, annealing temperature for each primer combination as mentioned in Table 5.2 for 15 s, extension at 72 °C for 1 min 20 s, and a final extension at 72 °C for 7 min.

5.3.6.2 Cloning of PCR Products Obtained from Thallus Extraction and DNA Purification

The PCR products were cloned using the TOPO® XL PCR Cloning Kit (Invitrogen by Life Technologies, Austria) according to the manufacturer's protocol. Colony PCRs were performed after the cloning procedure to check the success rate and the quality of the products. Sufficient colony PCR products were cleaned by Wizard® SV Gel and PCR Clean-Up System (Promega, Austria), whereas products with low signals in gel electrophoresis were enhanced by overnight liquid cultures and then cleaned with Wizard® *Plus* SV Minipreps DNA Purification System (Promega, Austria).

5.3.6.3 Sequencing Approach to Obtain Full Length Thallus PKS Gene

The sequencing was performed by Eurofins MWG Operon (Germany) using primer walking technique with primers of Table 5.1 and one further designed primer XsT5for_1 (5′-CTC TTT CAA AAA GCT TCG ATT TTG AC-3′). The obtained sequences were edited and assembled with Geneious version R6 (Biomatters) and aligned with the PKS sequence of the mycobiont culture for comparison and domain determination.

5.3.7 HPLC Analyses of Mycobiont Culture and Lichen Thallus from *X. Substrigosa*

Secondary metabolites of the subcultured mycobiont and the lichen thallus from *X. substrigosa* were analyzed by high-performance liquid chromatography (HPLC) using a Merck–Hitachi system with two pumps, a DAD (photodiode array detector; 190–800 nm wavelength) and a computer system. Subcultured fungal hyphae (around 1 cm diameter), which were vacuum dried by Univapo 150 ECH programmable (Uniequip, Germany), or dry thallus fragments (around 5 mm diameter) were transferred in glass vials under a laminar flow bench. The extractions of the mycobiont and thallus material were performed with methanol for four hours. After this period of time, the extracts were pipetted to HPLC vials and an aliquot of 5–20 µl of every sample was injected by a needle connected to an autosampler. The HPLC analyses were done with retention index values (RI) calculated from benzoic acid and solorinic acid controls (Elix 1996; Elix and Wardlaw 2000; Feige and Lumbsch 1993). Two solvent systems, A and B, were applied for these analyses: 1 % aqueous orthophosphoric acid and methanol in the ratio 7:3 (A) and methanol (B). The run started with 100 % solvent A and was raised to 58 % B within 15 min, then to 100 % B within a renewed 15 min, followed by isocratic elution in 100 % B for further 10 min. The HPLC chromatograms were recorded and the obtained spectra computer matched against a library of ultraviolet spectra of metabolites analyzed under identical conditions (Brunauer et al. 2007; Zocher and Stocker-Wörgötter 2005).

5.4 Specification of the Type I PKS Gene of *Xanthoparmelia Substrigosa*

The amplification of the KS domain with the degenerate primers LC1-Im and LC2c-Im (Schmitt et al. 2005) resulted in one single KS sequence from the Ainslie mycobiont and thus in the characterization of a single PKS gene.

5.4.1 Domain Formation of PKS Gene XsmPKSI

The complete sequence of the investigated PKS gene from the isolated and cultured mycobiont *Xanthoparmelia substrigosa* (XsmPKSI **KJ501919**) consists of 6,819 bp. The BLAST search (NCBI GenBank) revealed that this PKS gene belongs to the fungal non-reducing PKS type I enzymes, at which the sequence XsmPKSI encoded the following conserved protein domains: one KS domain, one AT domain, two PP-binding positions (phospho-pantetheine binding)/ACP domains, and one TE domain (Fig. 5.3). By dint of Scan Prosite tool and comparisons of other fungal PKS genes with XsmPKSI, it turned out that the KS domain (Fig. 5.4) contains the typical active site with the conserved cysteine (DTA**C**SSSAA) and two strongly conserved histidine motifs (VEM**H**GT, NVG**H**GEA) as described by Keating and Walsh (1999) and Wang et al. (2012). Active residues are also shown in the AT domain (GH**S**LG) with a pantetheine-binding serine, in the first ACP (GVD**S**LMS) and the second ACP (GMD**S**LMS) domain, both comprising a phosphopantetheine-binding serine (Brunauer et al. 2009; Serre et al. 1995; Wang et al. 2012; Wong et al. 2002). The TE domain has two active sites, the serine residue (GW**S**AGGVI) and a histidine motif at the C-terminal end (MPGN**H**FSMM), to initiate the Claisen-type cyclization (Schneider and Marahiel 1998; Wang et al. 2012; Watanabe and Ebizuka 2004).

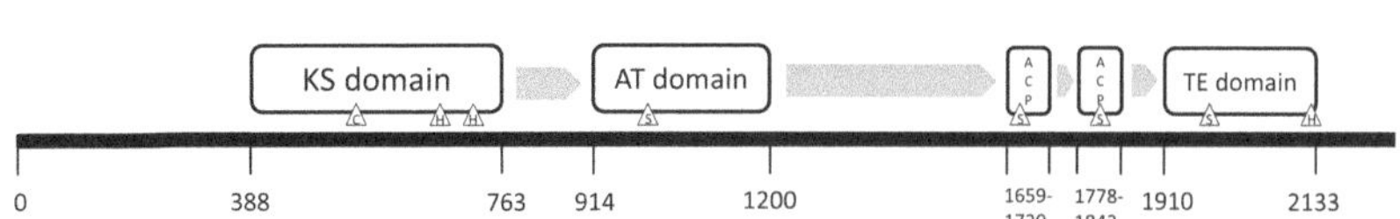

Fig. 5.3 Schematic domain organization of the fungal PKS gene XsmPKSI, which starts with the β-ketoacyl synthase (KS), followed by the acyltransferase (AT), two acyl carrier protein (ACP), and the thioesterase (TE) domains. The numbers indicate the amino acid position of the translated coding sequence, and the triangles with the characters represent the active sites of the domains

Ketoacyl synthase with the conserved cysteine

XsmPKSI	G	P	S	F	S	V	D	T	A	C	S	S	S	A	A	A	M	Q	L	A	C	T	S	L	W
Pseudomonas protegens AAC38075	G	P	A	M	T	V	T	T	A	C	S	S	S	L	V	A	M	H	L	A	C	R	A	L	Q
Mycobacterium tuberculosis Q10977	G	P	S	V	A	V	D	T	A	C	S	S	S	L	V	A	I	H	L	A	C	Q	S	L	R
Microbulbifer degradans ZP_00316393	G	P	S	M	S	V	D	T	A	C	S	S	S	L	V	S	V	H	L	A	C	Q	S	L	A
Botryotinia fuckeliana AAR90238	G	P	S	M	T	V	D	T	G	C	S	G	S	L	V	A	L	H	Q	A	C	Q	G	L	R
Magnaporthe grisea EAA48052	G	P	S	V	C	L	D	T	A	C	S	S	S	L	V	A	L	H	L	A	C	Q	D	L	R
Streptomyces globisporus AAL06699	G	G	G	Y	T	V	D	G	A	C	S	S	S	L	L	S	I	T	T	A	A	T	S	L	Q
	*	.				.	.	.	.	*	*		*	.		.				*	.			*	

Ketoacyl synthase with the histidine motif 1

XsmPKSI	D	V	N	Y	V	E	M	H	G	T	G	T
Pseudomonas protegens AAC38075	D	I	D	Y	V	E	A	H	G	T	G	T
Mycobacterium tuberculosis Q10977	E	V	D	Y	V	E	A	H	G	T	G	T
Microbulbifer degradans ZP_00316393	E	L	Q	Y	I	E	A	H	G	T	S	T
Botryotinia fuckeliana AAR90238	K	I	N	Y	V	E	A	H	G	T	G	T
Magnaporthe grisea EAA48052	R	C	Q	Y	F	E	A	H	G	T	G	T
Streptomyces globisporus AAL06699	T	V	P	L	F	E	G	H	G	T	G	T
				.		*		*	*	*	.	*

Ketoacyl synthase with the histidine motif 2

XsmPKSI	K	A	N	V	G	H	G	E	A	V	S	K
Pseudomonas protegens AAC38075	K	A	N	M	G	H	L	D	A	A	A	K
Mycobacterium tuberculosis Q10977	K	T	N	L	G	H	T	E	A	A	A	K
Microbulbifer degradans ZP_00316393	K	A	N	I	G	H	T	E	A	A	A	K
Botryotinia fuckeliana AAR90238	K	S	N	I	G	H	L	E	S	A	S	K
Magnaporthe grisea EAA48052	D	G	R	L	E	D	D	D	G	K	Q	D
Streptomyces globisporus AAL06699	K	G	M	I	G	H	T	K	A	A	A	K
	.				.	*						.

Acyl transferase with a pantetheine binding serine

XsmPKSI	P	D	V	V	I	G	H	S	L	G	E	Y	A	A	L
Mycobacterium tuberculosis Q10977	P	D	L	V	I	G	H	S	M	G	E	V	A	A	A
Mycobacterium bovis Q02251	P	G	A	V	V	G	H	S	M	G	E	S	A	A	A
Penicillium griseofulvum P22367	P	Q	A	V	I	G	H	S	V	G	E	I	A	A	S
Aspergillus nidulans Q00706	Q	A	I	F	A	G	H	S	L	G	E	Y	S	S	L
Saccharopolyspora erythraea Q03131	P	D	A	V	V	G	H	S	I	G	E	L	A	A	A
Saccharopolyspora erythraea Q03132	P	A	A	V	V	G	H	S	Q	G	E	I	A	A	A
	.			.		*	*	*		*	*		.	.	

Acyl Carrier Protein phosphor-pantetheine binding serine

XsmPKSI	M	G	V	D	S	L	M	S	L	S	I	S
Aspergillus oryzae Q2UGE1	I	G	F	D	S	L	M	M	I	E	L	S
Bacillus amyloliquefaciens Q1RS43	L	G	L	E	S	S	Q	L	L	A	I	I
Bipolaris maydis Q6RKE6	L	G	I	D	S	L	M	S	L	V	L	S
Marinobacter algicola ZP_01892962	V	G	L	D	S	L	L	L	T	Q	V	A
Ostreococcus lucimarinus XP_001416177	H	G	L	D	S	L	G	I	A	Y	F	F
Aspergillus niger XP_001390425	I	G	V	D	S	L	L	A	L	T	I	V
		*		.	*	.						

Acyl Carrier Protein phosphor-pantetheine binding serine

XsmPKSI	S	A	L	G	M	D	S	L	M	S	L	S
Mycobacterium sp. YP_001071136	F	Q	C	G	M	D	S	L	M	S	V	T
Trichodesmium erythraeum YP_723338	F	N	L	G	M	D	S	L	M	A	L	D
Saccharopolyspora spinosa AAS00421	T	E	L	G	F	D	S	L	T	A	V	E
Streptomyces antibioticus AAZ77693	K	D	L	G	F	D	S	L	T	A	V	D
Streptomyces parvulus CAE45670	R	E	L	G	F	D	S	L	S	A	V	K
Streptomyces ambofaciens CAJ88185	R	D	L	G	A	D	S	L	T	A	V	A
			.	*		*	*	*		.	.	

Thioesterase serine motif

XsmPKSI	G	G	W	S	A	G	G	V	I	S	Y	E
Bacillus subtilis 2K2Q_B	F	G	H	S	M	G	G	M	I	T	F	R
Streptomyces griseus P33586	F	G	H	S	M	G	A	A	L	A	F	E
Mycobacterium tuberculosis P63460	F	G	H	S	M	G	G	M	L	A	F	E
Brevibacillus brevis O30412	F	G	H	S	M	G	G	L	V	A	F	E
Streptomyces caelestis O30768	A	G	H	S	A	G	A	N	V	A	Y	A
Lysobacter lactamgenus P94873	F	G	W	S	F	G	G	V	V	S	L	E
		*	.	*		*	.					

Thioesterase histidine motif

XsmPKSI	R	Q	M	P	G	N	H	F	S	M	M
Bacillus subtilis 2K2Q_B	H	Q	F	D	G	G	H	M	F	L	L
Streptomyces griseus P33586	R	V	F	T	G	G	H	F	F	L	T
Mycobacterium tuberculosis P63460	R	V	F	P	G	D	H	F	Y	L	N
Brevibacillus brevis O30412	F	P	V	S	G	G	H	F	F	T	H
Streptomyces caelestis O30768	T	D	V	P	G	N	H	F	T	M	L
Lysobacter lactamgenus P94873	H	F	S	W	I	H	E	K	T	T	V
					.		.				

Fig. 5.4 XsmPKSI (KJ501919) aligned with closely related PKS genes of different organisms. Active residues are highlighted in *gray*. Conserved amino acids among all sequences are marked with an *asterisk*, and variations of two amino acids are marked with a *dot*

5.4.2 Secondary Metabolites of *X. Substrigosa*

HPLC analyses revealed that the six-month-old culture of the Ainslie mycobiont was not producing any detectable secondary metabolites. The lichen thallus showed the expression of the polyketides salazinic acid as major lichen substance, and norstictic acid and usnic acid as additional metabolites (Fig. 5.5).

5.5 Conclusion

Mycobiont cultures of *Xanthoparmelia substrigosa* were accomplished to facilitate the RNA isolation together with the SMARTer RACE cDNA amplification for the screening of PKS genes, since lichen thalli have been shown to be "microecosystems" composed of various organisms which probably also produce secondary metabolites (Grube et al. 2009; Miao et al. 2001; Nash 1996; Stocker-Wörgötter 2008). Grube et al. (2009) have accomplished culture experiments on 12 bacterial genera isolated from the lichens *Cladonia arbuscula* and *Umbilicaria cylindrica* to prove antagonistic activity of the cultured bacteria. It turned out that some bacterial strains have the ability to reduce the growth of different microorganisms by production of specific secondary compounds, e.g., *Botrytis cenerea* and *Verticillium dahlia*, both plant pathogenic ascomycetes, or acted against *Erwinia carotovora*, a plant pathogenic bacterium. These results raise more interests on bacterial communities living on or in lichen thalli and their impact against harmful organisms, which may enhance the survival of lichens in habitats with high competition or in extreme local conditions (e.g., cracks in rocks) and play an important role in colonization ability.

The PCRs with degenerate primers designed in the study of Schmitt et al. (2005) for the amplification of KS domains showed that only a single KS domain of the mycobiont culture was detected. The primer combination LC1-Im and

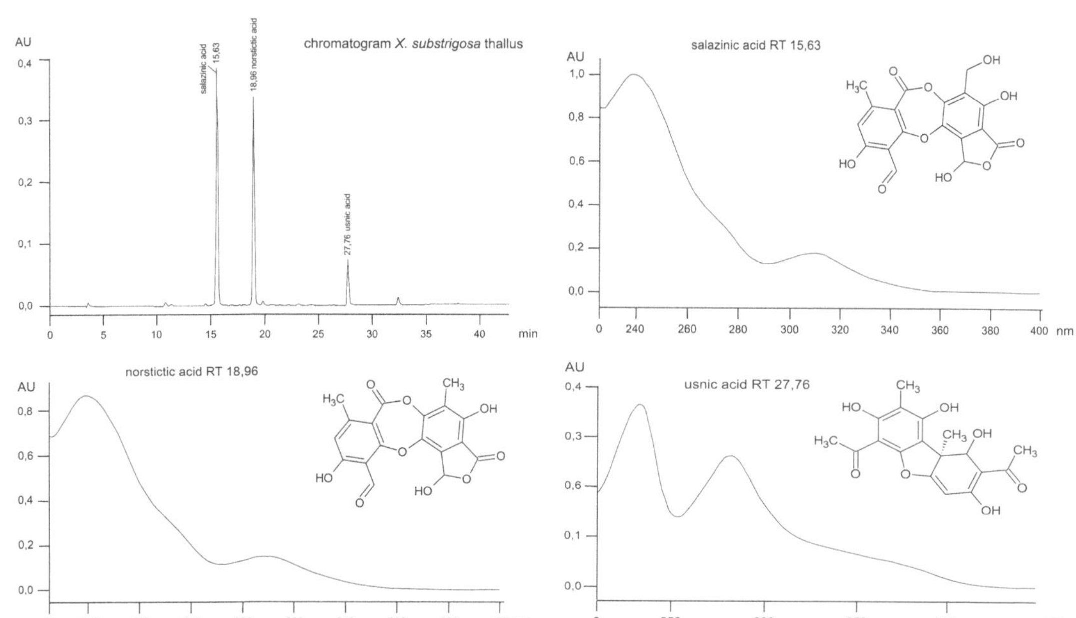

Fig. 5.5 HPLC chromatogram and UV spectra of secondary metabolites detected in the lichen thallus of *Xanthoparmelia substrigosa* collected in Australia

LC2c-Im was generated to target non-reducing PKSs, and the amplification performed in the study of Schmitt et al. (2005) yielded 1–5 different KS sequences in Pertusariales. This fact lead to the assumption that not all PKS genes of *X. substrigosa* were detected, as well as the HPLC analysis revealed the production of several secondary compounds with different origin in the polymalonate pathways. Furthermore, due to the results of HPLC and the consideration of domain formation from lichen PKS genes detected in previous studies (Armaleo et al. 2011; Brunauer et al. 2009; Chooi et al. 2008; Gagunashvili et al. 2009; Valarmathi et al. 2009), the presumption of production of several metabolites from a single gene could be possible depending on the biosynthetic origin of the secondary compounds. Having regard to these circumstances and to the low knowledge about lichen PKSs, the whole-genome sequencing of *X. substrigosa* would be a further option to complete the retrieval of PKS genes.

The resulting PKS gene XsmPKSI obtained via RACE-PCRs using specific primers of the *X. substrigosa*-KS domain had a high nucleotide analogy with the PKS gene of *Usnea longissima* strain CH050148 (Wang et al. 2012) according to the NCBI BLAST search. XsmPKSI and the *U. longissima* PKS gene (*UlPKS1*) belong to the type I non-reducing polyketide synthases, but XsmPKSI contained no starter unit of ACP transacylase (SAT domain) and product template domain (PT), whereas *UlPKS1* does. The molecular analysis of the PKS gene found in the thallus of *X. substrigosa* asserted the accuracy of amino acid configuration and thus the domain organization of XsmPKSI. Additionally, aligning genomic DNA to cDNA (complimentary DNA) revealed five introns along nucleotide positions 288–339, 638–689, 1359–1408, 1763–1815, and 2057–2112 with typical consensus splicing sites of the thalline PKS gene (data not shown).

Compared to other already detected and published lichen type I PKS genes, XsmPKSI showed a higher structural resemblance in the domain organization with the PKS gene found in *Dirinaria applanata* (Valarmathi et al. 2009) and the one found in *Xanthoria elegans* (Brunauer et al. 2009). The only difference between *Dn*PKS, XePKS1, and XsmPKSI is that the polyketide synthases of *Dirinaria applanata* and *Xanthoria elegans* possessed a SAT domain, which induces the transfer of C_6 fatty acids to one of the ACP domains to initialize the synthase activity (Brunauer et al. 2009; Crawford et al. 2006; Valarmathi et al. 2009). The PKS gene located in *Xanthoparmelia semiviridis* (XsePKS1) exhibited a varying domain formation (Chooi et al. 2008). The enzyme complex is composed of a KS, AT, one ACP, and CMeT (C-methyltransferase) domain, while XsmPKSI is comprised of one more ACP domain and a TE domain instead CMeT. The voucher specimen *X. semiviridis* contained succinprotocetraric acid, fumarprotocetraric acid, and usnic acid as major lichen substances, as well as conprotocetraric acid, consuccinprotocetraric acid, confumarprotocetraric acid, protocetraric acid, and virensic acid in minor traces (Chooi et al. 2008). The HPLC analysis of the *Xanthoparmelia substrigosa* thallus used in this study confirmed the production of norstictic acid, salazinic acid, and usnic acid, whereas the satellite substances consalazinic and connorstictic acids were lacking. Chooi et al. (2008) suggested that the production of minor compounds derives from intermediate stages of the β-orsellinic acid biosynthesis. Norstictic acid and salazinic acid are categorized into β-orcinoldepsidones and are known as closely related lichen substances (Culberson et al. 1981). One of the investigated biological functions of salazinic acid was its antioxidant activity (Manojlovic et al. 2012); norstictic acid has the ability to form metal–lichen–acid complexes (Purvis et al. 1987; Purvis 2014). The formation of such complexes arises as avoidance of the toxicity originating from the metal, in this case of copper. The capability of intracellular uptake of copper was assigned to usnic acid when the lichen was growing in nutrient-poor habitats (Hauck et al. 2009), additionally photoprotection by absorbing incident light to preserve the viability of the photobiont (Molnár and Farkas 2010; Rao and LeBlanc 1965), antimicrobial activity (Lauterwein et al. 1995), and some other functions (Ingolfsdottir 2002).

Xanthoparmelia substrigosa was collected on rocky substrate, on Silurian volcanic rocks (Brown and Ollier 1975), and therefore exposed to high UV radiation. Considering the conditions of the substrate and light-exposed habitat of *X. substrigosa*, the production of norstictic acid and usnic acid exhibits a benefit for the lichen to ensure good growth when the capabilities of the substances mentioned before were taken into account. The conditions of the location were different compared to *X. semiviridis*, which was collected in Wyperfeld National Park (Australia), since *X. semiviridis* is known as vagrant lichen. The foliose thallus of this *Xanthoparmelia* species is rolled up like a ball when it is desiccated and is completely expanded in rehydrated state. The dry thallus lobes were moved by wind and dispersed in that way to other locations (Mereschkowsky 1918). Another dispersal strategy of *X. semiviridis* was observed, while small thallus fragments became entangled in animal fur and thus were transported over longer distances. It also has been found that vagrant species of *Xanthoparmelia* were eaten by sheep and pronghorn antelopes in North America (Rosentreter 1993), which suggests that *X. semiviridis* can serve as food additive for herbivore animals or microorganisms and thus the production of fumarprotocetraric acid, protocetraric acid, and usnic acid, which exhibit verified strong antimicrobial activity and antiherbivory effect as protection against parasites and pathogens will be initiated (Ingolfsdottir 2002; Lawrey 1986; Molnár and Farkas 2010; Nimis and Skert 2006; Pöykkö and Hyvärinen 2003; Solhaug and Gauslaa 2012). The environmental conditions and life strategies of *X. semiviridis* and *substrigosa* required complete different adaptations of the lichens to the respective surroundings.

According the HPLC analysis of the cultured mycobiont of *X. substrigosa*, no secondary compounds were detected in the young mycobiont. This fact was often observed as mentioned in Stocker-Wörgötter (2008) which could be affiliated due to the fact that the culture was not kept under sufficient stressful conditions and was grown in nutrient-rich medium to get more pure mycelia material for the genetic investigations. Previous studies (Brunauer et al. 2007; Stocker-Wörgötter et al. 2004; Timsina et al. 2013; Valarmathi et al. 2009; Wang et al. 2012) investigated the effect of media composition (especially the addition of different carbon sources like ribitol, mannitol, and sorbitol), the exposure of the cultures to UV radiation, of desiccation, the influence of variable pH of the media, etc., in relation to the production of polyketides. It turned out that for instance, the addition of mannitol can induce the production of increased variety of anthraquinones in *Xanthoria elegans* culture (Brunauer et al. 2007) or different amounts of sucrose in the medium enhance the polyketide synthase expression in *Usnea longissima* (Wang et al. 2012).

Chemical analyses performed by Armaleo et al. (2011), Brunauer et al. (2009), Gagunashvili et al. (2009), and Valarmathi et al. (2009) revealed various types of secondary metabolites (e.g., atranorin, grayanic acid, gyrophoric acid, and parietin), which makes accurate statements about the identity of a particular polyketide metabolite and its production by a certain PKS at the present stage of knowledge difficult and still very speculative. For future investigations, it would be recommendable to use lichen specimens with similar substance composition for the screening of further PKS genes. These results and subsequent alignments of each PKS gene would facilitate the understanding of the functions and interactions between PKS enzymes and their metabolites produced via polymalonate pathways.

Acknowledgments The study was supported by the Austrian Science Foundation, FWF Grant No. 23570 to EST-W. We thank Jack Elix (Canberra, Australia) for identification of lichen substances. We are also grateful to Georg Brunauer (Salzburg, Austria) and Marion C. Höpflinger (Salzburg, Austria) for technical support.

References

Armaleo D, Sun X, Culberson C (2011) Insights from the first putative biosynthetic gene cluster for a lichen depside and depsidone. Mycologia 103:741–754. doi:10.3852/10-335

Bingle LE, Simpson TJ, Lazarus CM (1999) Ketosynthase domain probes identify two subclasses of fungal polyketide synthase genes. Fungal Genet Biol 26:209–223. doi:10.1006/fgbi.1999.1115

Brader G, Compant S, Mitter B, Trognitz F, Sessitsch A (2014) Metabolic potential of endophytic bacteria. Curr Opin Biotechnol 27:30–37. doi:10.1016/j.copbio.2013.09.012

Brown MC, Ollier CD (1975) Geology and scenery of Canberra. Aust Geogr 13:97–103. doi:10.1080/00049187508702684

Brunauer G, Hager A, Grube M, Türk R, Stocker-Wörgötter E (2007) Alterations in secondary metabolism of aposymbiotically grown mycobionts of Xanthoria elegans and cultured resynthesis stages. Plant Physiol Biochem PPB Soc Fr Physiol Veg 45:146–151

Brunauer G, Muggia L, Stocker-Wörgötter E, Grube M (2009) A transcribed polyketide synthase gene from Xanthoria elegans. Mycol Res 113:82–92. doi:10.1016/j.mycres.2008.08.007

Chooi Y-H, Stalker DM, Davis Ma, Fujii I, Elix Ja, Louwhoff SHJJ, Lawrie AC (2008) Cloning and sequence characterization of a non-reducing polyketide synthase gene from the lichen *Xanthoparmelia semiviridis*. Mycol Res 112:147–161. doi:10.1016/j.mycres.2007.08.022

Crawford JM, Dancy BCR, Hill EA, Udwary DW, Townsend CA (2006) Identification of a starter unit acyl-carrier protein transacylase domain in an iterative type I polyketide synthase. Proc Natl Acad Sci U S A 103:16728–16733. doi:10.1073/pnas.0604112103

Croteau R, Kutchan TM, Lewis NG (2000) Natural products (secondary metabolites). In: Buchanan B, Gruissem W, Jones R (eds) Biochemistry and molecular biology of plants. American Society of Plant Physiologists, pp 1250–1318

Culberson W (1969) The use of chemistry in the systematics of the lichens. Taxon 18:152–166

Culberson C, Culberson W, Johnson A (1981) A standardized TLC analysis of β-Orcinol Depsidones. Bryologist 84:16–29

Demain A (1999) Pharmaceutically active secondary metabolites of microorganisms. Appl Microbiol Biotechnol 52:455–463

Doyle JJ, Doyle JL (1987) A rapid DNA isolation procedure for small quantities of fresh leaf tissue. Phytochem Bull 19:11–15

Elix JA (1993) Progress in the generic delimitation of parmelia sensu lato lichens (Ascomycotina: Parmeliaceae) and a synoptic key to the Parmeliaceae. Bryologist 96:359–383

Elix JA (1994) Xanthoparmelia. Flora Australia, 55th edn. pp 201–308

Elix JA (1996) Biochemistry and secondary metabolites. In: Nash III T (ed) Lichen Biology. Cambridge University Press, Cambridge, pp 155–180

Elix JA, Wardlaw JH (2000) Lusitanic acid, peristictic acid and verrucigeric acid. Three new beta-orcinol depsidones from the lichens *Relicina sydneyensis* and *Xanthoparmelia verrucigera*. Aust J Chem 53:815–818

Fehrer J, Slavikova-Bayerova S, Orange A (2008) Large genetic divergence of new, morphologically similar species of sterile lichens from Europe (Lepraria, Stereocaulaceae, Ascomycota): concordance of DNA sequence data with secondary metabolites. Cladistics 24:443. doi:10.1111/j.1096-0031.2008.00216.x

Feige GB, Lumbsch HT (1993) Identification of lichen substances by a standardized high-performance liquid chromatographic method. J Chromatogr 646:417–427. doi:10.1016/0021-9673(93)83356-W

Fox EM, Howlett BJ (2008) Secondary metabolism: regulation and role in fungal biology. Curr Opin Microbiol 11:481–487. doi:10.1016/j.mib.2008.10.007

Gagunashvili AN, Davídsson SP, Jónsson ZO, Andrésson OS (2009) Cloning and heterologous transcription of a polyketide synthase gene from the lichen *Solorina crocea*. Mycol Res 113:354–363. doi:10.1016/j.mycres.2008.11.011

Gardes M, Bruns TD (1993) ITS primers with enhanced specificity for basidiomycetes—application to the identification of mycorrhizae and rusts. Mol Ecol 2:113–118. doi:10.1111/j.1365-294X.1993.tb00005.x

Grube M, Blaha J (2003) On the phylogeny of some polyketide synthase genes in the lichenized genus Lecanora. Mycol Res 107:1419–1426. doi:10.1017/S0953756203008724

Grube M, Cardinale M, de Castro JV, Müller H, Berg G (2009) Species-specific structural and functional diversity of bacterial communities in lichen symbioses. ISME J 3:1105–1115. doi:10.1038/ismej.2009.63

Hale ME (1990) A synopsis of the lichen genus Xanthoparmelia (Vainio) Hale (Ascomycotina, Parmeliaceae), 24th edn. Smithsonian contributions to botany, USA, pp 1–254

Hauck M, Willenbruch K, Leuschner C (2009) Lichen substances prevent lichens from nutrient deficiency. J Chem Ecol 35:71–73. doi:10.1007/s10886-008-9584-2

Huneck S (1999) The significance of lichens and their metabolites. Naturwissenschaften 86:559–570

Huneck S, Yoshimura I (1996) Identification of lichen substances. Springer, Berlin, 493 pp

Ingolfsdottir K (2002) Usnic acid. Phytochemistry 61:729–736

Katsuyama Y, Ohnishi Y (2012) Type III polyketide synthases in microorganisms. Methods Enzymol 515:359–377. doi:10.1016/B978-0-12-394290-6.00017-3

Keating TA, Walsh CT (1999) Initiation, elongation, and termination strategies in polyketide and polypeptide antibiotic biosynthesis. Curr Opin Chem Biol 3:598–606

Lauterwein M, Oethinger M, Belsner K, Peters T, Marre R (1995) In vitro activities of the lichen secondary microorganisms. In Vitro Activities of the Lichen Secondary Metabolites Vulpinic Acid, (+)-Usnic Acid, and (−)-Usnic Acid against Aerobic and Anaerobic Microorganisms. doi: 10.1128/AAC.39.11.2541. Updated

Lawrey JD (1986) Biological role of lichen substances. Bryologist 89:111–122

Lillly VG, Barnett HL (1951) Physiology of the fungi, 1st edn. McGraw-Hill Book Company, New York

Manojlovic NT, Vasiljevic PJ, Maskovic PZ, Juskovic M, Bogdanovic-Dusanovic G (2012) Chemical composition, antioxidant, and antimicrobial activities of lichen *Umbilicaria cylindrica* (L.) Delise (Umbilicariaceae). Evid Based Complement Alternat Med 2012:452431. doi:10.1155/2012/452431

Merewschkowsky C (1918) Note sur une nouvelle forme de Parmelia vivant a l'état libre. Bull la Soc Bot Genève 10:26–34

Miao V, Coëffet-LeGal MF, Brown D, Sinnemann S, Donaldson G, Davies J (2001) Genetic approaches to harvesting lichen products. Trends Biotechnol 19:349–355

Molnár K, Farkas E (2010) Current results on biological activities of lichen secondary metabolites: a review. Z Naturforsch C 65:157–173

Nash T (1996) Lichen biology. Cambridge University Press, Cambridge

Nicholson TP, Rudd BAM, Dawson M, Lazarus CM, Simpson TJ, Cox RJ (2001) Design and utility of oligonucleotide gene probes for fungal polyketide synthases. Chem Biol 8:157–178. doi:10.1016/S1074-5521(00)90064-4

Nimis PL, Skert N (2006) Lichen chemistry and selective grazing by the coleopteran *Lasioderma serricorne*. Environ Exp Bot 55:175–182. doi:10.1016/j.envexpbot.2004.10.011

Nylander W (1866) Les lichens des Jardins de Luxembourg. Soc Bot Fr 13:364–371

Pichersky E, Gang DR (2000) Genetics and biochemistry of secondary metabolites in plants: an evolutionary perspective. Trends Plant Sci 5:439–445

Piercey-Normore MD (2007) The genus Cladonia in Manitoba: exploring taxonomic trends with secondary metabolites. Mycotaxon 101:189–199

Pietra F (1997) Secondary metabolites from marine microorganisms: bacteria, protozoa, algae and fungi. Achievements and prospects. Nat Prod Rep 14:453–464

Pöykkö H, Hyvärinen M (2003) Host preference and performance of *lichenivorous Eilema* spp. larvae in relation to lichen secondary metabolites. J Anim Ecol 72:383–390

Purvis O (2014) Adaptation and interaction of saxicolous crustose lichens with metals. Bot Stud 55:23. doi:10.1186/1999-3110-55-23

Purvis OW, Elix JA, Broomheadj JA (1987) The occurrence of copper—Norstictic acid in Lichens from Cupriferous Substrata. The Lichenologist 19:193–203

Rao DN, LeBlanc F (1965) A possible role of atranorin in the lichen thallus. Bryologist 68:284. doi:10.1639/0007-2745(1965)68[284:APROAI]2.0.CO;2

Rosentreter R (1993) Vagrant lichens in North America. Bryologist 96:333–338. doi:10.2307/3243861

Savchuk L (2006) Versuche zur Etablierung eines Verfahrens zur Isolierung von RNA aus Eberspermien

Scharf DH, Heinekamp T, Brakhage AA (2014) Human and plant fungal pathogens: the role of secondary metabolites. PLoS Pathog 10:e1003859. doi:10.1371/journal.ppat.1003859

Schmitt I, Lumbsch HT (2004) Molecular phylogeny of the Pertusariaceae supports secondary chemistry as an important systematic character set in lichen-forming ascomycetes. Mol Phylogenet Evol 33:43–55

Schmitt I, Martín MP, Kautz S, Lumbsch HT (2005) Diversity of non-reducing polyketide synthase genes in the Pertusariales (lichenized Ascomycota): a phylogenetic perspective. Phytochemistry 66:1241–1253. doi:10.1016/j.phytochem.2005.04.014

Schmull M, Miadlikowsk J, Pelzer M, Stocker-Wörgötter E, Hofstetter V, Fraker E, Hodkinson B, Reeb V, Kukwa M, Lumbsch T, Kauff F, Lutzoni F (2011) Phylogenetic affiliations of members of the heterogeneous lichen-forming fungi of the genus Lecidea sensu Zahlbruckner (Lecanoromycetes, Ascomycota). Mycologia 103:983–1003

Schneider A, Marahiel MA (1998) Genetic evidence for a role of thioesterase domains, integrated in or associated with peptide synthetases, in non-ribosomal peptide biosynthesis in Bacillus subtilis. Arch Microbiol 169:404–410

Serre L, Verbree E, Dauter Z, Stuitje A, Derewenda Z (1995) The *Escherichia coli* Malonyl-CoA: acyl carrier protein transacylase at 1.5-Å Resolution. J Biol Chem 270:12961–12964

Solhaug KA, Gauslaa Y (2012) Secondary lichen compounds as protection against excess solar radiation and herbivores. In: Lüttge U, Beyschlag W, Büdel B, Francis D (eds) Progress in botany, 73rd ed. Springer, Heidelberg, pp 283–304

Stocker-Wörgötter E (2002) Resynthesis of photosymbiodemes. In: Kramer I, Beckett R, Varma A (eds) Protocols in lichenology-culturing, biochemistry, ecophysiology and use In biomonitoring. Springer Lab Manual, pp 47–60

Stocker-Wörgötter E (2008) Metabolic diversity of lichen-forming ascomycetous fungi: culturing, polyketide and shikimate metabolite production, and PKS genes. Nat Prod Rep 25:188–200. doi:10.1039/b606983p

Stocker-Wörgötter AE, Elix JA, Grube M (2004) Secondary chemistry of lichen-forming fungi: Chemosyndromic variation and DNA-analyses of cultures and chemotypes in the Ramalina farinacea complex. The Bryologist 107:152–162

Timsina B, Stocker-Wörgötter E, Piercey-Normore MD (2012) Monophyly of some North American species of Ramalina and inferred polyketide synthase gene function. Botany 90:1295–1307. doi:10.1139/b2012-097

Timsina BA, Sorensen JL, Weihrauch D, Piercey-Normore MD (2013) Effect of aposymbiotic conditions on colony growth and secondary metabolite production in the lichen-forming fungus *Ramalina dilacerata*. Fungal Biol 117:731–743. doi:10.1016/j.funbio.2013.09.003

Valarmathi R, Hariharan GN, Venkataraman G, Parida A (2009) Characterization of a non-reducing polyketide synthase gene from lichen *Dirinaria applanata*. Phytochemistry 70:721–729. doi:10.1016/j.phytochem.2009.04.007

Wang Y, Kim Ja, Cheong YH, Koh YJ, Hur J-S (2012) Isolation and characterization of a non-reducing polyketide synthase gene from the lichen-forming fungus *Usnea longissima*. Mycol Prog 11:75–83. doi:10.1007/s11557-010-0730-1

Watanabe A, Ebizuka Y (2004) Unprecedented mechanism of chain aromatic polyketide synthases. Chemistry & Biology 11:1101–1106. doi:10.1016/j

White TJ, Bruns T, Lee S, Taylor J (1990) Amplification and direct sequencing of fungal ribosomal RNA genes for phylogenetics. In: Innis MA, Gelfand DH, Sninsky JJ, White TJ (eds) PCR protocols: a guide to methods and applications. Academic Press, New York, pp 315–322

Wong HC, Liu G, Zhang Y-M, Rock CO, Zheng J (2002) The solution structure of acyl carrier protein from Mycobacterium tuberculosis. J Biol Chem 277:15874–15880. doi:10.1074/jbc.M112300200

Yamamoto Y (1990) Studies of cell aggregates and the production of natural pigments in plant cell culture. Nippon Paint, Singapore

Yu D, Xu F, Zeng J, Zhan J (2012a) Type III polyketide synthases in natural product biosynthesis. IUBMB Life 64:285–295. doi:10.1002/iub.1005

Yu NH, Kim JA, Jeong MH, Cheong YH, Jung JS, Hur J-S (2012b) Characterization of two novel non-reducing polyketide synthase genes from the lichen-forming fungus Hypogymnia physodes. Mycol Prog 12:519–524. doi:10.1007/s11557-012-0858-2

Zocher B, Stocker-Wörgötter E (2005) Secondary chemistry of lichen-forming ascomycetes: culture experiments with isolates of Xanthoparmelia species from Australia. Folia Cryptogam Est 41:123–133

Endolichenic Fungi in Kumaun Himalaya: A Case Study

6

Manish Tripathi and Yogesh Joshi

Contents

Abstract

Endolichenic fungi represent an important ecological group of species that form associations with lichens, and to extend the knowledge of their diversity within macrolichens, we isolated and identified the endolichenic fungi from some healthy macrolichens of Kumaun Himalaya. The main goal of the research presented here was to examine the diversity and taxonomic composition of the endolichenic fungal community of 14 macrolichen species in Kumaun Himalaya. Identification of fungal isolates was mainly done morphologically. The majority of endolichenic fungi belonged to Hyphomycetes, and the lowest were obtained from Zygomycetes. A total of 10 species are reported as true endolichenic fungi. Species reported in the present paper as true endolichenic fungi were previously reported as decomposers, potential pathogens, biological control agents, and secondary metabolite producers, reflecting an interesting functional diversity and new host. This study will encourage further studies regarding endolichenic fungi including biotechnological aspects and bioprospection of endolichenic fungi. In this study, we enlighten the areas in endolichenic fungi in need of further study. We hope by calling attention to the largely unexplored biodiversity of endolichenic fungi, investigators will take up the study of these fascinating organisms.

M. Tripathi · Y. Joshi (✉)
Lichenology Division, Department of Botany, Kumaun University, S.S.J. Campus, Almora 263601, Uttarakhand, India
e-mail: dryogeshcalo@gmail.com

D.K. Upreti et al. (eds.), *Recent Advances in Lichenology*,
DOI 10.1007/978-81-322-2235-4_6

Keywords
Himalaya · Fungi · Lichens · Endolichenic fungi

6.1 Introduction

Symbiotic associations between fungi and photosynthetic organisms are both ancient and ubiquitous (Alexopoulos et al. 1996; Berbee 2001; Heckman et al. 2001). Such interactions led to permanent associations now recognized as lichens, mycorrhizas, mycophycobioses, and endophytes (Atsatt 1988, 1991; Selosse and LeTacon 1998). Hawksworth and Rossman (1997) identified three categories where undescribed species of fungi can be found: (1) fungi in tropical forests; (2) fungi in unexplored habitats; and (3) lost or hidden species. The second category (fungi in unexplored habitats) includes hypogeous fungi of Australia, fungi in guts of beetles and insects, lichenicolous fungi, and endophytic fungi.

Dreyfuss and Chapela (1994) have estimated that around 1 million endophytic fungi reside in plants. They represent an important and quantifiable component of fungal biodiversity and are known to affect plant community diversity and structure (Krings et al. 2007). Studies on endophytic fungi from tropical and temperate forests support the high estimates of species diversity (Rodrigues et al. 1993; Sánchez Márquez et al. 2007; Santamaria and Bayman 2005).

Endophytes are found in every major plant lineage known on the earth, such as marine algae (Cubit 1974; Hawksworth 1988), mosses and ferns (Schulz et al. 1993; Fisher 1996), palms (Rodrigues and Samuels 1994; Frohlich and Hyde 1999), liverworts (Boullard 1988), pteridophytes (Dhargalkar and Bhat 2009), leaves, bark, and xylem of almost all plants (Petrini 1986), from different regions of the world ranging from dry deserts to Arctic tundra and tropical rainforests. Lichens did not remain exceptional to endophytes (Stone et al. 2000; Arnold 2007; Kannangara et al. 2009; Li et al. 2007 and Suryanarayanan et al. 2005), and the endophytes residing in lichens (=endolichenic fungi) preferentially associate with green algal photobionts in lichen thalli.

The term "endolichenic fungi" was introduced by Miadlikowska et al. (2004) for lineages of Ascomycota that reside within the lichen thallus and which are quite distinct from lichen mycobionts (the primary fungal component of the lichen thallus), lichenicolous fungi (which fruit or are otherwise symptomatic on thalli), and incidental fungi on thallus surfaces (Arnold et al. 2009; Lawrey and Diederich 2003; Lutzoni et al. 2001). The high level of fungal diversity within lichen thalli may be the result of the highly porous and heterogeneous nature of the lichen thalli.

However, few studies pertaining to endolichenic fungal diversity are known across the world (Stone et al. 2000; Davis et al. 2003; Arnold 2007; Kannangara et al. 2009; Li et al. 2007) and in India the study has been started recently (Suryanarayanan et al. 2005). Suryanarayanan et al. (2005) performed their studies on endolichenic fungi of tropical lichens in South India, and till then, no further publications on endolichenic fungi came from India. Kumaun Himalaya was previously explored several times for its lichen diversity, but nothing is being mentioned about their endophytes. This led authors to work on endolichenic fungal diversity in Kumaun Himalaya, keeping following objectives in mind:

(a) Diversity of endolichenic fungi colonizing macrolichens of Kumaun Himalaya
(b) Do endophytes promote/have any role in lichen thallus development
(c) To develop best protocol for isolating endolichenic fungi with minimum occurrence of Mycelia Sterilia
(d) To check correlation between endophytes of substratum and lichen species colonizing that substratum

6.2 Collection and Washing Procedures

The macrolichen samples were collected from different substrates and taken in the laboratory in sterilized paper bags. Plastic bags were not used to avoid the moisture accumulation inside the bags and unwanted growth of fungi on the surface of the thallus. The lichen samples were processed within 24 h after the time of collection. The samples were subjected to surface sterilization to remove the surface contaminants by the modified procedure of Suryanarayanan et al. (2005). Surface sterilization procedures vary according to the species of host plant and host tissue type sampled. It has been noted earlier that different surface sterilization procedures (Petrini 1992; Schulz et al. 1993; Bissegger and Sieber 1994), isolation medium (Bills and Polishook 1991), and sample unit size (Carroll 1995) affect the isolation frequency of endophytes. To check the effectiveness of the sterilization procedure, thallus segments were tapped on to the PDA plates and incubated (Schulz et al. 1993).

When the surface of thallus segments were found well sterilized, they were cut randomly into segments and placed on petri plates having PDA medium supplemented with streptomycin sulfate and incubated at 25 °C for 28 days and checked in the alteration of 3 days. For the isolation and subculturing of endolichenic fungi, the routine mycological media (PDA, MEA, and YMEA) were used. The percentage of the media used varies in different methods. Colony-limiting agents and antibiotics also are often used for primary isolations. Many fungal species produce more diffuse, spreading, and less recognizable colonies on weak media, and to avoid this problem, water agar was used for isolations to reduce contamination. Identification of the isolated endolichenic fungi was done on the basis of cultural characteristics and morphology of fruiting bodies and spores with the help of published flora (Subramaniam 1971; Sutton 1980; Barnett and Hunter 1972; Ellis 1971, 1976; Chowdhry 2000; Gilman 1967). Most of the isolates did not sporulate and were treated as Mycelia sterilia. The fungal specimens are deposited in the herbarium of Kumaun University (ALM).

6.3 Isolation and Identification of Endolichenic Fungi

The present study documents the presence of 25 isolates (excluding mycelia sterilia) of endolichenic fungi belonging to 20 genera (Table 6.1) encompassing Hyphomycetes (52.0 %), Plectomycetes (16.0 %), Coelomycetes (12.0 %), Pyrenomycetes (12.0 %), Agaricomycetes (4.0 %) and Zygomycetes (4.0 %). Our study corroborates with results of previously conducted studies on endolichenic fungi across the world, mentioning that members of Hyphomycetes are the abundant colonizers of lichens as endophytes, while population of Zygomycetes and Agaricomycetes was very less. *Mucor racemosus* and *Rhizoctonia* sp. are the only species of Zygomycetes and Agaricomycetes encountered in the present study respectively.

Of these 25 endophytes, 10 fungal species (*Acremonium lichenicola*, *Bipolaris australiensis*, *Nigrospora sphaerica*, *Papulospora* sp., *Pestalotiopsis maculans*, *Rhizoctonia* sp., *Sordaria fimicola*, *Spegazzinia tessarthra*, *Trichophyton roseum*, *Xylaria hypoxylon*) (Figs. 6.1, 6.2 and 6.3) are reported for the first time across the world as true endolichenic fungi. Meanwhile, presence of *Xylaria hypoxylon* as an endophyte from Kumaun Himalaya indicates the shift in the geographical preference of this species, as members of Xylariaceae predominate as endophytes in tropical regions.

Besides this, the study also reveals that some species of fungi are lichen specific, i.e., restricted to a particular lichen (*Spegazzinia tessarthra*, *Nigrospora sphaerica*, *Nigrospora oryzae*, *Pestalotiopsis maculans*, *Sordaria fimicola*, *Rhizoctonia* sp.), while some species are more generalized and are reported from several or all target lichen species (*Alternaria alternata*, *Aspergillus flavus*, *Fusarium solani*) (Table 6.1).

Results also suggest that the endolichenic fungi may be related to the mycoflora of the substrate and surroundings where the host lichen thallus is

Table 6.1 Endolichenic fungi isolated from macrolichens of Kumaun Himalaya

Lichen species	Family	Endolichenic fungi	References
Bulbothrix meizospora (Nyl.) Hale	Parmeliaceae	*Alternaria alternata*, *Aspergillus flavus*, *Cylindrosporium* sp., *Fusarium solani*, *Gilmaniella humicola*, *Mycelia sterilia*, *Penicillium* sp.	Suryanarayanan et al. (2005), Tripathi et al. (2014c)
Flavoparmelia caperata (L.) Hale	Parmeliaceae	*Alternaria alternata*, *Aspergillus* cfr. *coremiiformis*, *Aspergillus flavus*, *Fusarium solani*, *Mycelia sterilia*	Tripathi et al. (2014c)
Heterodermia flabellata (Fée) D. D. Awasthi	Physciaceae	*Alternaria alternata*, *Aspergillus flavus*, *Aspergillus niger*, **Bipolaris australiensis**, *Fusarium solani*, *Pestalotiopsis* sp. 1, *Pestalotiopsis* sp. 2, ***Spegazzinia tessarthra***, *Trichoderma harzianum*	Suryanarayanan et al. (2005), Li et al. (2007), Tripathi et al. (2014a, b)
Heterodermia hypochraea (Vain.) Swinsc & Krog	Physciaceae	*Alternaria alternata*, *Aspergillus flavus*, *Aspergillus niger*, *Fusarium solani*, *Papulospora* sp.	Suryanarayanan et al. (2005), Tripathi et al. (2014c)
Leptogium burnetiae Dodge	Collemataceae	*Alternaria alternata*, *Aspergillus flavus*, *Fusarium solani*, *Gilmaniella humicola*	Tripathi et al. (2014c)
Parmelaria thomsonii (Stirton) D.D. Awasthi	Parmeliaceae	*Acremonium* sp., *Alternaria alternata*, *Aspergillus flavus*, *Aspergillus niger*, *Fusarium solani*, ***Nigrospora sphaerica***, *Pestalotiopsis* sp., *Trichoderma harzianum*	Suryanarayanan et al. (2005), Li et al. (2007), Tripathi et al. (2014c)
Parmotrema crinitum (Ach.) Choisy	Parmeliaceae	*Alternaria alternata*, *Aspergillus flavus*, *Fusarium solani*, *Mycelia sterilia*, *Trichoderma harzianum*	Suryanarayanan et al. (2005), Li et al. (2007), Tripathi et al. (2014c)
Parmotrema graynum (Hue) Hale	Parmeliaceae	*Alternaria alternata*, *Aspergillus flavus*, *Gilamniella humicola*, *Fusarium solani*, *Trichophyton roseum*	Tripathi et al. (2014c)
Parmotrema nilgherrense (Nyl.) Hale	Parmeliaceae	*Alternaria alternata*, *Aspergillus flavus*, *Chaetomella* sp., *Cladosporium* sp., *Gilmaniella humicola*, *Fusarium solani*, *Mycelia sterilia*	Suryanarayanan et al. (2005), Tripathi et al. (2014c)
Parmotrema praesorediosum (Nyl.) Hale	Parmeliaceae	*Alternaria alternata*, *Aspergillus flavus*, *Cladosporium* sp., *Fusarium solani*	Suryanarayanan et al. (2005), Tripathi et al. (2014c)
Parmotrema reticulatum (Taylor) Choisy	Parmeliaceae	*Acremonium lichenicola*, *Alternaria alternata*, *Aspergillus flavus*, *Fusarium solani*, ***Nigrospora oryzae***, **Mucor racemosus**, *Papulospora* sp., *Penicillium* sp., ***Pestalotiopsis maculans***, ***Sordaria fimicola***, *Xylaria hypoxylon*, **Acremonium lichenicola**	Suryanarayanan et al. (2005), Tripathi et al. (2014c)
Physcia dilatata Nyl.	Physciaceae	*Alternaria alternata*, *Aspergillus flavus*, *Aspergillus niger*, ***Bipolaris australiensis***, *Cladosporium* sp., *Fusarium solani*, *Trichoderma harzianum*	Suryanarayanan et al. (2005), Li et al. (2007), Tripathi et al. (2014b, c)
Punctelia subruducta (Nyl.) Krog	Parmeliaceae	*Gilmaniella humicola*, *Fusarium solani*, *Rhizoctonia* sp.	Tripathi et al. (2014c)
Usnea sp.	Parmeliaceae	*Fusarium solani*, *Trichoderma harzianum*	Tripathi et al. (2014c)

colonizing. For example, *Sordaria fimicola* isolated in the present study from *Parmotrema reticulatum* is a coprofillous fungus and is isolated from terricolous specimen of the host thallus; however, its presence from the corticolous specimen of the same lichen species was not reported (Table 6.2).

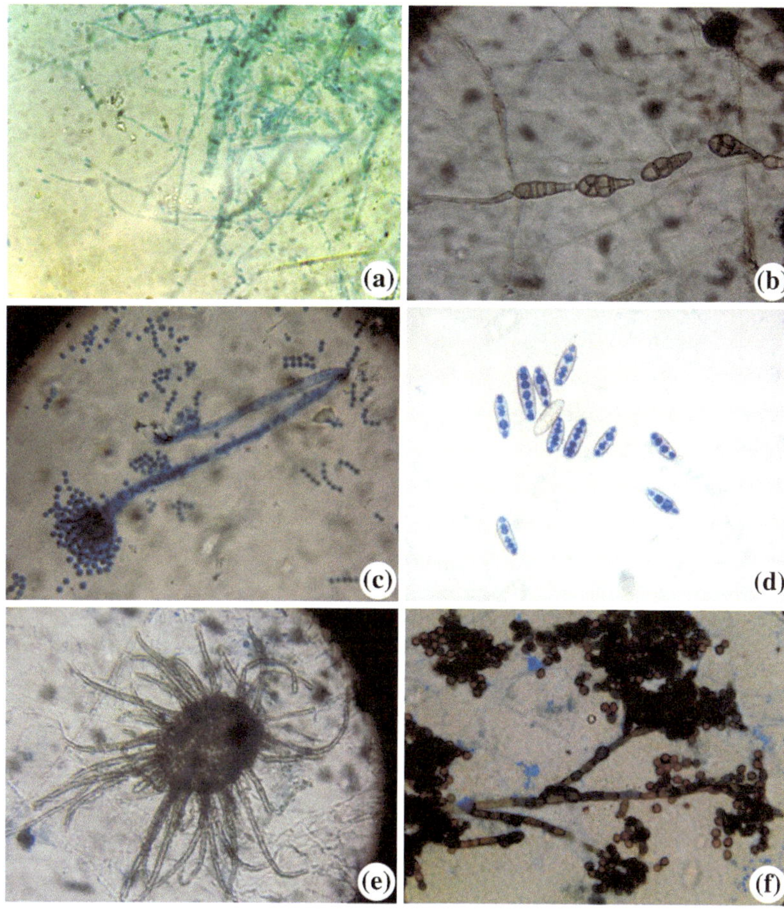

Fig. 6.1 Endolichenic fungi. **a** *Acremonium lichenicola*, **b** *Alternaria alternata*, **c** *Aspergillus flavus*, **d** *Bipolaris australiensis*, **e** *Chaetomella* sp., **f** *Cladosporium* sp.

In the present study, another hypothesis whether the diversity of endolichenic fungi colonizing a particular epiphyte is somewhat related to the diversity of endophytes of that host tree. For this, three lichen species (*Heterodermia flabellata*, *Parmotrema praesorediosum*, *Physcia dilatata*) growing on the bark of *Quercus* tree were tested to check this relation, and it was found that all the three lichen species have approximately the same endophytes as that of *Quercus* (Tables 6.3).

It was also noticed that there were no major changes in the diversity of endophytes of the same lichen species growing in different locations, indicating that endolichenic fungi occur in a random manner and no fixed pattern has been observed in their occurrence within lichen thallus, indicating that these endophytes are not responsible for lichen growth and development (Table 6.4).

It has been observed by various workers that the isolation of endolichenic fungi is generally a method-dependent process, i.e., depends on the surface sterilization procedure, sample size of the explant, and media to culture them. The more the number of thallus segments, the more the isolates we can found. To check the effectiveness of different surface sterilization procedures for our samples (Petrini 1986; Petrini and Carroll 1981; Rodrigues and Samuels 1994; bayman et al. 1975; Suryanarayanan et al. 2005 modified), we found that the isolation frequency of endolichenic fungi was maximum in the modified protocol of Suryanarayanan et al. (2005) (Table 6.5) and the most effective media to culture endolichenic fungi was PDA.

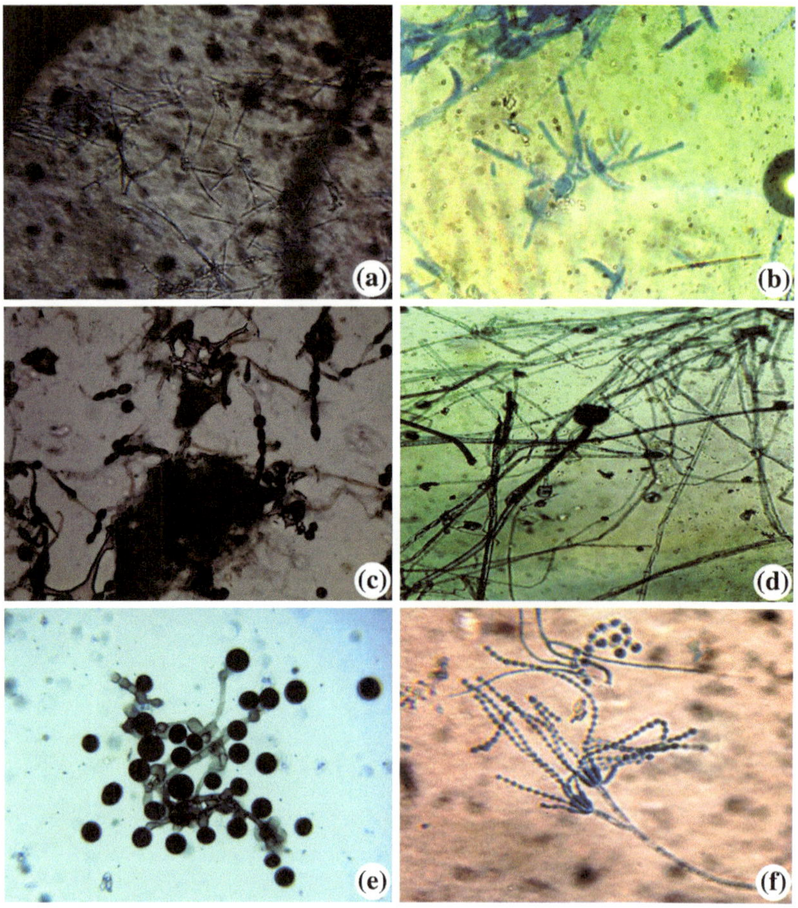

Fig. 6.2 Endolichenic fungi. **a** *Cylindrosporium* sp., **b** *Fusarium solani*, **c** *Gilmaniella humicola*, **d** *Mucor racemosus*, **e** *Nigrospora sphaerica*, **f** *Penicillium* sp.

6.4 Practical and Theoretical Considerations

If we accept Hawksworth (1991) hypothesis that a single tree hosts 6 fungal species, then the total fungal diversity will be around 1.5 million species in the world, and if we follow Frohlich and Hyde (1999), who gave a more realistic figure of 33 species of endophytic fungi per host, then the tally increases up to 5 times greater than that of 1.5 million of Hawksworth (1991). But, as far as Hawksworth and Rossman (1997) hypothesis of undescribed fungi is concerned, the lichens are the least explored among plants for their endophytic fungal diversity. Endolichenic fungi are highly specialized and successful group of organisms which develop inside the healthy lichen thallus. They are relatively inconspicuous and rarely studied by non-lichenologists. Lichenologists in the past were not much interested in these fungi, perhaps owing to their lack of knowledge of non-lichenized fungi. Professional mycologists who would have the knowledge of these groups rarely studied lichens, the very specialized substrate. For these reasons, studies on endolichenic fungi remain in dark for a long time and are poorly known.

Though such type of preliminary studies have to some extent explored the fungal diversity residing within lichen thallus, still, there remains a lot to be done in this field. The effect of incubation temperature and light cycles on emergence of endophytes and their sporulation is still unknown. Till now, abundance of mycelia sterilia in the cultures of endolichenic fungi is an unsolved mystery. How these microbial endophytes gain access to their host plants, their biology,

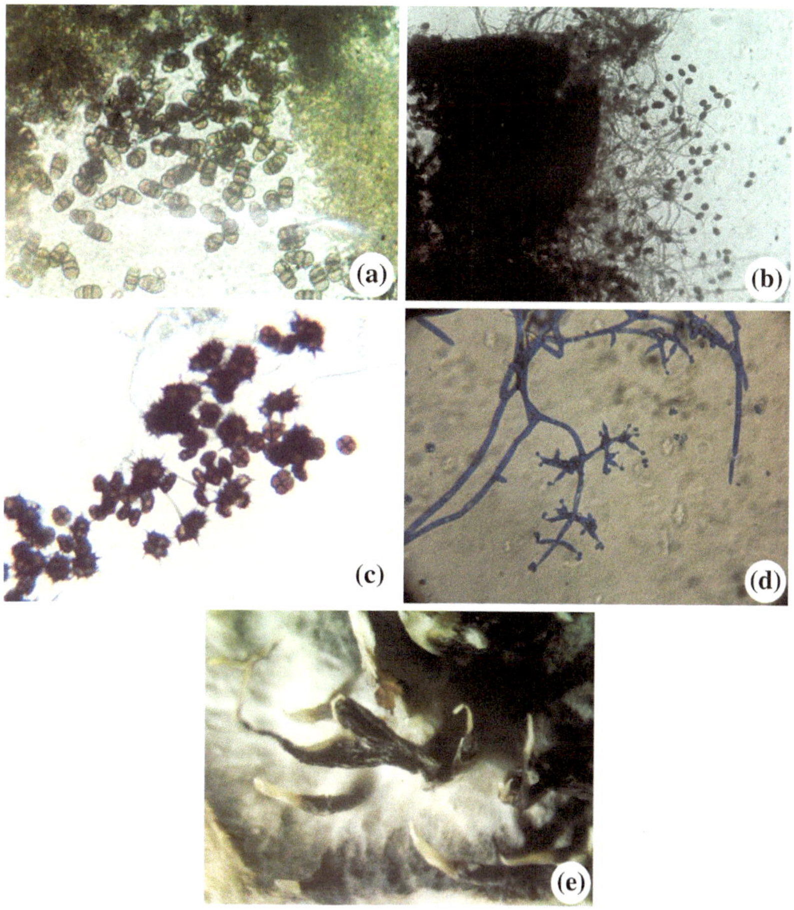

Fig. 6.3 Endolichenic fungi. **a** *Pestalotiopsis maculans*, **b** *Sordaria fimicola*, **c** *Spegazzinia tessarthra*, **d** *Trichoderma harzianum*, **e** *Xylaria hypoxylon*

Table 6.2 Endolichenic fungi isolated from lichen colonizing on two different substrates

Lichenized fungi	Soil	Bark
Parmotrema reticulatum	***Acremonium lichenicola***, *Alternaria alternata*, *Aspergillus flavus*, *Fusarium solani*, *Nigrospora oryzae*, *Papulospora* sp. 1, *Penicillium* sp., *Pestalotiopsis maculans*, ***Sordaria fimicola***, ***Xylaria hypoxylon***	***Acremonium* sp.**, *Alternaria alternata*, *Fusarium solani*, ***Trichoderma harzianum***, *Penicillium citrinum*, ***Gilmaniella humicola***, *Pestalotiopsis* sp. 1

Table 6.3 Endolichenic fungi isolated from different lichenized fungi growing on the same substrate

Lichenized fungi	Thallus	Bark
Heterodermia flabellata	***Mycelia sterilia***, ***Pestalotiopsis* sp. 1,** *Pestalotiopsis* sp. 2, *Spegazzinia tessarthra*, ***Trichoderma harzianum***	*Fusarium solani*, ***Mycelia sterilia***, ***Pestalotiopsis* sp. 1, *Trichoderma harzianum***, *Verticillium* sp.
Parmotrema praesorediosum	***Alternaria alternata***, *Aspergillus flavus*, *Cladosporium* sp., ***Mycelia sterilia***	***Alternaria alternata***, *Fusarium solani*, ***Mycelia sterilia***, *Penicillium* sp.
Physcia dilatata	***Alternaria alternata***, ***Aspergillus flavus***, *Bipolaris australiensis*, *Cladosporium* sp., ***Mycelia sterilia***	***Alternaria alternata***, ***Aspergillus flavus***, *Fusarium solani*, ***Mycelia sterilia***, *Trichoderma harzianum*

Table 6.4 Endolichenic fungi isolated from *Parmotrema reticulatum* from different locations

Lichenized fungi	Location	Endolichenic fungi
Parmotrema reticulatum	Syahi Devi Forest	*Acremonium lichenicola*, *Alternaria alternata*, *Aspergillus flavus*, *Fusarium solani*, *Nigrospora oryzae*, *Papulospora* sp., *Penicillium* sp., *Pestalotiopsis maculans*, *Sordaria fimicola*, *Xylaria hypoxylon*
	Banri Devi Forest	*Alternaria alternata*, *Aspergillus flavus*, *Fusarium solani*, *Nigrospora sphaerica*, *Penicillium citrinum*, *Pestalotiopsis* sp. 1
	Binsar Wildlife Sanctuary	*Acremonium* sp., *Alternaria alternata*, *Fusarium solani*, *Trichoderma harzianum*, *Penicillium citrinum*, *Gilmaniella humicola*, *Pestalotiopsis* sp. 1
	Jhakar Sem Forest	*Acremonium* sp., *Alternaria alternata*, *Aspergillus* sp., *Fusarium* sp., *Nigrospora sphaerica*, *Penicillium citrinum*, *Trichoderma harzianum*
	Dhaula Devi forest	*Acremonium* sp., *Alternaria alternata*, *Aspergillus* sp., *Penicillium citrinum*, *Nigrospora sphaerica*

Table 6.5 Effect of different surface sterilization procedures on the isolation of endolichenic fungi

Surface sterilization procedures	Lichenized fungi
Petrini (1986)	*Mycelia sterilia*
Petrini and Carroll (1981)	*Acremonium* sp.
Rodrigues and Samuels (1994)	*Aspergillus flavus*
Bayman et al. (1997)	*Mycelia sterilia*
Suryanarayanan et al. (2005) modified	*Alternaria alternata*, *Nigrospora sphaerica*, *Pestalotiopsis* sp.

preference for a particular host, omnipresence nature, and their relationship and communication with other endophytes of the same host are some questions which still need to be answered.

Prior to this, studies on the diversity of endolichenic fungi along an altitudinal and latitudinal gradient, within the different host individuals of the same species, same host species on different substrates, seasonal variation, and phytosociology of endolichenic fungi need to be done.

Besides this, lichenology is the most neglected branch of plant science in the context of screening their endophytes for novel compounds. Endophytes from different plant species have been screened out across the globe, but endolichenic fungi are at its elementary stage, and so far available literature concerning endolichenic fungi mentions about their fungal diversity, nothing about bioactive potential of these endophytes. The endophytes residing inside lichen thallus were considered to evolve from the pathogens of the host, and if they are residing inside a healthy lichen thallus without causing any symptoms, then there would have been a change in the genotype of the endophyte during this evolutionary process (pathogen to mutualist); hence, the same fungus which is acting as pathogen or mutualist has either the same genome or same bioactivity is needed to be checked out. For example, *Trichoderma harzianum* is a potent biocontrol agent and has been screened out for its activity, but as an endolichenic fungal strain, it has not been screened out. Previously, fungi, such as *Cladosporium*, *Cylindrosporium,* and *Pestalotiopsis* isolated from various plants, have been tested for their bioactivity and were found as potent inhibitor of pathogenic microbes; hence, presence of these fungal species in lichens shows the significance of lichens and associated endolichenic mycoflora as a source of bioactive compounds and can be proven as endless source of secondary metabolites acting as novel compounds and will also resolve to some extent the fungal diversity debate.

The present work is a preliminary step toward an alluring field of endolichenic fungi and opens new opportunities for the researchers who are working in the field of lichens (Fig. 6.4).

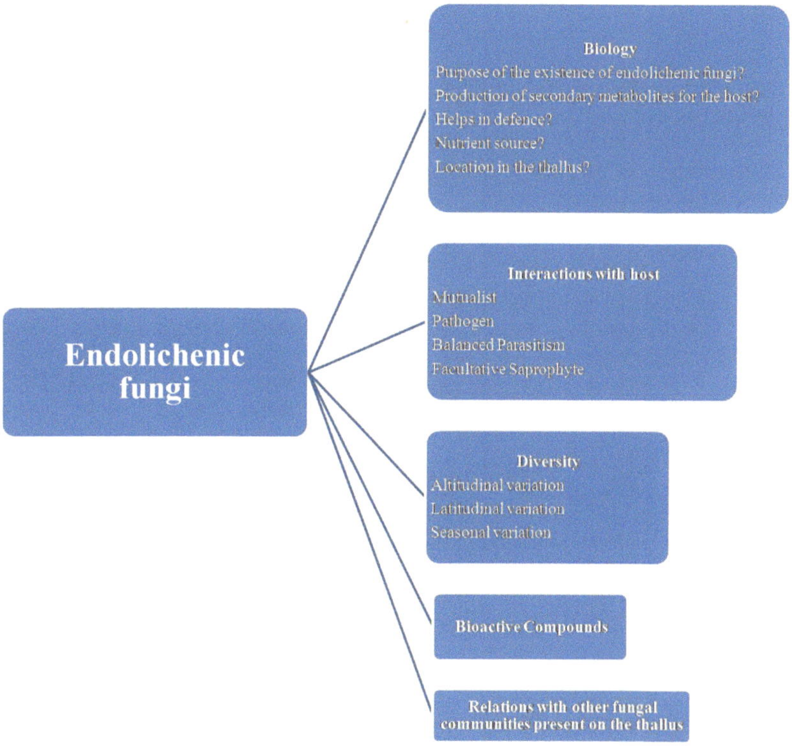

Fig. 6.4 Future prospects of endolichenic fungi

Acknowledgment Authors would like to thank University Grants Commission [File No. 41-488/2012 (SR)] for the financial assistance.

References

Alexopoulous CJ, Mims CW, Blackwell M (1996) Introductory Mycology. Willey, New York

Arnold AE (2007) Understanding the diversity of foliar endophytic fungi, progress, challenges, and frontiers. Fungal Biol Rev 21:51–66

Arnold AE, Miadilkowska J, Higgins KL, Sarvate SD, Gugger P, Way A, Hofstter V, Kauff F, Lutzoni F (2009) A phylogenetic estimation of trophic transition networks for ascomycetous fungi: are lichens cradles of symbiotrophic fungal diversification? Syst Biol 58:283–297

Atsatt PR (1988) Are vascular plants inside-out lichens? Ecology 69:17–23

Atsatt PR (1991) Fungi and the origin of land plants. In: Margulis L, Fester R (eds) Symbiosis as a source of evolutionary innovation: speciation and morphogenesis. MIT Press, Cambridge, pp 301–315

Barnett HL, Hunter BH (1972) Illustrated Genera of imperfect fungi, 3rd edn. Burgess Publishing Co., Minneapolis, p 234

Bayman BP, Lebron LL, Tremblay RL, Lodge DJ (1997) Variation in endophytic fungi from roots and leaves of lepanthes (orchidaceae). New Phytol 135:143–9

Berbee ML (2001) The phylogeny of plant and animal pathogens in the Ascomycota. Physiol Mol Plant Pathol 59:165–187

Bills GF, Polishook JD (1991) Microfungi from *Carpinus caroliana*. Can J Bot 69:1477–1482

Bissegger M, Sieber TN (1994) Assemblages of endophytic fungi in coppice shoots of *Castanea sativa*. Mycologia 86:648–655

Boullard B (1988) Observations of the coevolution of fungi with hepatics. In: Pirozynski KA, Hawksworth DL (eds) Coevolution of fungi with plants and animals. Academic Press, London, UK, pp 107–124

Carroll GC (1995) Forest endophytes: pattern and process. Can J Bot 73(Suppl. 1):S1316–S1324

Chowdhry PN (2000) Manual on identification of plant pathogenic and biocontrol fungi of agricultural importance. IARI, New Delhi

Cubit JD (1974) Interactions of seasonally changing physical factors and grazing affecting high intertidal communities on a rocky shore. PhD dissertation, Universityu of Oregon

Davis EC, Franklin JB, Shaw AJ, Vilgalys R (2003) Endophytic xylaria (xylariaceae) among liverworts and angiosperms: phylogenetics, distribution and symbiosis. Am J Bot 90:1661–1667

Dhargalkar S, Bhat DJ (2009) Echinosphaeria pteridis sp. nov. and its Vermiculariopsiella anamorph. Mycotaxon 108:115–122

Dreyfuss MM, Chapela IH (1994) Potential of fungi in the discovery of novel, low-molecular weight pharmaceuticals. In: Gullao VP (ed) The Discovery of natural products with therapeutic potential. Butterworth-Heinemann, London, pp 49–80

Ellis MB (1971) Dematiaceous hyphomycetes. CABI Publications, Commonwealth Mycological Institute, p 608

Ellis MB (1976) More dematiaceous hyphomycetes. CABI Publications, Commonwealth Mycological Institute, p 507

Fisher PJ (1996) Survival and spread of endophyte *Stagonospora pteridocola* in *Pteridium aquilinum*, other ferns and some flowering plants. New Phytol 132:119–122

Frohlich J, Hyde KD (1999) Biodiversity of palm fungi in the tropics: are global fungal diversity estimates realistic? Biodivers Conserv 8:977–1004

Gilman JC (1967) Manual of soil fungi. Oxford and IBH Publishing Co., New Delhi, p 450

Hawksworth DL (1988) The variety of fungal-algal symbioses, their evolutionary significance, and the nature of lichens. Bot J Linn Soc 96:3–20

Hawksworth DL (1991) The fungal dimension of biodiversity: magnitude, significance, and conservation. Mycol Res 95:641–655

Hawksworth DL, Hill DJ (1984) The lichen-forming fungi. Blackie, Glasgow

Hawksworth DL, Rossman AY (1997) Where are all the undescribed fungi? Phytopathology 87:888–891

Heckman DS, Geiser DM, Eidell BR, Stauffer RL, Kardos NL, Hedges SB (2001) Molecular evidence for the early colonization of land by fungi and plants. Science 293:1129–1133

Kannangara BTSDP, Rajapaksha, RSCG, Paranagama PA (2009) Nature and bioactivities of endolichenic fungi in *Pseudocyphellaria* sp., *Parmotrema* sp. and *Usnea* sp. at Hakgala montane forest in Sri Lanka. Lett Appl Microbiol 48:203–209

Krings M, Taylor TN, Hass H, Kerp H, Dotzler N, Hermsen EJ (2007) Fungal endophytes in a 400-million-yrold land plant, infection pathways, spatial distribution, and host responses. New Phytol 174:648–657

Lawrey JD, Diederich P (2003) Lichenicolous fungi: interactions, evolution and biodiversity. Bryologist 106:80–120

Li WC, Zhou J, Guo SY, Guo LD (2007) Endophytic fungi associated with lichens in Baihua mountain of Beijing, China. Fungal Diversity 25:69–80

Lutzoni FL, Pagel M, Reeb V (2001) Major fungal lineages are derived from lichen symbiotic ancestors. Nature 411:937–940

Miadlikowska A, Arnold A, Lutzoni F (2004) Diversity of cryptic fungi inhabiting healthy lichen thalli in a temperate and tropical forest. Ecol Soc Am Annu Meet 89:349–350

Petrini LE (1992) *Rosellinia* species of the temperate zones. Sydowia 44:169–281

Petrini O (1986) Taxonomy of endophytic fungi of aerial plant tissues. In: Fokkema NJ, van den Huevel J (eds) Microbiology of the phyllosphere. Cambridge University Press, Cambridge, pp 175–187

Petrini O, Carroll GC (1981) Endophytic fungi in foliage of some cupressaceae in oregon. Can J Bot 59:629–36

Rodrigues KF, Leuchtmann A, Petrini O (1993) Endophytes species of Xylaria, cultural and isozymic studies. Sidowia 45:116–138

Rodrigues KF, Samuels GJ (1994) *Letendraeposis palmarum*, a new genus and species of loculascomycetes. Mycologia 86:254–258

Sánchez Márquez S, Bills GF, Zabalgogeazcoa I (2007) The endophytic mycobiota of the grass *Dactylis glomerata*. Fungal Divers 27:171–195

Santamaria J, Bayman P (2005) Fungal epiphytes and endophytes of coffee leaves (*Coffea arabica*). Microbial Ecol 50:1–8

Schulz B, Wanke U, Draeger S, Aust HJ (1993) Endophytes from herbaceous plants and shrubs-effectiveness of surface sterilization methods. Mycol Res 97:1447–1450

Selosse MA, Le Tacon F (1998) The land flora: a phototroph-fungus partnership? Trends Ecol Evol 13:15–20

Stone JK, Bacon CW, White JF Jr (2000) An overview of the endophytic microbes: endophytism defined. In: Bacon CW, White JF (eds) Microbial endophytes. Marcel Dekker, New York, pp 3–19

Subramaniam CV (1971) Hyphomycetes: an account of Indian species except Cercosporae. ICAR, p 930

Suryanarayanan TS, Thirunavukkarasu N, Hariharan GN, Balaji P (2005) Occurrence of non-obligate microfungi inside lichen thalli. Sydowia 57:120–130

Sutton BC (1980) The Coelomycetes: fungi imperfecti with pycnidia, acervuli and stromata. CABI Publications, Commonwealth Mycological Institute, p 696

Tripathi M, Gupta RC, Joshi Y (2014a) *Spegazzinia tessarthra* isolated as a true endophyte from lichen *Heterodermia flabellata*. Indian Phytopathol 67 (1):109–110

Tripathi M, Gupta RC, Joshi Y (2014b) *Physcia dilatata* Nyl. (lichenized fungi, *Physciaceae*); a new host of *Bipolaris australiensis* (M.B. Ellis) Tsuda and Ueyama from Kumaun Himalaya, India. Proc Nat Acad Sci Lett 37(5):477–479

Tripathi M, Joshi Y, Gupta RC (2014c) Assessment of endolichenic fungal diversity in some forests of Kumaun Himalaya. Curr Sci 107(5):745–748

The Diversity of Lichenised Fungi: Ecosystem Functions and Ecosystem Services

7

Luciana Zedda and Gerhard Rambold

Contents

Abstract

Biodiversity supports ecosystem functioning and productivity, and makes ecosystems more resilient and resistant to changes. Ecosystem functions are due to the biological, geochemical and physical processes occurring within an ecosystem. They relate to the structural components (e.g. water, soil, atmosphere, and biota) and their interactions within and across ecosystems. Functions that are useful to human well-being are defined as 'ecosystem services'. Lichenised fungi are complex and form diverse 'functional organismic communities'. They can be regarded as individuals as well as microhabitats comprising a huge variety of coexisting fungal, algal and bacterial taxa or genotypes, pertaining to most different domains of life. Lichens and their symbionts underpin a great number of ecosystem functions (i.e., rock decomposition, soil formation, carbon, and nitrogen fixation), support the diversity of numerous organisms, e.g. through the provision of food, habitat, shelter, camouflage, or nesting material. Furthermore, they provide numerous direct and indirect ecosystem services, which are presented in detail in this overview. Examples are the provision of lichen secondary metabolites and other compounds for medicinal and other purposes, the use of lichens as bioindicators of environmental changes, and as inspiration source in the context of culture, arts and design. The aim of the present review is to give insight in the current knowledge on ecosystem functions provided by lichens, as well as to point out which of these are, directly or indirectly, of benefit for human beings. Lichens are often neglected in ecosystem service analyses and nature conservation management, mostly due to underestimation of their role and difficulties in identification. The primary agents and lichen

L. Zedda (✉)
BIO-Diverse, Ließemer Straße 32a, 53179 Bonn, Germany
e-mail: luciana.zedda@bio-diverse.de

G. Rambold
Mycology Department, University of Bayreuth, Universitätsstraße 30, 95447 Bayreuth, Germany
e-mail: gerhard.rambold@uni-bayreuth.de

D.K. Upreti et al. (eds.), *Recent Advances in Lichenology*,
DOI 10.1007/978-81-322-2235-4_7

traits involved in ecosystem processes are analysed, and possible approaches on how to quantify, estimate the value, model and map lichen ecosystem services are discussed.

Keywords
Lichenised ascomycetes • Biodiversity • Ecosystem services • Ecosystem functions

7.1 Introduction

According to the definition of the Convention on Biological Diversity (CBD 1992), biological diversity means the variability among living organisms and includes diversity at inter- and intraspecific (genes, individuals, functional types) as well as community and ecosystem levels. This concept is therefore wide and includes different levels of biological organization. Biodiversity entails a wide range of key functional roles in terrestrial, fresh water and marine ecosystems, which have been deeply analysed for the first time in term of ecosystem services and benefits for human well-being by the Millennium Ecosystem Assessment (MEA 2005; Norris et al. 2011).

Many studies have confirmed that the diversity and complexity of ecosystems support biodiversity and that increased biodiversity improves ecosystem functioning and productivity, and makes ecosystems more resilient and resistant to perturbation and sudden environmental change. Not only species diversity is relevant for ecosystem functioning, but all levels of biodiversity, i.e. functional group diversity. More biodiverse ecosystems are also likely to be more stable and efficient (Sekercioglu 2010).

Ecosystem functions are based on and define the biological, geochemical and physical processes occurring within an ecosystem. They relate to the structural components of an ecosystem (e.g. water, soil, atmosphere, and biota) and their interactions within and across ecosystems. The most relevant exchanges within ecosystems, for instance, concern energy and nutrients (i.e. carbon and nitrogen), as well as production and decomposition of biomass (Jax 2010). The set of ecosystem functions that are useful to humans are defined as **ecosystem services**. They concern the direct or indirect services, provided by the biota of natural ecosystems to human well-being. Although people have long been aware that natural ecosystems help support human societies, the explicit recognition of ecosystem services is relatively recent (Sekercioglu 2010). Numerous ecosystem services have been assessed by the MEA in an analysis carried out in the period 2001–2005. Its major aim was to identify and quantify the consequences of ecosystem change for human well-being due to increasing economies, and to establish a scientific basis for actions required to enhance conservation and a sustainable use of ecosystems (MEA 2005). As relevance and stability of ecosystem services tend to improve with increasing biodiversity, ecosystems hosting a wide range of phylogenetic or functional groups of organisms and life-form diversity or 'biotic diversity' provide multiple and diverse services and associated values (Norris et al. 2011).

Ecosystem services include so called **supporting services**, such as soil formation, photosynthesis, and nutrient cycling. These are preconditioned for the accomplishment of all other ecosystem services; **provisioning services** concern products of ecosystems such as food and fodder, construction and bulk materials, and fibers; **regulating services** relate to the benefits obtained from the regulation of ecosystem processes for instance influencing climate, water flow, availability and quality, air quality, pollution and health; **cultural services** concern aesthetic, spiritual and recreational benefits (MEA 2005).

Lichens as obligate symbioses of fungi and algae and/or cyanobacteria are highly relevant as providers of numerous ecosystem functions and services (Rambold et al. 2013). Compared to non-lichenised fungi, lichens provide additional functions and services, such as carbon assimilation and nitrogen fixation (cyanolichens) (Norris et al. 2011). Due to identification difficulties and underestimation of their ecosystem roles, lichens are often neglected in analyses of ecosystem services. Only recently and sparsely, multiple stakeholders have started to include these mutualistic associations of organisms in their conservation plans and monitoring studies, and to recognize the importance of taking threats on lichen functions in ecosystems into consideration (Will-Wolf and Scheidegger 2002). Just few studies on this topic have been hitherto carried out. A first short analysis on the provision of ecosystem services by lichens for Great Britain was given by Dighton (2003) and Ellis (in: Norris et al. 2011).

The aim of the present publication is to review the current knowledge on the ecosystem functions and services of lichens, and to point out which of these are, directly or indirectly, of benefit for human beings. Primary agents (mycobiont, photobiont, entire lichen) and lichen traits involved in ecosystem processes are analysed. Possible research approaches for quantifying, modeling and mapping lichen ecosystem services are presented.

7.2 Lichen Diversity

About 20 % of all fungi and 40 % of the ascomycetes (13,500 species) are lichen-forming fungi (Blackwell 2011). According to Boonpragob et al. (2012), the species number of lichens worldwide may be estimated of about 28,000 taxa. When considering the range of intraspecific genetic diversity and cryptic species, which is not included in this estimation, this number is even much higher (Leavitt et al. 2011; Lumbsch and Leavitt 2011).

Traditionally, different lichen morphotypes have been regarded as distinct species, based on typical phenotypic and chemical traits, as well as on anatomical characters of the fungal partner. Lichens may be however regarded not only as an individual, but also as a 'functional organismic community' or as a microhabitat with a huge variety of coexisting fungal, algal and bacterial genotypes (Boonpragob et al. 2012). The lichen microhabitat consists of symbionts pertaining to most different domains of life and major phylogenetic clades within eukaryotes and prokaryotes.

The genetic diversity of the mycobiont within one and the same thallus has been hitherto scarcely studied. The diversity of mycobionts attributed to the same morphologically defined species in thalli growing side by side, and in geographically distant populations is comparatively better known and occasionally linked with certain ecological or environmental factors (Crespo et al. 1997; Domaschke et al. 2012).

Knowledge of lichen photobiont diversity is still rather limited due to the lack of diagnostically relevant traits at species and subspecies levels (Gärtner 1985; Friedl 1989). Since recently, DNA marker gene data increasingly contribute to a better knowledge of the distribution of the various taxa. To date, most studies on photobiont genotype variation were focused on certain lichen taxa from species to family level (Blaha et al. 2006; Grube and Muggia 2010; Helms et al. 2001; Muggia et al. 2008; Piercey-Normore and De Priest 2001). A certain degree of photobiont diversity may even occur within one lichen species, or within one and the same thallus of a species, including different photobiont strains (Friedl 1989). For instance, Blaha et al. (2006) recorded unrelated lineages of the green algal genus *Trebouxia* in the saxicolous, crustose lichenised species aggregate of *Lecanora rupicola* s.l.

A review on studies concerning the genetic diversity of symbiotic cyanobacteria was recently presented by Rikkinen (2013). More than 1,500 species of lichen-forming fungi are known to have cyanobacteria as primary or accessory photobionts. However, the genetic diversity within several important lineages of lichen symbiotic cyanobacteria still is largely unknown and also applies to many biomes being rich in asymbiotic and symbiotic cyanobacterial species. Remarkably, few square millimetres of a lichen thallus may host several different cyanolichen genotypes. Such diversity reflects most probably an adaptative strategy to habitat variation and environmental change, which may allow some fungal hosts to associate in parallel with several cyanobacterial genotypes that are optimally adapted to the prevailing habitat conditions (Rikkinen 2013).

Lichens occur in a wide range of micro- and macro-ecosystems and form diverse community types. They are found in temperate, extreme cold or heat conditions as well as in natural and managed habitats, where they colonize most types of substrata. Lichens can show high species richness even within few square millimetres, as in the case of foliicolous lichens in tropical forests. Species diversity per hectare may range from around ten to several hundred (Nimis et al. 2002).

7.3 Ecosystem Functions and Ecosystem Services Provided by Lichens

In the present article, we explore the functional role of lichen diversity in ecosystems and present a review of ecosystem functions and ecosystem services provided by lichens as entity or by single lichen symbionts. Ecosystem functions are analysed according to the target group, abiotic (rock, soil, water air) or biotic (bacteria, viruses, non-lichenised fungi, plants, animals). Ecosystem services are categorized in accordance to the classification of the MEA (2005). Furthermore, we emphasize the 'primary agents' (mycobiont = mb, photobiont = pb, cyanobiont = cb, lichen = lic) and the lichen traits involved in given ecosystem functions or services (Table 7.1).

7.3.1 Ecosystem Functions

7.3.1.1 Rock, Soil, Water, Air, Climate

Rock Decomposition

Many epilithic lichen species are known to affect their substrates of growth (Seaward 1997). The lichen-rock interface has been demonstrated to be a location of considerable physical and chemical activities, where lichens play an important role in accelerating weathering processes of rock surface minerals (Chen et al. 2000). The hyphal penetration of rock crevices can induce mechanical erosion of rocks, with disaggregation and exfoliation of the rock surface (Nimis et al. 1992; Chen et al. 2000). The mechanical disruption of rock is also linked to the expansion and contraction of lichen thalli due to wetting and drying processes. Particularly, the medullar layer of the thalli has hygroscopic properties and a water holding capacity up to 300 % of the dry weight (Chen et al. 2000). Lichens can even incorporate separated or disaggregated rock and mineral particles, i.e. quartz, feldspar and mica, in their thalli (Chen et al. 2000). Chemical weathering is induced by lichens through different mechanisms: (1) the dissolution of respiratory CO_2 in water resulting in the production of carbonic acid, which supports solubilisation of minerals by increased acidity; (2) the mycobiontal secretion of oxalic acid, which can be transformed into metal oxalates at the lichen-rock interface and within the lichen thallus, as well as of other acids taking acidic effect and having chelation potential by their hydrogen ions; (3) lichen secondary compounds with chelating abilities, forming metal-organic-complexes with silicate-originating cations. Secondary weathering products are metal oxalates, i.e. calcium and magnesium oxalates (i.e., weddellite and whewellite), frequently accumulated in the lower layers of the lichen thalli, iron oxides, amorphous alumino-silicate gels, clay minerals (e.g., kaolinite and halloysite) (Nimis et al. 1992; Chen et al. 2000).

Primary agent: mb. **Lichen traits involved**: see Table 7.1. **Ecosystem services**: *Direct:* (cultural services) scientific value; lichens as destroyers of rock monuments. *Indirect:* (supporting, provisioning, cultural services) mineral reallocation and reformation in soil; soil nutrient enrichment, entailing benefits for food and feed provision; support of other organisms.

Primary Colonization and Soil Formation

Lichen photobionts assimilate carbon and in certain cases also nitrogen from the atmosphere through photosynthesis and N-fixation. Surplus nutrients are forwarded to the mycobiont, which in return provides the photobiont with minerals, mobilized, for instance, from rock by the

Table 7.1 Phenotypic lichen traits (morphological, photobiont, chemical and ecological) involved in the provision of the different ecosystem functions and ecosystem services (marked with "+")

Lichen trait (phenotype level)	Ecosystem functions (EF) and ecosystem services (ES)																				
	Rock decomposition	Primary colonization and soil formation	Nitrogen fixation	Photosynthesis and carbon fixation	Denitrification	Metal chelating	Erosion prevention	Water flow regulation	Water and substrate quality regulation	Air quality regulation	Influence on substrate temperature (albedo)	Influence on ozone content	Food provision	Provision of lichen secondary metabolites and other compounds	Provision of habitat, shelter, camouflage and nesting material	Plant germination and growth support	Bedding material (fibber)	Fuel and tinder	Biodindication	Scientific and education value	Recreational value; inspiration for culture, art and design
Morphology																					
thallus growth habit	++	++	+	++			++	++	++	++	+		++		++	++	++	+	+	+	++
thallus growth rate	+	++		+			+	+	+	+	+		++		+	+	+	+	+	+	
thallus size	+	+	+	++			+	++	++	++	++		++	+	++	++	++	+	+	+	+
thallus upper layers morphology		++		+			+	+							++	++				+	+
rhizinoid structure formation	++	++				?	++	+	?						?	?				+	
reproductive structures presence		+											+	+	+					+	+
Photobiont																					
photobiont type		+	++	++	++	?		+	?	?			+	++	+				+	+	
cyanobiont presence		+	++	+	++	?		?	?	?	+			+	+				+	+	
Chemistry																					
secondary metabolism presence	++	+			+	++			++	++	+		+	++	++	+		?	+	+	+
secondary metabolite type	++					++							+	++	+					+	
release of volatile compounds												++								+	
thallus and other structures pigmentation presence											++			+	+				+	+	++
chelation capacity	+	++				++	+		+	+				+		+				+	
crystal formation	+													+						+	
Ecology																					
substrate preference	++	++			+	?	++	+	+		+		+	+	++	++			+	+	+
association in communities	+	+					+	+	+	+	+			+	++	+			+	+	++
extremophily	+	+	+	+	+	+	+	+						+		+			+	+	
pollution tolerance		+		+		+			+	++				+	+				+	+	
pollution by nitrogen tolerance		+		+	+				+	+				+	+				+	+	
heavy metal tolerance	+	+				+			+					+					+	+	
heavy metal accumulation	+					+			+	+				+					+	+	
water-retaining capacity	+	+		+	?		+	++	+			+				+	+			+	
adaptation by high thallus water content				+	?		?	+				++								+	

Particularly relevant traits are marked with "++". The role played by some traits in given EF and ES is not clear in some cases and therefore marked with "?"

production of organic acids. In that way, lichens are capable to build up significant amounts of biomass and, in addition, various kinds of extracellular secondary compounds and polymers (Nash 1996). Lichen biomass may be decomposed by destruents like fungi and bacteria and consumers like animals of the surrounding ecosystem or be removed due to erosion and runoff. This kind of primary biomass production is particularly important in arid regions and other terrestrial environments with low abundance of organic nutrients (Elbert et al. 2012). Many lichen species have evolved to live under nutrient- and water-limited environment conditions, and are therefore excellent primary colonizers of bare substrates, such as rock and soil. As primary colonizers, they play a pioneering role in soil formation and stabilization, supporting initial vegetation growth and plant succession. Dead lichen thallus plectenchyma form together with mineral material the very first organic-matter additions to soil, thereby increasing its organic matter content (Ashman and Puri 2002). Organic matter derived from lichen decomposition, together with detached particles of the substratum and atmospherically derived dust trapped by the thalli, may contribute all to the accumulation of further fine material and the subsequent establishment of a primitive soil layer (Seaward 1997; Chen et al. 2000).

Primary agent: mb. **Lichen traits involved**: see Table 7.1. **Ecosystem services**: *Direct:* (cultural services) scientific value. *Indirect:* (supporting, provisioning, cultural services) soil nutrient enrichment and stabilization, entailing benefits for food and feed provision; support of other organisms.

Nitrogen Fixation

This process involves the conversion of atmospheric nitrogen to ammonia. The photobionts of cyanolichens from various ecosystems and phytogeographical regions contribute significantly to the global biological nitrogen fixation rate. This impact is most relevant, for instance, in moist forest biota (Pike 1978). Numerous studies in temperate and boreal as well as in desert areas have been carried out to quantify the contribution of cyanolichens to nitrogen fixation. Lichen species of the genera *Lobaria, Collema, Leptogium, Peltigera* and *Sticta* occurring in forests of Northern European and North American, for instance, exhibited annual N-fixation rates of 2–11 kg ha^{-1} (Becker et al. 1977). In deserts and semi-desert regions of the World, lichens are particularly frequent in biological soil crusts, (BSC) where their coverage may approach 100 % of the plant interspaces. Total rates of nitrogen input by BSCs (including also non-lichenised organisms) vary according to crust species composition, abundance and seasonality, and may reach values up to ca. 12 kg N ha^{-1} a^{-1}. Nitrogen fixation by BSCs has been demonstrated to be the predominant form of N input in arid ecosystems (Evans and Lange 2001), the cyanolichen crusts exhibiting more than six-fold higher N-fixation activity compared to cyanobacterial crusts in other arid habitats like in the southeastern regions of Utah (Barger et al. 2013). Elbert et al. (2012) reported on a global nitrogen uptake by ‘cryptogamic’ covers (comprising variable proportions of cyanobacteria, algae, fungi, lichens and bryophytes) of around 49 tg $year^{-1}$, which accounts for nearly half of the biological nitrogen fixation rate on land. Nitrogen input from BSCs may influence N-cycles, not only through leakage of N from the crust-forming organisms, but also due to the decomposition of crust material. Where soil N content is greater, plants can profit of higher nitrogen concentrations having a positive effect on pasture. Man-made disturbance limiting growth of BSCs, particularly lichens, which require more stable growth conditions, causes nitrogen loss and in consequence also loss of pasture quality (Evans and Lange 2001).

Primary agent: cb. **Lichen traits involved**: see Table 7.1. **Ecosystem services**: *Direct:* (provisioning, cultural services) lichens for food and nitrogen consumption; scientific value. *Indirect:* (supporting, provisioning, regulating, cultural services) soil nutrient enrichment, entailing benefits for food and feed provision; support of other organisms; regulation of environmental nitrogen.

Photosynthesis and Carbon Fixation

Lichens contribute to the carbon cycle through fixation of carbon dioxide from the atmosphere and therefore play an important role in carbon exchange between atmosphere and terrestrial biota through photosynthesis and respiration (Insarov and Schroeter 2002). Elbert et al. (2012) estimate that the global cover by soil- and plant-inhabiting 'cryptogams' takes up around 3.9 Pg year^{-1}, corresponding to around 7 % of the net primary production of the terrestrial vegetation. In a study assessing the annual net carbon deposition related to BSC growth in a Mediterranean shrubland, Wilske et al. (2009) demonstrated that a BSC-related net carbon deposition may range from 7 to 51 kg ha^{-1} year^{-1}. Maximum rates of net CO_2-assimilation of soil crust lichens were reported in the same order of magnitude as those of concomitant vascular plants or 10 plants of temperate regions (Lange 2001; Lange et al. 1990). Lange et al. (1994) extrapolated from 13 days of field measurements in the coastal fog zone of the Namib desert an average of 250 days with fog, which resulted in a first estimate of the annual carbon balance of 16 g C m^{-2} year^{-1} for soil crust lichens. Carbon gains of soil crust lichens monitored with an automatic cuvette system under temperate habitat conditions resulted in a C balance of 25.8 and 21.7 g C m^{-2} year^{-1} for *Collema cristatum* and *Lecanora muralis*, respectively (Lange 2001). Beymer and Klopatek (1991) showed that approximately 28 kg C ha^{-1} was fixed by lichen crust communities in forests of a semiarid environment in the Grand Canyon, of which about 34–35 % became incorporated in organic soil matter. Epiphytic lichens are able to utilize the mineral nutrients that are intercepted by tree canopies and drip down branches and trunks. This also supports photosynthesis by the lichens photobionts (Dighton 2003).

Primary agent: pb. **Lichen traits involved**: see Table 7.1. **Ecosystem services**: *Direct:* (supporting, provisioning, regulating and cultural services) atmospheric oxygen due to photosynthesis contributes to greater oxygen availability in the atmosphere, to air cleaning and to UV ray protection; lichen biomass may serve as nutrition for human, bedding material, fuel and tinder; scientific value. *Indirect:* (supporting, provisioning, regulating and cultural services) soil nutrient enrichment (organic matter added to soil), entailing benefits for food and feed provision; reduction of atmospheric CO_2 (influence on climatic conditions); support of other organisms; regulation of environmental nitrogen.

Denitrification

Both, lichen mycobionts and cyanobionts may contribute to denitrification processes, in particular in subsequence to rain events (Evans and Lange 2001). N_2O as denitrification products has been demonstrated in cyanobacterial and lichen crusts in Oman, were denitrification rates for lichens reach 58 ± 20 μmol N m^{-2} h^{-1} (Abed et al. 2013). Estimates of denitrification from cyanolichen crusts constituted 4 % of the N fixation rates in cool desert environments like in the southeastern part of Utah, suggesting that a loss of N due to denitrification is insignificant compared to N input by fixation (Barger et al. 2013).

Primary agent: mb, cb. **Lichen traits involved**: see Table 7.1. **Ecosystem services** *Direct:* (regulating and cultural services) changes in atmospheric composition and contribution to nitrogen reduction processes; scientific value. *Indirect:* (supporting, provisioning, regulating and cultural services) nutrient cycling, food and feed provision, partial impact on global climate change; support of other organisms.

Metal Chelating

Many secondary metabolites of lichens are known to possess metal chelator functions by forming complexes with ions of polyvalent metals such as Mg, K, P, Ca, Fe. These substances maintain metals in soil which would otherwise precipitate in solution. The secreted chelators are water-soluble and capable of being translocated in soil and water. In this way metal ions become also available to non-chelating plants or microbes (Belnap et al. 2001).

Primary agent: mb, pb. **Lichen traits involved**: see Table 7.1. **Ecosystem services**: *Direct:* (provisioning, regulating and cultural services) potential use of lichen metabolites for soil fertilization in agriculture and for soil bioremediation; scientific value. *Indirect:* (supporting, provisioning, regulating and cultural services) soil nutrient enrichment and consequent benefits for food and feed provision; support of other organisms.

Erosion Prevention

Loss of vegetation cover usually increases the erosional impact of rain and wind, resulting in the removal of the top soil layers, loss of valuable nutrients, and finally desertification. Erosion also lowers soil productivity and reduces the amount of organic carbon returned into soil as plant residue (Sekercioglu 2010). Soil lichens, together with other organisms forming BSCs, protect soil surface against water and wind erosion by binding soil particles, sealing and stabilizing soils through aggregates. In particular, soil surface stabilization is considerably higher by crustose lichen thalli and in presence of rhizines (Belnap et al. 2001; Belnap and Gillette 1997). Lichen crusts also modify the morphology of the soil surface, which becomes rougher and by that enables the accumulation of organic substances, fine particles and water (Belnap et al. 2001). Due to stabilization by lichens and other BSC organisms, also loss of soil particles as dust into the atmosphere is reduced (Lazaro et al. 2008). By decreased soil erosion, also the movement of surface runoff is slowed, thus allowing excess water to infiltrate (Egoh et al. 2012).

Primary agent: lic. **Lichen traits involved**: see Table 7.1. **Ecosystem services**: *Direct:* (cultural services) scientific value. *Indirect:* (supporting, provisioning, regulating and cultural services) soil protection and stabilization entailing benefits for food and feed provision; support of other organisms.

Regulation of Water Flow

Lichens are poikilohydric organisms exhibiting specific mechanisms of rapid liquid water absorption and water conduction into the thallus. Water fills the intercellular capillary and air spaces between hyphae and algal cells entailing inflation of the hyphal plectenchyma. In contrary, water-vapour absorption processes are considerably slower (Ahmadijian 2012). In so-called 'gelatinous' thalli of cyanobacterial lichens, the polysaccharide sheaths of the cyanobionts significantly contribute to water retention (Karunaratne et al. 2012). Similarly to higher vegetation, also lichen crusts intercept water raindrops gradually releasing absorbed water into soil, and thus contribute to limitation of erosion and floods (Sekercioglu 2010). In this way, water available for plants and other organisms is increased and soil water evaporation decreased (Belnap 2006). Water holding capacity of soils and water quality are also a function of water infiltration, which mainly depends from soil surface characteristics. Natural plant vegetation and soil surface lichen cover slow down the runoff, thereby retaining nutrients and allowing water to infiltrate into the ground efficiently compared to eroded or sealed soil surfaces (Mills and Fey 2004; Nash 1996). Not only in arid areas, but also in the Arctic and subarctic regions, where ground is covered by mats of lichen thalli, water evaporation from the soil is strikingly reduced. Also epiphytic lichens may intercept precipitation in form of rain, fog and dew (Nash 1996). All these types of 'hydrological regulation' also contribute to buffering seasonal extremes in water flows (Sekercioglu 2010).

Primary agent: lic. **Lichen traits involved**: see Table 7.1. **Ecosystem services**: *Direct:* (regulating and cultural services) water availability, scientific value. *Indirect:* (supporting, provisioning, regulating and cultural services) protection of soils from water erosion; greater water availability for plants and animals; benefits for food and feed provision; support of other organisms.

Water and Substrate Quality Regulation

Lichens, especially terricolous ones, play an important role for filtering contaminants and purifying water and growth substrata thanks to their ability of accumulating substratum, water and airborne nutrients in the thallus. Lichens possess highly efficient mechanisms for the accumulation of a range of major and minor essential elements by cation exchange and soluble cation uptake over the entire thallus surface (Ahmadijian 2012). These processes involve the reversible binding to negatively charged anionic sites on the cell walls or to the exterior plasma membrane surface. Soil or atmospheric particles adsorbed to the thallus surface or trapped in the loose hyphal weft of the medulla contribute to the total of element concentrations in the thallus as well (Bargagli et al. 1999). The majority of metals are deposited in insoluble and inactive form (as metallic oxides), and for this reason lichens are able to accumulate large amounts of such metals. Metal-tolerant species occur on metalliferous substrates with considerable concentrations of metals, for instance in abandoned mine sites or vineyards polluted by copper-based solutions (Nash 1996). Uptake of elements from water and substrata, along with metal and mineral transformation processes may lead to metal leaching and thereby to detoxification, and xenobiotic pollutant degradation, which may support vegetation recovery (Gadd 2007). Lichen growth on asbestos-rich substrata, as observed by Favero-Longo (2005), may be made use as potential tool for bioremediation, since, when colonized by lichens, fibres appear to have been modified in their toxicity.

Primary agent: lic. **Lichen traits involved**: see Table 7.1. **Ecosystem services**: *Direct:* (regulating and cultural services) availability of water with better quality and less polluted substrates; scientific value. *Indirect:* (supporting, provisioning, regulating and cultural services) water and substratum treatment and detoxification; benefits for health, food and feed provision; support of other organisms.

Air Quality Regulation

Lichens exhibit air cleaning activity by capturing pollutants and their accumulation in intracellular spaces. Many species are capable of trapping airborne dust particles even at concentrations that vastly exceed their physiological requirements (Seaward 1997; Nash 1996). In particular in the surroundings of pollution sources, the occurrence and accumulation of such particles in the thalli has been frequently demonstrated. Depositions may concern metals, but also anions associated with acid rain and organics from agricultural and industrial activities. Thanks to this property, lichens are used bioindicators for atmospheric deposition (Nash 1996).

Primary agent: lic. **Lichen traits involved**: see Table 7.1. **Ecosystem services**: *Direct:* (regulating and cultural services) better air quality, scientific value. *Indirect:* (supporting, provisioning, regulating and cultural services) detoxification and improvement of air quality with benefits for plants and animals, benefits for food and feed provision; support of other organisms.

Influence on Substrate Temperature (Albedo)

Mechanical destruction of soil surface covered by lichens and/or other organisms exposes the underlying substrate layers of mostly lighter colour, which implicates an increased albedo, i.e. reflectance, by up to 50 % at most of the measured wavelengths. Associated energy loss involves corresponding temperature reduction at the soil surface. Therefore, large-scale changes of surface albedo may entail changes of the micro- and macroclimatic conditions at a regional level (Belnap and Eldridge 2001).

Primary agent: lic. **Lichen traits involved**: see Table 7.1. **Ecosystem services**: *Direct:* (regulating, cultural services) regulation of climatic changes, scientific value. *Indirect:* (supporting, provisioning, regulating and cultural services) modification of rock and soil temperature, climate regulation, influence on food and feed provision; support of other organisms.

Influence of the Ozone Content

Lichens may influence the ozone content in the lower atmosphere by release of volatile organic compounds (Insarov and Schroeter 2002). They also play a role in the exchange of organic acids and aldehydes between atmosphere and biosphere, as demonstrated by Wilske and Kesselmeier (1999) for the boreal zone. Ethanolic fermentation can occur within thalli of various boreal lichen species as demonstrated in experiments where the co-emission of acetaldehyde and ethanol due to intrathalline oxygen deficiency at high thallus water contents was measurable. It is, however, not clear, whether ethanolic fermentation is carried out just by one of the symbionts or by both (Wilske et al. 2001).

Primary agent: Unknown. **Lichen traits involved**: see Table 7.1. **Ecosystem services**: *Direct:* (regulating, cultural services) regulation of climatic changes, scientific value. *Indirect:* (supporting, provisioning, regulating and cultural services) impact on climate, influence on food and feed provision; effects on the biological diversity of other organisms.

7.3.1.2 Organisms (Bacteria, Viruses, Non-lichenised Fungi, Plants, Animals)

Food Provision

Lichens provide biomass as a food resource for numerous animal species and human beings. A functional relationship between lichen biomass and the diversity of food sources for different animals was pointed out by several authors (Bokhorst et al. 2007; Llano 2012; Liu et al. 2013; Norris et al. 2011). Common intracellular products occurring in lichens include amino acids and proteins, polysaccharides and polyols, carotenoids and vitamins (Elix 1996). Lichens also produce different kinds of lipids in response and as adaptation to environmental abiotic factors such as temperature, light, radiation, and chemical and physical peculiarities of the substratum. Ascomata for instance may comprise numerous ascospores containing lipids at high amounts (Bychek-Guschina 2002). Plectenchyma of high nutritive value for animals are the hymenial layer of the ascomata as well as the photobiont layers of the vegetative thallus. The outer layers of the hyphal walls of the mycobionts are mainly built up by polysaccharides and glycoproteins, but also the photobionts accumulate polysaccharides. In gelatinous cyanolichens, the photobionts produce polysaccharidic sheaths being capable of water retention (Karunaratne et al. 2012). Terricolous mat-forming lichens of the genera *Cladonia* (*Cladina*), *Cetraria, Stereocaulon* and *Alectoria* form extensive ground cover in subarctic lichen woodland, forest tundra and tundra heathlands, where they represent the principal food source for reindeer and caribou during winter time (Crittenden 2000). Some monkey species like the endangered golden snub-nosed monkey (*Rhinopithecus roxellana*) from China is known to be primarily herbivore on lichens as the main food source (Guo et al. 2007; Liu et al. 2013). The northern flying squirrel (*Glaucomys sabrinus yukonensis*) (Rosentreter et al. 1997), as well as the Eurasian red squirrel (*Sciurus vulgaris*) include lichens in their diet (Lurz et al. 2005). Beard lichens make up a considerable part of the diet of the bank vole (*Myodes glareolus*), in particular in autumn and winter (Nybakken et al. 2010; Viro and Sulkava 1985). Gazelles feed on *Ramalina duriaei* in Oman (Hawksworth et al. 1984). Also many insects are reported to feed on lichen. The polyphagus coleopteran *Lasioderma serricorne* selectively feeds on different lichen species stored in herbaria, but avoids the ones containing certain lichen substances such as atranorin or fumarprotocetratic acid (Nimis and Skert 2006). Larvae of various Lepidopteran species (e.g., several genera of the Arctiidae) feed on lichens (Hesbacher et al. 1995b; Rambold 1985). The same applies to Attine ants which seasonally feed on lichen thalli and other fungi in the Neotropical savannas (Brazil), most notably during the dry season (Leal and Oliveira 2000). Fungus-feeding thrips (Thysanoptera) are particularly diverse in the Neotropics and some species are known to feed on lichens (Mound 2002). Numerous lichen-grazing snails are known as well (Gauslaa 2005; Hesbacher et al. 1995a). Also many species of micro-invertebrates (water bears, rotifers and roundworms) seem to depend more or less obligatorily from lichen thalli and algae as a food source (German and Foster 2011). Lichen palatability depends on the investments in herbivore defence. Selective feeding by avoiding deterrent secondary metabolites has been observed in different groups of vertebrates and invertebrates (Gauslaa 2005; Hesbacher et al. 1995b; Nimis and Skert 2006; Nybakken et al. 2010; Rambold 1985). Continuing research and detailed observations on this kind of animal-lichen interaction find their expression in an increasing number of publications.

Primary agent: lic. **Lichen traits involved**: see Table 7.1. *Direct:* (provisioning and cultural services) use of lichens for human nutrition; economic, culinary and scientific value. *Indirect:* (supporting, provisioning, regulating and cultural services) soil nutrient enrichment (organic matter added to soil), use of domestic animals feeding on lichens (i.e. reindeer and caribou); support of the biological diversity of other organisms.

Provision of Lichen Secondary Metabolites and Other Compounds

Lichens produce unique chemical substances, many of which have pharmaceutical relevance. More than 1000 lichen secondary compounds are known so far (Huneck and Yoshimura 1996). They belong to various groups and most of them are phenolic compounds, anthraquinones, dibenzofurans, depsides, depsidones, depsones, γ-lactones and pulvinic acid derivates (Muggia et al. 2009). In nature, lichen secondary metabolites have multiple functions as light filters to protect photobiont cells from excessive radiation, as deterrent against herbivorous animals, as antibiotics against microbial degradation or as weathering agents for degrading rock substrates (Muggia et al. 2009). Lichen metabolites exert a wide variety of biological activities including antibiotic, antimycobacterial, antiviral, anti-inflammatory, analgesic, antipyretic, antiproliferative and cytotoxic effects (Malhotra et al. 2013). Lichens have therefore been used in many countries in folk medicine for the treatment of various diseases like arthritis, alopecia, constipation, and leprosy (Malhotra et al. 2013). Some of the most commonly used lichens are *Cetraria islandica* (L.) Ach. and *Usnea* spp. as remedy for coughing and as expectorants, *Lobaria pulmonaria* (L.) Hoffm. for lung diseases (Huneck and Yoshimura 1996; Sharnoff 1998; Rogers 2011), and *Peltigera canina* (L.) Willd. as remedy for liver ailments (Nash 1996). Biological activities have also been reported for lichen polysaccharides (Karunaratne et al. 2012). Those of lichen secondary compounds have been investigated since the beginnings of the last century and their activities can be divided into:

- Antibiotic activity. 50 % of all lichens are estimated to have antibiotic properties (Malhotra et al. 2013). Depisides, depsidone and usnic acid are active in particular against gram-positive microorganisms (Huneck and Yoshimura 1996). Antibiotic activity has been shown especially for usnic, evernic and vulpinic acids, as well as for derivates of diploicin with antitubercolous activity (Muggia et al. 2009).
- Antitumor and antimutagenic activity. Numerous lichen secondary compounds like usnic acid, protolichesterinic acid, neprosteranic acids, polyporic acid and derivates, physodalic acid, lichen glucans and lichenin derivates exhibit antitumor activity (Huneck and Yoshimura 1996). The most active lichen substances are water-soluble polysaccharides because of their stimulation of the immune system (Nash 1996). Recently, acetone extracts of *Lecanora atra (=Tephromela atra), L. muralis* and *Cladonia furcata* were shown to induce significant cytotoxic effects on tested cancer cell lines (Ranković et al. 2011). Strong anticancer activity was also demonstrated for *Xanthoria parietina* (Triggiani et al. 2009). Traditionally, *Cetraria, Parmelia* and *Usnea* spp. have been reportedly used as folk remedies against cancer and tumors in various regions of the World (Azenha et al. 1998).
- Antiviral activity. Is has been demonstrated an activity against the Human Immunodeficiency Virus (HIV) by a partially acetylated β (1 → 6)-glucan from the lichen *Umbilicaria esculenta* (Miyoshi) Mink. (Huneck and Yoshimura 1996). Anthraquinones, bianthrones and hypericin derivatives from lichens exhibited strong inhibitory activity against the herpes simplex virus type 1 (HSV-1) (Cohen et al. 1996).

- Antifungal activity. Some lichen substances such as usnic and haematommic acids inhibit the growth of moulds (Nash 1996). Antifungal activity in particular for the anthraquinone parietin has been reported by Manojlovic et al. (2005). Aqueous and acetone extracts of lichens *Ramalina* spp. and *Stereocaulon* spp. from India showed strong efficacy against plant pathogenic fungi as *Alternaria alternata, Aspergillus flavus* and *Penicillium italicum* (Shukla et al. 2011).
- Antioxidant activity. Lichen extracts from many lichens, in particular *Lecanora atra (=Tephromela atra), L. muralis* and *Cladonia furcata* were found to have various antioxidant activities (DPPH radical scavenging, superoxide anion radical scavenging and reducing power), similar or even greater than the standard antioxidants (ascorbic acid, BHA and α-tocopherol) (Ranković et al. 2011).
- Immunomodulator activity. This kind of activity has been reported for heteroglycans and a beta-glucan isolated from *Thamnolia vermicularis* var. *subuliformis*. Such compounds have various influences on the immune system (Malhotra et al. 2013).
- Plant growth inhibitory activity. Many lichen substances affect the growth of higher plants (Huneck and Yoshimura 1996).
- Antiherbivore activity. Certain lichen compounds may inhibit feeding of lichens by animals. Nimis and Skert (2006) report that the coleopteran *Lasioderma serricorne* avoid lichens in herbaria, if they contain atranorin, calcium oxalate, fumarprotocetraric acid, gyrophoric acid, lecanoric acid, skyrin, usnic acid and zeorin. Asplund et al. (2008) found that the depsidones of *Lobaria pulmonaria* deter grazing molluscs.
- Poison and allergenic activity. A number of lichen substances are allergens (i.e. atranorin, barbatic, evernic, usnic, fumarprotocetraric and stictic acids) (Huneck and Yoshimura 1996) and may cause contact dermatitis and skin photosensitizing (Nash 1996).
- Enzyme inhibitory activity. Some lichen secondary compounds, such as lecanoric and usnic acids, were found to be enzyme inhibitors (Huneck and Yoshimura 1996).
- Other activities. Cardiotonic, phlogistic, anti-inflammatory, analgesic and antipyretic activities were recorded for several lichen substances (Huneck and Yoshimura 1996).

Sequestration of lichen compounds by lichen-feeding members of the Arctiidae (Lepidoptera) has been demonstrated in some studies. This is regarded as a defense mechanism by which lichen compounds may be utilized for the chemical defense or against microbial pathogens (Hesbacher et al. 1995a). Lichen-grazing snails are also known to sequester lichen compounds like parietin and atranorin, when feeding on the crustose and foliose lichens. Sequestered parietin may even be translocated from the tissue of the mother individual to the eggs in the reproductive tract (Hesbacher et al. 1995a).

Primary agent: mb. **Lichen traits involved**: see Table 7.1. **Ecosystem services**: *Direct:* (regulating, provisioning, cultural services) use of lichen secondary metabolites for their biological activities in biotechnology, medicine, pharmacy, cosmetics, dyeing etc.; metal chelating and bioremediation; scientific value. *Indirect:* (supporting, regulating, provisioning, cultural services) benefits from the support of the biological diversity of other organisms.

Provision of Habitat, Shelter, Camouflage and Nesting Material

Lichens offer habitat and shelter for other organisms thus supporting biological diversity. Lichen thalli and their reproductive structures are for long known being persistent habitat for a wide range of inhabiting groups of organisms. They create shelter for micro- and mesoorganisms and correspondingly enhance the structural complexity of microhabitats on trees, soil and rock (Itzhak Martinez et al. 2014). Epiphytic lichens are known to contribute to the overall biodiversity of ecosystems (Lindo and Stevenson 2007).

Recently, it has been shown that lichens are reservoirs for various types of **viruses**, like the plant cytorhabdovirus and the Apple Mosaic Virus (ApMV), a pathogen on apple and pear plants. It is not known, whether lichen thalli hosting rhabdovirus or ApMV can be considered either as long-term or accidental hosts (Petrzik et al. 2013).

Lichens provide unique habitats for **bacteria**, which may form communities being different from those of adjacent substrates. Furthermore, lichen species of different communities and growth habit may house distinct bacterial assemblages (Bates et al. 2011; Mushegian et al. 2011). They often comprise non-photosynthetic nitrogen-fixing bacteria, which may provide benefit to the host lichen by their metabolic activities. The most common groups belong to Alphaproteobacteria. Several studies suggest that different groups of the lichen-inhabiting bacterial communities are highly structured and not evenly or casually distributed across the centre and edges of the lichen thallus, the central parts mostly exhibiting the highest species richness. Indications exist for some kind of 'climax communities' in the central parts of the thalli, whereas the edges house more or less random assemblages (Mushegian et al. 2011). In consequence, the traditional concept of symbiotic relationship between lichenised fungi and algae or cyanobacteria is in need to include multiple bacterial partners, as these have been shown to be also involved in the nutrient cycling of lichens (Grube et al. 2009; Bates et al. 2011).

Not only green algae and cyanobacteria, but also **diatoms** may inhabit lichen thalli, most probably as accessory photobionts. The lichen complex *Coenogonium linkii* Ehrenb., for instance, was shown to host 18 species of terrestrial diatoms in specimens from Panama. These were found between the thallus filaments on extracellular material of the mycobiont, and could potentially benefit from lichen symbiosis in relation to irradiance, and water and nitrogen availability (Lakatos et al. 2004).

High diversity is known for **lichenicolous and endolichenic fungi**, being accessory mycobionts. Among these, various taxa form obligate associations with the host lichens, either with saprotrophic or biotrophic life habit, exhibiting a balanced mutualistic to antagonistic relationship with the primary mycobionts and photobionts. More than 1,800 species of lichenicolous fungi (forming reproductive structures) are known so far, with an estimate of over 3,500, most of them being Ascomycetes. The host specificity of these fungi is usually high, at least at lichen genus level (Hawksworth 2004; Molina et al. 2005). Endolichenic fungi live asymptomatically in lichenised thalli and usually have an extremely high diversity. They appear to be very selective with respect to habitat preference but are still poorly known. The fungal community associated with the lichen community *Letharietum vulpinae* from the European Alps and California was found to host groups of fungi exclusively associated with lichens (obligate inhabitants) as well as ubiquitous, mostly endophytic and rock-inhabiting taxa (Peršoh and Rambold 2012).

Animal diversity often depends on the presence of lichens. Various corticolous lichens form unique aquatic micro-habitats, which undergo frequent desiccation. These allow only specifically adapted organisms to survive. Aquatic micro-invertebrate phyla commonly occurring on and in the thalli of such lichens belong to Tardigrada (**water bears**), Rotifera (**rotifers**), and Nematoda (**roundworms**). Their abundance in such ephemeral aquatic micro-habitat is correlated with lichen biomass and humidity. However, diversity of these aquatic organisms is still poorly investigated (German and Foster 2011; Kaczmarek et al. 2011). The lichens' dwelling microfauna, consisting of protists and micrometazoa, has been studied in lichen *Xanthoria parietina* by Šatkauskienė (2012), who reports 24 taxa living on this lichen. More than 4,600 individuals of bdelloid rotifers were counted in epilithic lichen samples of 10 cm^2 (Erdoğan and Kaya 2013).

Mesofauna elements like **mites** (Acari), suborder Oribatida, often colonize lichens with high species diversity. They may represent more than 80 % of the lichen-associated arthropods and seem to be host lichen species-specific (Behan-Pelletier et al. 2008). Altogether, 104 oribatid species were recorded from saxicolous lichen and moss communities (Materna 2000).

Thrips (Thysanoptera) are found not only on living plants, dead branches or twigs and leaf, but also on and in lichens, where they feed and reproduce (Chiarini Monteiro 2002).

Certain **collembolans** (Collembola) have been reported to feed on lichens, i.e. in Antarctic ecosystems (Longton 1988). Soil collembolans are feeders on lichenised and non-lichenised fungi, as well as on fragmented litter and bacteria, and may preferentially select certain fungal taxa (Jørgensen et al. 2005). However, according to Aptroot and Berg (2004), the collembolan *Anurophorus laricis* only seeks protection on lichens but seems to feed on algal crusts only in the surrounding of the lichens. Also springtail species like *Pachytullbergia scabra* are lichen-associated, do not feed on the thallus, but profit from shelter from climatic extreme conditions only (Messuti and Kun 2007).

Numerous **insects** use lichens for shelter and are supposed to feed on them. They belong to very different phylogenetic groups and include beetles (Coleoptera), butterflies and moths (Lepidoptera), psocids (Psocoptera), termites (Isoptera), aphids (Hemiptera), stoneflies (Plecoptera), earwigs (Dermaptera) and embiopterans (Embioptera) (Gerson 1973; Seaward 1988). In extreme habitats, such as the Namib Desert, lichen-dominated soil crusts are important supporters of the arthropod diversity, and it has been demonstrated that crustose lichens and lichen cover as a whole play a key role in driving the variance in arthropod assemblages (Lalley et al. 2006). Itzhak Martinez et al. (2014) recently observed higher abundance of cecidogenic aphid species on branches of *Pistacia atlantica* trees covered by *Xanthoria parietina* compared to branches lacking lichens. The lichen thalli seem to be used for protection and oviposition.

Lichen species richness also promotes **spider** diversity due to structural heterogeneity and occurrence of prey arthropods such as mites or bark lices (Gunnarsson et al. 2004). Species richness and abundance of spiders (170 species) investigated on Robben Island (South Africa) was found to be greater in lichen habitats than in bushes (Mukherjee et al. 2010). **Avian species** using lichens might be important in the control of insect herbivores causing domestic plant damage, with consequent economic benefit. Promoting natural predators through the preservation of their native habitat like lichen thalli may have positive effects on crop yields and improve food security (Sekercioglu 2010).

Camouflage is a widespread strategy among the animal kingdom and lichens are used in many ways by various animal species for camouflaging themselves or for protecting their nests, by using lichen thalli directly or by imitating lichen community patterns (Cannon 2010). Among the animals attaching lichen fragments to their bodies many insect species exist (Cannon 2010; Gerson and Seaward 1977). Lacewing larvae (Neuroptera: Chrysopidae), for instance, collect small lichen bits and form some sort of lichen bug or shelter on their back to escape potential predators. Lacewing larvae themselves are important predators of eggs and larvae of many soft-bodied pest insects, and therefore contribute to plant pest protection (Cannon 2010; Skorepa and Sharp 1971). Bagworm larvae (Lepidoptera: Psychidae) construct cases out of silk and lichens, and often also feed on lichens (Cannon 2010).

The nymph of a new genus of katydid (Tettigoniidae), *Lichenodraculus matti*, described a few years ago from the Ecuadorian Andes, perfectly mimics epiphytic lichens and exclusively feeds on various lichen species (Braun 2011). *Gymnopholus lichenifer* (Curculionidae), a beetle species endemic to Papua New Guinea, covers its vulnerable back with living lichens (Gressit 1977). Birds may selectively harvest particular lichens for incorporating them into the nest design (Gerson and Seaward 1977). Some species of spiders are known to mimic lichen communities and this gives them a strong advantage both in capturing prey and evading predators, as the Lichen Huntsman spider (*Pandercetes gracilis*) from Australasia. Crab spiders construct nests using lichens to protect their eggs (Botsford Comstock 1986). Lichens also are important compositions of bird nests (i.e. hummingbirds). They may serve as camouflage, insulation, waterproofing, and may even have antibiotic effects.

Geckos, lizards and frogs also comprise species showing camouflage with lichens. Noteworthy are the Kuhl's Flying Gecko (*Ptychozoon kuhli*) from the Philippines (Brown et al. 1997), the Parachute Gecko (*Ptychozoon kaengkrachanense*) from

Thailand (Sumotha et al. 2012), the gray tree frog (*Hyla versicolor*), most of them being vulnerable and most sensitive to habitat manipulations (Hocking and Semlitsch 2007). Lizards and grasshoppers camouflaging on soil and rock lichen communities have been observed in South Africa by Zedda and Rambold (2011). One well-known example of natural selection linked to lichen communities is the peppered moth (*Biston betularia*) which mainly in the last century developed industrial melanismic forms in response to pollutants (Brodo et al. 2001; Majerus 2009).

Primary agent: lic. **Lichen traits involved**: see Table 7.1. **Ecosystem services**: *Direct:* (cultural services) scientific value. *Indirect:* (supporting, regulating, provisioning, cultural services) pest control; benefits from the support of the biological diversity of other organisms.

Plant Germination and Growth

The presence of terricolous lichens can positively affect vascular plant germination, due to changes in soil surface morphology, facilitating seed capture and the accumulation of organic substances, fine particles and water on soil surface. Lichen-moss crusts in cool and cold deserts of the World have been demonstrated for instance to increase perennial vascular plant seed entrapment, germination, establishment, survival, biomass, and nutritional status (Belnap and Eldridge 2001). In arid areas, highly structured phycolichen-moos crusts trap much more seeds, especially of perennial plants compared to smooth cyanobacterial crusts (Belnap et al. 2001). However, it has been reported that the germination and establishment of exotic annual plants (such as *Schimus* sp. and *Bromus tectorum*) may be inhibited by BSCs. Also temperature, moisture and soil acidity shifts, as well as nitrogen fixation by cyanobacterial lichens may have a strong impact on seed germination of plant species. Due to metal chelating and correlated soil acidity shifts by lichens, availability of minerals such as Mg, K, CuS and Zn is greater in crusted soil surface, and increased uptake for plants possible. Some authors even report a positive correlation between vascular plant species richness, biomass and cover and development of lichen-moss soil crusts (Belnap et al. 2001).

Primary agent: lic. **Lichen traits involved**: see Table 7.1. **Ecosystem services**: *Direct:* (cultural services) scientific value. *Indirect:* (supporting, regulating, provisioning, cultural services) influence on food and feed provision; benefits from the support of the biological diversity of other organisms.

7.3.2 Ecosystem Services

In this part of the work we attempt to categorize ecosystem functions which are particularly important, directly or indirectly, for human beings.

7.3.2.1 Supporting Services

Direct: People usually benefit only indirect and over a very long time from supporting ecosystem services (see below).

Indirect: primary colonization of bare substrates, soil formation (pedogenesis) and rock decomposition, reallocation and reformation of minerals, nutrient cycling (carbon and nitrogen fixation, denitrification, chelation of metals), soil enrichment by organic matter and contribution to humus formation, and biomass production, provisioning of clean drinking water and the decomposition of wastes. These are services necessary for the production of all other ecosystem services and people benefit from the indirect support of the diversity of other organisms and for food and feed provision.

7.3.2.2 Provisioning Services

Direct: Direct benefits for human being are the use of lichens as food, raw material, dyes, fuel, medicinal and ornamental resources, as well as the use of lichen extracts in cosmetics and in the fragrance industry. **Food**: Traditionally, lichens have been important to some extent for human nutrition. They were used to escape epochs and periods of starvation and survive famines (i.e. the lichen manna). Lichens were just boiled or used for flavouring soups (Japan and North America) and for preparing bread. The most frequently used species in North America, Russia and Scandinavia were *Bryoria, Cetraria, Cladina*, and *Nephroma* spp., *Evernia prunastri* and *Pseudevernia furfuracea*, in Egypt and Turkey. In India, the Kubal Garam Masala, a curry additive, includes a high proportion of Parmeliacean and Ramalinacean species. Iunuits are known to consume caribou stomach (rumen) together with digested lichen thalli content. Several lichens were used for brewing of alcoholic beverage, especially in Russia and Siberia (i.e. *Lobaria pulmonaria*) (Sharnoff 1998; Redzic et al. 2010; Llano 2012). In France, lichens were formerly used in the manufacture of chocolates and some pastries. **Primary agent**: lic. **Lichen traits involved**: see Table 7.1.

Lichens have been also used to a small extent as **bedding material** (fibber), in particular in North America and India, for absorbent qualities in wound dressing, baby diapers, sanitary napkins for women etc. or in bags and pillows when feathers were unavailable (Sharnoff 1998). **Primary agent**: lic. **Lichen traits involved**: see Table 7.1. Some lichens (i.e. *Cetraria*) were also used as **fuel and tinder** for priming wood fires or gathered for cooking fuel by eskimos or inuits respectively, and Indians (Sharnoff 1998). **Primary agent**: lic. **Lichen traits involved**: see Table 7.1.

Direct use of **lichen secondary metabolites** for their biological activities concern their antibiotic, antimycobacterial, antiviral, anti-inflammatory, analgesic, antipyretic, antiproliferative and cytotoxic effects in biotechnology, (folk) medicine, pharmacy, and cosmetics. Usnic acid, for instance, is used in commercially available antiseptic creams and in other topical antiseptic products (i.e. shaving treatments), and is very effective in the treatment of external wounds and burns (Nash 1996; Muggia et al. 2009). The insecticidal activity of lichen extracts was proven against the Maize Weevil (Yildirim et al. 2012). Lichens have been exploited in the past also for **dyeing** textiles as wool, silk, and cotton, by the extraction of lichen pigments. Particularly in the Mediterranean area and in the Canary Islands, lichens of the genus *Roccella* were collected at high amounts to extract a purple pigment called orchil, especially from the fifteenth to seventeenth centuries (Zedda 1996; Muggia et al. 2009). In North America, the most significant lichen dye is *Letharia vulpina*, from which indigenous people extract yellow dye based on vulpinic acid. In Scotland brown lichen dyes from *Parmelia omphalodes* and/or *P. saxatilis* (both generally called crottle) and red-purple lichen dyes from *Ochrolechia tartarea, Lasallia*, and *Umbilicaria* spp. were extensively used for dying wool and Harris tweed (Diadick Casselman 2003). Lichen extracts are used in the **fragrance industry**, the most used species being the oakmoss (*Evernia prunastri*) and tree moss (*Pseudevernia furfuracea*). About 700 tons of oakmoss are processed every year in France. Two other species of lichens are exploited in China under the name 'Chinese oakmoss', being *Evernia mesomorpha* and *Cetrariastrum nepalense*. Lichen species exploited in perfumery are known to contain more than 170 relevant compounds in their extracts (Joulain and Tabacchi 2009). However, oak moss extracts are considered as cause of perfume allergy (Johansen et al. 2003). Lichen extracts also have certain importance in the cosmetic industry for the production of crème, soap and toothpaste with bleaching and antibacterial effects (Sharnoff 1998). Lichens were formerly used as hallucinogens in shamanistic ritual, especially in the American Continent (Sharnoff 1998). The online bibliographical database of the human uses of lichens can be consulted for further information on given species and uses (Sharnoff 1998). **Primary agent**: lic. **Lichen traits involved**: see Table 7.1.

Indirect: Support of the biological diversity of other organisms through food provision for numerous animals, and provision of lichen metabolites, habitat, shelter and nesting material for other organisms.

7.3.2.3 Regulating Services

Direct: purification and detoxification of water, air and substrates; pest and disease control. Metal and mineral transformations by lichens have the potential for counteracting environmental pollution, in particular by bioremediation and revegetation of abandoned mines and heavy metal-polluted landscapes. Lichens are among the organisms which are most active in the uptake, immobilization or detoxification of metallic and radionuclide pollutants (Gadd 2009). However, in only few studies, lichens have been investigated as agents for possible bioremediation (Kumar et al. 2012).

Indirect: Climate regulation and moderation of extreme events, erosion prevention, regulation of water flow, water and substrate quality regulation: fresh water (Egoh et al. 2012).

7.3.2.4 Cultural Services

Direct: Lichens are subject of numerous **scientific investigations** and rich number of papers and scientific lichenology societies worldwide demonstrate their importance for science. Many areas of the world still need to be lichenologically explored and the genetic variability of lichenised fungi and photobionts still reserve discovery opportunities. Interactions among lichenised fungi, photobionts, non-lichenised fungi, bacteria, free-living algae, animals etc. need to be further investigated. **Primary agent**: mb, pb, cb, lic.**Lichen traits involved**: see Table 7.1.

Lichens provide information on environmental change and are valuable **bioindicators** of air pollution, climate change and land use intensity, and can provide important information on habitat alteration and fragmentation, valuable for the management of natural resources. They are perennial, slow-growing organisms, lack a waxy cuticle and stomata and can therefore absorb contaminants over the entire lichen surface. Furthermore, many species have high habitat specificity and broad geographic distribution and their spatial pattern are mostly documented well (Nash 1996; Nimis et al. 2002). Environmental changes induce change in lichen diversity, morphology, physiology, chemistry and accumulation of pollutants (Nimis et al. 2002). During the past years they have been mainly used for estimating air quality and indirectly pollution by sulphur dioxide. Nowadays, the reduced influence of sulphur dioxide has been accompanied by an increase in environmental nitrogen inputs, which favour the distribution of nitrogen tolerant lichens. Also atmospheric pollution by heavy metals is still considerable and a number of lichen species are regarded as valuable biomonitors for the assessment of environmental quality. The rate of absorption and the accumulation of heavy metals depend on morphological and chemical features of lichen thalli (Garty 2001).

Cultural services are mostly also related to non-material benefits, for example to spiritual or aesthetic value (Egoh et al. 2012). Lichens provide **inspiration** for culture, art and design and significantly contribute to the allover aesthetic landscapes appeal. In natural and agricultural habitats lichens often determine the appearance of trees, rocks and soils by forming specific colour mosaics. In urban environments, lichens colonize artificial substrates as cement, asphalts, walls and gravestones, as well as wood and roadside trees, colourising their surfaces. The beauty of lichens has inspired major works of art, photography, literature, and music. Some examples are poetries dedicated to lichens by Sbarbaro (1948), an Italian poet and lichenologists, and by the American poet Bishop (1955), as well as a science fiction romance by Wyndham (2008), dealing with the discovery of a rare lichen with anti-aging effect. Impressive images of lichens exist on internet sites, cured by lichenologists and professional photographers. Lichens inspire diverse types of figurative arts: painting, drawing, sculptures and jewellery, and a surprising number of crafts replicating or imitating lichen species or communities are also found. Examples are glazes with crawling and reticulation patterns in appearance similar to lichens, crochet art with lichen-shaped objects, lichen necklaces etc. For ornamental purposes lichens are used in cemetery floristry as well as for other kind of decoration or floristry creation and miniatures, e.g. in terrariums and mini gardens. Lichen wall art is a new trend to create natural-inspired edges (vertical gardens) and visual indoor effects. Furthermore, slurries of lichen thalli are sometimes used as paint for exterior walls of urban buildings. Species of the genus *Cladina* are mostly used as decorative florist 'moss', and are sometimes dyed in different colours. Reindeer moss (*Cladina stellaris*) is one of the most important decorative materials in the Nordic countries. About 500 tonnes of lichens are collected in peak years in Finland and mainly exported to Central Europe and the USA (Kettunen et al. 2012). Traditionally, several folks have used lichens for decorating traditional costumes, as artificial hair of masks, worn in rituals and ceremonies, for instance in Papua New Guinea, North America and some villages in Austria (Sharnoff 1998). Lichens have considerable **educational and recreational value** and numerous workshops, excursions, citizen science and education projects for schools on lichen are offered worldwide (e.g. the Opal project, UK), and are often organized by the different lichenological societies. In some places of the world lichens are even a tourist attraction, like the famous lichen fields in the Namib Desert. Online- and offline identification guides and resources are increasingly established for use by professional lichenologists as well as non-specialists (Rambold et al. 2014). **Primary agent**: lic. **Lichen traits involved**: see Table 7.1.

Indirect: Human beings profit from cultural services underpinned by other organisms whose diversity is supported by lichen ecosystem functions.

7.4 Discussion and Conclusions

From the present review and analysis of ecosystem functions and ecosystem services underpinned by lichens it emerges that lichenised fungi play a crucial role in ecosystem processes and in the support of the diversity of many other organisms. Also human beings profit in different ways, directly or indirectly from lichen diversity, and should preserve it from destruction due to man-made activities.

The major drivers of change in lichen diversity are climate change, deforestation, agricultural intensification, growing urbanisation and industrialisation, in particular causing pollution and loss of growth substrates. Many lichen traits give information on ecological changes and can be regarded as biodindicators. An important trait with specific adaptive function, likely to shift with regard to geographic distribution or with modified ecological conditions, is the occurrence of given lichen secondary metabolites. Also lichen community compositions demonstrate a strong correlation with climatic factors. Climatic shifts may have impact on lichen biomass, cover, frequency, species diversity and community, causing often species extinction (Insarov and Schroeter 2002).

The decline of natural ecosystems and consequent significant biodiversity loss during the last decades has been documented for many regions of the world. Here many ecosystem functions are often lost and services unsustainably used. The recognition of an urgent need to safeguard ecosystem functions and services has led to the establishment of new policies like the 'EU Biodiversity Strategy for 2020'. Interest in defying ecosystem services provided by various organismic groups, and the need of corresponding indicators is increasing (Norris et al. 2011; Egoh et al. 2012). The Intergovernmental Platform on Biodiversity and Ecosystem Services (IPBES; www.ipbes.net) was recently launched with the aim of facilitating the flow of scientific information related to biodiversity and ecosystem services to stakeholders like governments and practitioners. Ecosystem services are increasingly included in conservation policies and by now also the business sector is interested in understanding how to manage its dependence and impact on ecosystem services (Egoh et al. 2012). Research on ecosystem functions and services has grown substantially in the last years and many institutions are presently involved in detecting and modeling them at a global level. As the consideration of ecosystem services is a relatively new approach, concept and data underpinning for metrics and indicators are still underdeveloped compared to other fields, but there have been an increasing number of publications on ecosystem services quantification, modelling and mapping in the last years (Egoh et al. 2012; Crossman et al. 2013).

Concerning lichenised fungi, it is essential to gather available information, to collect new biodiversity data, and to establish work and data flows for knowledge generation on provided ecosystem functions and services.

7.4.1 Data Gathering and Knowledge Management

According to Norris et al. (2011), understanding of the quantitative links between biodiversity, traits and ecosystem services is presently still poor as biodiversity monitoring data have rarely been linked to ecosystem service criteria. Future research programs should therefore address such knowledge gaps and support careful documentation and data archiving of long-term monitoring biodiversity projects. Different kinds of data and information on lichens should be collected in databases and linked to environmental information for obtaining knowledge on biodiversity distribution and shifts. With such data and information it will be possible to quantify losses and gains of ecosystem functions and services provided by lichens for given case areas.

Relevant lichenological data for this purpose are:

- Occurrence and distribution data of lichen taxa (i.e. genotypes, species), of functional groups, and communities, as well as data on ecosystems.
- Monitoring data on biodiversity. As to be seen in the Global Biodiversity Information Facility (GBIF) (www.gbif.org), for instance, data on taxa distribution are still poor or missing for a number of biodiversity groups. Furthermore, assessments of lichen diversity status and trends are limited for many regions and more monitoring programs should be started to assess biodiversity changes and ecosystem functions. Biodiversity should be assessed at different levels, from genes to communities, and on different time and space scales (Norris et al. 2011).
- Data on lichen traits, considered at different level (genome, proteome, phenotype and ecological) and ecological data from given geographical regions. As shown in Table 7.1, numerous lichen traits play an important role in ecosystem functions and in the provision of ecosystem services. A database of selected lichen traits, 'LIAS light' (Rambold et al. 2001, onwards; Rambold et al. 2014), is maintained at the Botanische Staatssammlung München, Germany. The correlation of lichen traits with ecological data will surely reserve important information on the dependence of lichen traits on ecological conditions, and in particular on ecosystem functions and ecosystem services underpinned by lichens.
- Data and information on ecosystem functions and ecosystem services of lichens (i.e. interactions of lichens with biotic and abiotic factors; ethnobotanical knowledge).

7.4.2 Mapping and Modelling Ecosystem Services

Mapping is useful to visualize ecosystem services and their spatio-temporal distribution at different scale (from local to global). However, a map can only contain limited information and most mapping studies have therefore to focus on selected ecosystem services. Such maps are a prerequisite for ecosystem planning and management and for the sustainable use of natural resources and ecosystem services (Kandziora et al. 2013). Standardised methodological approaches to quantify and map ecosystem services are still going to be developed (Crossman et al. 2013). Presently, different methods are typically used to map and quantify ecosystem services: (1) collection of primary data through direct observation or measurements; (2) proxy methods in which single or combined indicators are used to define ecosystem services; (3) process models in which indicators are used as variables in the equation (Egoh et al. 2012). Of the different ecosystem service categories, observed primary data are often used for the quantification of provisioning services, whereas process models are predominantly used to quantify regulating services, especially with regard to climate, regulation of water flows, erosion prevention, and moderation of extreme events (Egoh et al. 2012).

Information from mapping and modelling analyses can be used to estimate biophysical quantities, establish trends, estimate costs and trade-offs, and place monetary value on biophysical quantities (Egoh et al. 2012). According to Cowling et al. (2008), the biophysical quantification of ecosystem services is an essential step towards successful implementation of actions to safeguard them. This step most often precedes monetary valuation as well as evaluation of trends and trade-offs. Unfortunately, many ecosystem services cannot be directly quantified, thus making the use of indicators indispensable. While ecosystem services providing goods can be directly quantified, most regulating, supporting, and cultural services are not easy to be quantified and researchers must rely on indicators or proxy data for their quantification (Egoh et al. 2012). An important step also for valuating some kind of ecosystem service provided by lichens will be to develop reliable and feasible indicators for mapping and modelling, as well as for bridging current data gaps.

One important limiting factor for mapping ecosystem services is data availability (Kandziora et al. 2013). Only few databases of ecosystem services and their indicators are available at the present, but none for lichens. These databases aim to integrate ecosystem services concepts into public and private sector decisions. One example is the 'Ecosystem Service Indicators Database' (ESID) (http://www.esindicators.org/), which seeks to support the application of ecosystem service metrics and indicators and consists of nine elements grouped in four categories (Ecosystems, Services, and Benefits; Human Well-being; Policy Strategies and Interventions; Drivers and Pressures). The quantification of ecosystem services supports biodiversity conservation activities and sustainable development. Especially developing countries heavily rely on ecosystem services for the well-being of their citizens; therefore, their inclusion in policies on biodiversity and development may have impacts on local livelihoods.

7.4.3 Calculating the Value of Ecosystem Services by Lichens

Valuing ecosystem services is not an end in itself, but is the first step towards integrating these services into public decision making to ensure the continuity of ecosystems. The scientific community can help develop the necessary tools to calculate the value of ecosystem services and to present them to decision makers (Sekercioglu 2010; Crossman et al. 2013). The 'Economics of Ecosystems and Biodiversity' (TEEB) is a global initiative focused on drawing attention to the economic benefits of biodiversity and of its ecosystem services. Its objective is to highlight the growing cost of biodiversity loss and ecosystem degradation. TEEB presents an approach that can help decision-makers recognize, demonstrate and capture the values of ecosystems and biodiversity, including how to incorporate these values into decision-making (TEEB 2010; Crossman et al. 2013). The integration of ecosystem services provided by lichens in such calculations and monetary evaluations should be supported.

In the face of decreasing biodiversity and ongoing global changes, maintaining ecosystem functioning is seen both as a means to preserve biological diversity as well as for safeguarding human well-being by securing the services ecosystems provide (Jax 2010).

Acknowledgments Dr. B. Wilske (University of Gießen, Germany) is kindly thanked for providing literature and useful information.

References

Abed RM, Lam P, de Beer D, Stief P (2013) High rates of denitrification and nitrous oxide emission in arid biological soil crusts from the Sultanate of Oman. ISME J 7(9):1862–1875

Ahmadijian V (2012) The lichens. Elsevier Verlag, Amsterdam

Aptroot A, Berg MP (2004) Collembola help lichens in competition with algae. Lichenologist 36(2):167–169

Ashman M, Puri G (2002) Essential soil science: a clear and concise introduction to soil science. Blackwell Science Ltd., Oxford

Asplund J, Solhaug KA, Gauslaa Y (2008) Fungal depsidones—an inducible or constitutive defence against herbivores in the lichen *Lobaria pulmonaria*? Basic Appl Ecol 10:273–278

Azenha G, Iturriaga T, Michelangeli FI, Rodriguez E (1998) Ethnolichenology, biological activity, and biochemistry of Amazonian lichen species. Emanations Rainforest 1(1). http://bhort.bh.cornell.edu/nih-mirt/emanations1-1/ethnolichenology.htm

Bargagli R, Sanchez-Hernandez JC, Monaci F (1999) Baseline concentrations of elements in the antartic macrolichen *Umbilicaria decussata*. Chemosphere 38(3):475–487

Barger NN, Castle SC, Dean GN (2013) Denitrification from nitrogen-fixing biologically crusted soils in a cool desert environment, southeast Utah, USA. Ecol Processes 2:16

Bates ST, Cropsey GW, Caporaso JG, Knight R, Fierer N (2011) Bacterial communities associated with the lichen symbiosis. Appl Environ Microbiol 77 (4):1309–1314

Becker VE, Reeder J, Stetler R (1977) Biomass and habitat of nitrogen fixing lichens in an oak forest in the North Carolina Piedmont. Bryologist 80(1):93–99

Behan-Pelletier VM, John MGSt, Winchester N (2008) Canopy Oribatida: Tree specific or microhabitat specific? Eur J Soil Biol 44:220–224

Belnap J (2006) The potential roles of biological soil crusts in dryland hydrologic cycles. Hydrol Process 20:3159–3178

Belnap J, Eldridge D (2001) Disturbance and recovery of biological soil crusts. In: Belnap J, Lange OL (eds) Biological Soil crusts: structure, function, and management. Springer, Berlin, pp 363–383

Belnap J, Gillette DA (1997) Disturbance of biological soil crusts: impacts on potential wind erodibility of sandy desert soils in Southeastern Utah. Land Degrad Dev 8:355–362

Belnap J, Prasse R, Harper KT (2001) Influence of biological soil crusts on soil environments and vascular plants. In: Belnap J, Lange OL (eds) Biological soil crusts: structure, function, and management. Springer, Berlin, pp 281–300

Beymer RJ, Klopatek JM (1991) Potential contribution of carbon by microphytic crusts in pinyon-juniper woodlands. Arid Soil Res Rehabil 5:187–198

Bishop E (1955) The shampoo—a cold spring. Houghton Mifflin Harcourt, Boston

Blackwell M (2011) The Fungi: 1, 2, 3 ... 5.1 million species? Am J Bot 98(3):426–438

Blaha J, Baloch E, Grube M (2006) High photobiont diversity associated with the euryoecious lichen-forming ascomycete *Lecanora rupicola* (Lecanoraceae, Ascomycota). Biol J Linn Soc 88:283–293

Bokhorst S, Ronfort C, Huiskes A, Convey P, Aerts R (2007) Food choice of Antarctic soil arthropods clarified by stable isotope signatures. Polar Biol 30:983–990

Boonpragob K, Crittenden PD, Lumbsch TH (2012) Lichens: from genome to ecosystems in a changing world. MycoKeys 6:1–2. doi:10.3897/mycokeys.6.4829

Botsford Comstock A (1986) Handbook of nature study. Comstock Publishing, Sacramento

Braun H (2011) The little lichen dragon—an extraordinary katydid from the Ecuadorian Andes (Orthoptera, Tettigoniidae, Phaneropterinae, Dysoniini). Zootaxa 3032:33–39

Braun RF, Ferner JW, Diesmos AC (1997) Definition of the Philippine parachute gecko, *Ptychozon intermedium* Taylor 1915 (Reptilia: Squamata: Gekkonidae): redescription, designation of a neotype, and comparison with related species. Herpetologica 53(3):357–373

Brodo I, Duran Sharnoff S, Sharnoff S (2001) Lichens of North America. Yale University Press, New Haven

Bychek-Guschina IA (2002) Analysis of lipids in lichens. In: Protocols in Lichenology. Springer Lab Manuals, pp 332–347

Cannon P (2010) Lichen camouflage and lichen mimicry. Br Lichen Soc Bull 106:39–41

Chen J, Blume HP, Beyer L (2000) Weathering of rocks induced by lichen colonization—a review. Catena 39:121–146

Chiarini Monteiro R (2002) The Thysanoptera fauna of Brazil. In: Thrips and Tospoviruses: proceedings of the 7th international symposium on Thysanoptera, pp 325–340

Cohen PA, Hudshon JB, Towers GH (1996) Antiviral activities of anthraquinones, bianthrones and hypericin derivatives from lichens. Experientia 52(2):180–183

Convention on Biological Diversity (CBD) (1992) http://www.cbd.int/convention/text/

Cowling RM, Egoh B, Knight TA, O'Farrel P, Reyers B, Rouget M, Roux D, Welz A, Wilhelm-Rechman A (2008) An operational model for mainstreaming ecosystem services for implementation. Proc Natl Acad Sci 105:9483–9488

Crespo A, Bridge PD, Cubero OF, Hawksworth DL (1997) Determination of genotypic variability in the lichen-forming fungus *Parmelia sulcata*. Bibliotheca Lichenologica. In: Türk R, Zorer R (eds) Progress and problems in Lichenology in the Nineties. J. Cramer, Berlin, pp 73–79

Crittenden PD (2000) Aspects of the ecology of mat-forming lichens. Rangifer 20(2–3):127–139

Crossman ND, Burkhard B, Nedkov S, Willemen L, Petz K, Palomo I, Drakou EG, Martín-Lopez B, McPhearson T, Boyanova K, Alkemade R, Egoh B, Dunbar MB, Maes J (2013) A blueprint for mapping and modelling ecosystem services. Ecosyst Serv 4:4–14

Diadick Casselman K (2003) Lichen dyes: the new source book. Dover Publications Inc., Mineola

Dighton J (2003) Fungi in ecosystem processes. CRC Press, Boca Raton

Domaschke S, Fernández-Mendoza F, García MA, Martín MP, Printzen C (2012) Low genetic diversity in Antarctic populations of the lichen-forming ascomycete *Cetraria aculeata* and its photobiont. Polar Res 31:17353. doi:10.3402/polar.v31i0.17353

Egoh B, Drakou EG, Dunbar MB, Maes J, Willemen L (2012) Indicators for mapping ecosystem services: a review. The European Commission's Joint Research Centre (JRC) Scientific and Policy. http://publications.jrc.ec.europa.eu/repository/bitstream/111111111/26749/1/lbna25456enn.pdf

Elbert W, Weber B, Burrows S, Steinkamp J, Büdel B, Andreae MO, Pöschl U (2012) Contribution of cryptogamic covers to the global cycles of carbon and nitrogen. Nat Geosci 5:459–462. doi:10.1038/ngeo1486

Elix JA (1996) Biochemistry and secondary metabolites. In: Nash III TH (ed) 1996. Lichen biology. Cambridge University Press, Cambridge, pp 155–180

Erdoğan S, Kaya M (2013) The biodiversity of Bdelloid rotifers on a single rock affected by cardinal directions. J Earth Sci Res 1(1):21–24

Evans RD, Lange OL (2001) Biological soil crusts and ecosystem nitrogen and carbon dynamics. In: Belnap J, Lange OL (eds) Biological soil crusts: structure, function, and management. Springer, Berlin, pp 263–279

Favero Longo SE (2005) Rischio amianto nelle Alpi occidentali: licheni come agenti di bioremediation: un'analisi ecologica, microbiologica e chimica delle interazioni fra organismi vegetali e fungini e substrati asbestiferi applicata alla ex-miniera di crisotilo di Balangero e Corio (Torino, Italia). Ph.D. thesis, University of Torino (Italy)

Friedl T (1989) Systematik und Biologie von *Trebouxia* (Microthamniales, Chlorophyta) als Phycobiont der Parmeliaceae (lichenisierte Ascomyceten). Ph.D. thesis, University of Bayreuth, Germany

Gadd GM (2007) Geomycology: biogeochemical transformations of rocks, minerals, metals and radionuclides by fungi, bioweathering and bioremediation. Mycol Res 111(1):3–49

Gadd GM (2009) Metals, minerals and microbes: geomicrobiology and bioremediation. Microbiology 156 (3):609–643

Gärtner G (1985) Die Gattung *Trebouxia* Puymaly (Chlorellales, Chlorophyceae). Algol Stud 41:495–548

Garty J (2001) Biomonitoring atmospheric heavy metals with lichens: theory and application. Crit Rev Plant Sci 20(4):309–371. doi:10.1080/20013591099254

Gauslaa Y (2005) Lichen palatability depends on investments in herbivore defence. Oecologia 143(1):94–105

German PJ, Foster JR (2011) Difference in the aquatic micro-invertebrate fauna of two common foliose epiphytic lichens. www.oneonta.edu/academics/biofld/PUBS/ANNUAL/2011/16%20Lichen%20Paper.pdf

Gerson U (1973) Lichen-arthropod associations. Lichenologist 5:434–443

Gerson U, Seaward MRD (1977) Lichen-invertebrate associations. In: Seaward MRD (ed) Lichen ecology. Academic Press, London, pp 69–119

Gressit JL (1977) Symbiosis runs wild on the backs of high-living weevils. Smithsonian 7:135–136

Grube M, Muggia L (2010) Identifying algal symbionts in lichen symbioses. In: Nimis PL, Vignes Lebbe R (eds) Tools for identifying biodiversity: progress and problems, pp 295–299

Grube M, Cardinale M, de Castro JV Jr, Müller H, Berg G (2009) Species-specific structural and functional diversity of bacterial communities in lichen symbioses. ISME J 3(9):1105–1115

Gunnarsson B, Hake M, Hultengren S (2004) A functional relationship between species richness of spiders and lichens in spruce. Biodivers Conserv 13(4):685–693

Guo S, Li B, Watanabe K (2007) Diet and activity budget of *Rhinopithecus roxellana* in the Qinling Mountains, China. Primates 48(4):268–276

Hawksworth DL (2004) Fungal diversity and its implications for genetic resource collections. Stud Mycol 50:9–18

Hawksworth DL, Lawton RM, Martin PG, Stanley-Price K (1984) Nutritive value of *Ramalina duriaei* grazed by gazelles in Oman. Lichenologist 16:93–94

Helms G, Friedl T, Rambold G, Mayrhofer H (2001) Identification of photobionts from the lichen family Physciaceae using algal-specific ITS rDNA sequencing. Lichenologist 33:73–86

Hesbacher S, Giez I, Embacher G, Fiedler K, Max W, Trawöger A, Türk R, Lange OL, Proksch P (1995a) Sequestration of lichen compounds by lichen-feeding members of the Arctiidae (Lepidoptera). J Chem Ecol 21(12):2079–2089

Hesbacher S, Baur B, Baur A, Proksch P (1995b) Sequestration of lichen compounds by three species of terrestrial snails. J Chem Ecol 21(2):233–246

Hocking DJ, Semlitsch RD (2007) Effects of timber harvest on breeding-site selection by gray treefrogs (*Hyla versicolor*). Biol Conserv 138:506–513

Huneck S, Yoshimura I (1996) Identification of lichen substances. Springer, Berlin

Insarov G, Schroeter B (2002) Lichen monitoring and climate change. In: Nimis PL, Scheidegger C, Wolseley PA (eds) Monitoring with lichens—monitoring lichens. Kluwer Academic Publishers, Berlin, pp 183–201

Itzhak Martinez JJ, Raz R, Mgocheki N, Álvarez R (2014) Epiphytic lichen is associated with species richness of gall-inducing aphids but not with niche differentiation among them. Arthopod-Plant Interact 8:17–24

Jax K (2010) Ecosystem functioning. Cambridge University Press, Cambridge

Johansen JD, Andersen KE, Svedman C, Bruze M, Bernard G, Giménez-Arnau E, Rastogi SC, Lepoittevin JP, Menné T (2003) Chloroatranol, an extremely potent allergen hidden in perfumes: a dose-response elicitation study. Contact Dermatitis 49(4):180–184

Jørgensen HB, Johansson T, Canbäck B, Hedlund K, Tunlid A (2005) Selective foraging of fungi by collembolans in soil. Biol Lett 1(2):243–246

Joulain D, Tabacchi R (2009) Lichen extracts as raw materials in perfumery. Part 1: oakmoss. Flavour Fragrance J 24(2):49–61

Kaczmarek L, Goldyn B, Welnicz W, Michalczyk L (2011) Ecological factors determining tardigrada distribution in Costa Rica. J Zool Syst Evol Res 49(Supp 1):78–83

Kandziora M, Burkhard B, Müller F (2013) Mapping provisioning ecosystem services at the local scale using data of varying spatial and temporal resolution. Ecosyst Serv 4:47–59

Karunaratne DN, Jayalal RG, Karunaratne V (2012) Lichen polysaccharides. In: Karunaratne DN (ed) The complex world of polysaccharides, Chap. 8. InTech Open Access Publishers, pp 215–226

Kettunen M, Vihervaara P, Kinnunen S, D'Amato D, Badura T, Argimon, M, Ten Brink P (2012) Socio-economic importance of ecosystem services in the Nordic Countries. Norden Publications

Kumar A, Singh N, Gaurav SS (2012) Bioremediation of metal contaminated sites by natural growing lichens found in hilly areas of Himachal Pradesh. Int J Curr Res 4(11):167–168

Lakatos M, Lange-Bertalot H, Büdel B (2004) Diatoms living inside the thallus of the green algal lichen *Coenogonium linkii* in Neotropical lowland rain forests. J Phycol 40:70–73

Lalley JS, Viles HA, Henschel JR, Lalley V (2006) Lichen-dominated soil crusts as arthropod habitat in warm deserts. J Arid Environ 67(4):579–593

Lange OL (2001) Photosynthesis of soil-crust biota as dependent on environmental factors. In: Belnap J, Lange OL (eds) Biological soil crusts: structure, function, and management. Springer, Heidelberg, pp 217–241

Lange OL, Meyer A, Zellner H, Ullmann I, Wessels DCJ (1990) Eight days in the life of a desert lichen: water relations and photosynthesis of *Teloschistes capensis* in the coastal fog zone of the Namib Desert. Modoqua 17:17–30

Lange OL, Meyer A, Zellner H, Heber U (1994) Photosynthesis and water relations of lichen 25 soil crusts: field measurements in the coastal fog zone of the Namib Desert. Funct Ecol 8:253–264

Lazaro R, Canton Y, Sole-Benet A, Bevan J, Alexander R, Sancho LG, Puigdefabregas J (2008) The influence of competition between lichen colonization and erosion on the evolution of soil surfaces in the Tabernas badlands (SE Spain) and its landscape effects. Geomorphology 102(2):252–266. doi:10.1016/j.geomorph.2008.05.005

Leal IR, Oliveira PS (2000) Foraging ecology of attine ants in a Neotropical savannah: seasonal use of fungal substrate in the cerrado vegetation of Brazil. Insectes Soc 47:376–382

Leavitt SD, Johnson LA, Goward T, St. Clair LL (2011) Species delimitation in taxonomically difficult lichen-forming fungi: an example from morphologically and chemically diverse *Xanthoparmelia* (Parmeliaceae) in North America. Mol Phylogenet Evol 60(3):317–332

Lindo Z, Stevenson SK (2007) Diversity and distribution of oribatid mites (Acari: Oribatida) associated with arboreal and terrestrial habitats in interior cedar-hemlock forests, British Columbia, Canada. Northwest Sci 81(4):305–315

Liu X, Stanford CB, Yang J, Yao H, Li Y (2013) Foods eaten by the Sichuan snub-nosed Monkey (*Rhinopithecus roxellana*) in Shennongjia National Nature Reserve, China, in relation to nutritional chemistry. Am J Primatol 75:860–871. doi:10.1002/ajp.22149

Llano G (2012) Utilization of lichens in the arctic and subartict. Econ Bot 10:367–392

Longton RE (1988) Biology of polar bryophytes and lichens. Studies in polar research. Cambridge University Press, Cambridge

Lumbsch HT, Leavitt SD (2011) Goodbye morphology? A paradigm shift in the delimitation of species in lichenized fungi. Fungal Divers 50:59–72

Lurz PWW, Gurnell J, Magris L (2005) Sciurus vulgaris. Mamm Species 769:1–10

Majerus MEN (2009) Industrial melanism in the peppered moth, *Bistonbetularia*: an excellent teaching example of darwinian evolution in action. Evo Edu Outreach 2:63–74

Malhotra S, Subban R, Singh A (2013) Lichens-role in traditional medicine and drug. Internet J Altern Med 5(2). http://ispub.com/IJAM/5/2/4012

Manojlovic NT, Solujic S, Sukdolak S, Milosev M (2005) Antifungal activity of *Rubia tinctorum, Rhamnus frangula* and *Caloplaca cerina*. Fitoterapia 76:244–246

Materna J (2000) Oribatid communities (Acari: Oribatida) inhabiting saxicolous mosses and lichens in the Krkonoše Mts. (Czech Republic). Pedobiologia 44 (1):40–62

Messuti MI, Kun M (2007) The occurrence of *Pachytullbergia scabra* (Collembola: Pachytullbergiidae) on *Pseudocyphellaria granulate* (lichenized Ascomycota). Revista de la Sociedad Entomológica Argentina 66(1–2):177–179

Millennium Ecosystem Assessment (MEA) (2005) Ecosystems and human well-being: synthesis. Island Press, Washington, DC. http://millenniumassessment.org/documents/document.356.aspx.pdf

Mills AJ, Fey MV (2004) A simple laboratory infiltration method for measuring the tendency of soils to crust. Soil Use Manag 20:8–12

Molina MC, DePriest PT, Lawrey JD (2005) Genetic variation in the widespread lichenicolous fungus *Marchandiomyces corallines*. Mycologia 97(2):454–463

Mound LA (2002) Thysanoptera biodiversity in the Neotropics. Revista de Biología Tropical 50(2):477–484

Muggia L, Grube M, Tretiach M (2008) Genetic diversity and photobiont associations in selected taxa of the *Tephromela atra* group (Lecanorales, lichenised Ascomycota). Mycol Prog 7:147–160

Muggia L, Schmitt I, Grube M (2009) Lichens as treasure chests of natural products. SIM NEWS, 85–97. http://www.uni-graz.at/~grubem/treasure.pdf

Mukherjee A, Wilske B, Navarro RA, Dippenaar-Schoeman A, Underhill LG (2010) Association of spiders and lichen on Robben Island, South Africa: a case report. JoTT Commun 2(4):815–819

Mushegian AA, Peterson CN, Baker CCM, Pringle A (2011) Bacterial diversity across individual lichens. Appl Environ Microbiol 77(12):4249–4252

Nash III TH (ed) (1996) Lichen biology. Cambridge University Press, Cambridge

Nimis PL, Skert N (2006) Lichen chemistry and selective grazing by the coleopteran *Lasioderma serricorne*. Environ Exp Bot 55(1–2):175–182

Nimis PL, Pinna D, Salvadori O (1992) Licheni e conservazione dei monumenti. Editori CLUEB Bologna

Nimis PL, Scheidegger C, Wolseley PA (2002) Monitoring with lichens—monitoring lichens. Kluwer Academic Publishers, Berlin

Norris K, Bailey M, Keith A, Maskell L, Reading C, Turner S, Vanbergen A, Watt A et al (2011) Biodiversity in the context of ecosystem services (Chap. 4). UK National Ecosystem Assessment. Understanding nature's value to society. Technical report. UNEP-WCMC, Cambridge, pp 63–104

Nybakken L, Helmersen AM, Gauslaa Y, Selås V (2010) Lichen compounds restrain lichen feeding by bank voles (*Myodes glareolus*). J Chem Ecol 36(3):298–304

Peršoh D, Rambold G (2012) Lichen-associated fungi of the *Letharietum vulpinae*. Mycol Prog 11(3):753–760

Petrzik K, Vondrák J, Barták M, Peksa O, Kubešová O (2013) Lichens—a new source or yet unknown host of herbaceous plant viruses? Eur J Plant Pathol. doi:10.1007/s10658-013-0246-z

Piercey-Normore M, De Priest PT (2001) Algal switching among lichen symbioses. Am J Bot 88(8):1490–1498

Pike LH (1978) The importance of epiphytic lichens in mineral cycling. Bryologist 81:247–257

Rambold G (1985) Fütterungsexperimente mit den an Flechten fressenden Raupen von *Setina aurita* Esp. (Lepidoptera, Arctiidae). Nachrichtenblatt der Bayerischen Entomologen 34:82–90

Rambold G, Davydov E, Elix JA, Nash III TH, Scheidegger C, Zedda L (eds.) (2001 onwards) (last visited: 2014-07-13). LIAS light—a database for rapid identification of lichens. http://liaslight.lias.net/

Rambold G, Stadler M, Begerow D (2013) Mycology should be recognized as a field in biology at eye level with other major disciplines—a memorandum. Mycol Prog. doi:10.1007/s11557-013-0902-x

Rambold G, Elix JA, Heindl-Tenhunen B, Köhler T, Nash TH III, Neubacher D, Reichert W, Zedda L, Triebel D (2014) LIAS light—towards the ten thousand species milestone. MycoKeys 8:11–16. doi:10.3897/mycokeys.8.6605

Ranković BR, Kosanić MM, Stanjković T (2011) Antioxidant, antimicrobial and anticancer activity of the lichens *Cladonia furcata, Lecanora atra* and *Lecanora muralis*. BMC Complement Altern Med. doi:10.1186/1472-6882-11-97

Redzic S, Barudanovic S, Pilipovic S (2010) Wild mushrooms and lichens used as human food for survival in war conditions; Podrinje—Zepa region (Bosnia and Herzegovina, W. Balkan). Res Hum Ecol 17:175–187

Rikkinen J (2013) Molecular studies on cyanobacterial diversity in lichen symbioses. MycoKeys 6:3–32. doi:10.3897/mycokeys.6.3869

Rogers R (2011) The fungal pharmacy: the complete guide to medicinal mushrooms and lichens of North America. North Atlantic Books, Berkeley

Rosentreter R, Hayward GD, Wicklow-Howard M (1997) Northern flying squirrel seasonal food habits in the interior conifer forests of Central Idaho, USA. http://hdl.handle.net/2376/1250

Šatkauskienė I (2012) Microfauna of lichen (*Xanthoria parietina*) in Lithuania: diversity patterns in polluted and non-polluted sites. Baltic Forestry 18(2):255–262

Sbarbaro C (1948) Trucioli. Mondadori, Milano

Seaward MRD (1988) Contribution of lichens to ecosystems. In: Galun M (ed) CRC handbook of lichenology, vol II. CRC Press Inc., Boca Raton, pp 107–129

Seaward MRD (1997) Major impact made by lichens in biodeterioration processes. Int Biodeterior Biodegrad 40:269–273

Sekercioglu CH (2010) Ecosystem functioning and services. In: Sodhi NS, Ehrlich PR (eds) Conservation biology for all. Oxford University Press, Oxford, pp 45–72

Sharnoff SD (1998) Bibliographic database of the human uses of lichens. http://www.lichen.com/usetaxon.html

Shukla AC, Chinlampianga M, Verma A, Dikshit A, Upreti DK (2011) Efficacy and potency of lichens of Mizoram as antimycotic agents. Indian Phytopath 64 (4):367–370

Skorepa AC, Sharp AJ (1971) Lichens in "packets" of lacewing larvae (Chrysopidae). Bryologist 74(3): 363–364

Sumotha M, Pauwels OSG, Kunya K, Limlikhitaksorn C, Ruksue S, Taokratok A, Ansermet M, Chanhome L (2012) A new species of Parachute Gecko (Squamata: Gekkonidae: genus *Ptychozoon*) from Kaeng Krachan National Park, western Thailand. Zootaxa 3513:68–78

TEEB (2010) The economics of ecosystems and biodiversity: ecological and economic foundations, London. Earthscan, Washington DC

Triggiani D, Ceccarelli D, Tiezzi A, Pisani T, Munzi S, Gaggi C, Loppi S (2009) Antiproliferative activity of lichen extracts on murine myeloma cells. Biologia 64:59–62

Viro P, Sulkava S (1985) Food of the bank vole in Northern Finnish Spruce Forests. Acta Theriol 30 (15):259–266

Will-Wolf S, Scheidegger C (2002) Monitoring lichen diversity and ecosystem function. An introduction. In: Nimis PL, Scheidegger C, Wolseley PA (eds) Monitoring with lichens—monitoring lichens. Kluwer Academic Publishers, Berlin, pp 143–145

Wilske B, Kesselmeier J (1999) First measurements of the C_1-and the C_2-organic aids and aldehydes exchange

between boreal lichens and the atmosphere. Phys Chem Earth (B) 24(6):725–728
Wilske B, Holzinger R, Kesselmeier J (2001) Evidence for ethanolic fermentation in lichen during periods of high thallus water content. Symbiosis 31:95–111
Wilske B, Burgheimer J, Maseyk K, Karnieli A, Zaady E, Andreae MO, Yakir D, Kesselmeier J (2009) Modeling the variability in annual carbon fluxes related to biological soil crusts in a Mediterranean shrubland. Biogeosciences Discuss 6:7295–7324
Wyndham J (2008) Trouble with Lichen. Penguin, UK
Yildirim E, Emsen B, Aslan A, Bulak Y, Ercisli S (2012) Insecticidal activity of lichens against the Maize Weevil, *Sitophilus zeamais* Motschulsky (Coleoptera: Curculionidae). Egypt J Biol Pest Control 22(2):151
Zedda L (1996) Note etnobotaniche sui licheni della Sardegna. Not Soc Lich Ital 9:35–37
Zedda L, Rambold G (2011) Lichens and their importance for the monitoring of environmental changes in southern Africa. http://publikationen.ub.uni-frankfurt.de/frontdoor/index/index/docId/23334

8 In Vitro Culture of Lichen Partners: Need and Implications

Neeraj Verma and Bhaskar C. Behera

Contents

Abstract

In the world, lichens are slowest growing symbiotic organism associations between fungi and a photosynthetic alga and/or cyanobacteria. Lichens produce a wide array of secondary metabolites which are unique and forms under lichenized conditions. Secondary metabolites of lichens demonstrated significant inhibition of various biological activities at very low concentrations. Although lichens are reservoir for various biologically active compounds, only few lichens and their compounds have been tested for their biological significance and still there are many more yet to be tested. Hence, there is clearly an urgent need for expanding research in this area including in-depth studies of those compounds which have shown promising results. A strong focus is also needed on the most promising lichen-based drug therapies followed by large scale production of the best of those compounds. One of the main issues related to the limited use of lichen compounds in modern medicine is the slow growth rate of lichen thalli which remains as a challenge to enhance their biomass with in vitro culture. Advance research in the field of cultivating lichens and their symbionts will enable the mass production of lichen substances and their pharmaceutical applications. The present review on tissue culture of lichens of various groups, biosynthesis of their unique secondary compounds, physiological conditions required for their synthesis, and biological activities will significantly contribute to the present knowledge in the field of experimental lichenology and will also attract the attention of industry/society/nation as a whole.

N. Verma (✉) · B.C. Behera
Mycology and Plant Pathology Group, Plant Science Division, Agharkar Research Institute, G.G. Agarkar Road, Pune 411004, India
e-mail: neerajverma2310@gmail.com

D.K. Upreti et al. (eds.), *Recent Advances in Lichenology*,
DOI 10.1007/978-81-322-2235-4_8

Keywords
Lichen · Tissue culture · Bioreactor · Secondary metabolites

8.1 Introduction

Lichen is a symbiotic association of a fungus (mycobiont) and a photosynthetic partner (photobiont), which may be an alga or a cyanobacterium. This association is not a simple mixture, however, but one in which the fungus produces a thallus, or body, within which the photobionts are housed (Ahmadjian 1993). Lichens are an outstandingly successful group, exploiting a wide range of habitats throughout the world and dominating about 8 % of terrestrial ecosystems. In nature, lichens grow very slowly. The growth rate of lichen species has been generalized by Hale (1973) as 0.5–5 mm yr^{-1} (Fig. 8.1). Lichens are the integral part of all ecosystems and are often responsible for either fixing or capturing essential nutrients from the air and returning them to the environment. The differential sensitivity of lichens to sulfur dioxide, nitrogen dioxide, ozone, and their ability to absorb/accumulate heavy metals and radionuclides have made them valuable pollution indicators around the urban and industrial sites of the world (Ahmadjian 1993). Irrespective of the advances in medical sciences, the tribal peoples still utilize these organisms for different purposes depending on their nutritive, medicinal, decorative, brewing, distilling, dying, cosmetic, and perfumery properties (Upreti and Chatterjee 2007).

Fig. 8.1 Various growth forms of lichens; foliose, fruticose and crustose lichens growing on tree bark

Lichens may grow under rather diverse and sometimes extreme ecological environments, either at polar latitudes or at extreme altitudes (up to 7,400 m) (Boustie and Grube 2005), but wherever they grow in nature, they often add a colorful aspects to their habitat. The distinct colors of many lichens are due to the massive accumulation of diverse secondary compounds, the “lichen substances”. These substances are stress compounds, which help the organism for self-defence and to grow in very adverse and harsh environmental conditions (Huneck 1999). These lichen substances or metabolites are unique with respect to those of higher plants. Lichen metabolites include radical scavengers, antibiotics, pigments, enzyme inhibitors, immune modulating agents, toxins, pesticides, and antitumor agents. They have major role on human health and nutrition. For the past hundred years, many lichenologists have studied lichen chemistry and have found over 1,050 compounds considered as secondary metabolites (Stocker-Wörgötter 2008).

Lichen substances represent comparatively a small group, but chemically they are diverse complex molecules. There are two main groups of lichen compounds: primary metabolites (intracellular) and secondary metabolites (extracellular). The common intracellular product occurs in lichens are proteins, amino acids, polyphenols, carotenoids, polysaccharides, and vitamins, which are bound in the cell walls and the protoplasts. As the lichen thallus is composite structure, it is not always possible to decide where a particular compound is biosynthesized. Most of the intracellular compounds are water soluble and nonspecific. These compounds also occur in free-living fungi, algae, and in higher green plants (Hale 1983).

The majority of organic compounds found in lichens are secondary metabolites of the fungal component, which are exported outside the fungal cells to be found on cell surfaces as crystals in different parts of the thallus. They often accumulate in the upper cortex or in specialized structures such as fruiting bodies (Fahselt 1994; Elix 1996; Oksanen 2006). These compounds are usually insoluble in water and can only be extracted with organic solvents.

A number of secondary metabolites produced by lichens through their diverse biosynthetic pathways, mainly, polyketide, shikimic acid, and mevalonic acid pathways. Most of the lichen substances are phenolic compounds. Polyketide-derived aromatic compounds such as depsides, depsidones, dibenzofurans, xanthones, and naphthaquinones are of great interest. Compounds from other pathways are esters, terpenes, steroids, terphenylquinones, and pulvinic acid (Fahselt 1994; Cohen and Towers 1995; Elix 1996; Müller 2001; Brunauer and Stocker-Wörgötter 2005; Stocker-Wörgötter 2005; Oksanen 2006).

8.2 Limitations and Advancement in Lichens Biotechnological Approaches

To date, many lichens and lichen products have proved to be a source of important secondary metabolites for food and pharmaceutical industries (Huneck 1999; Oksanen 2006; Shukla et al. 2010; Paudel et al. 2011) and still holds a considerable interest as alternative treatments in various parts of the world (Richardson 1991). Although immense biological potential of lichens reported by various authors but they have been neglected by pharmaceutical industries (Crittenden and Porter 1991; Yamamoto et al. 1998; Behera et al. 2003, 2004). Industrial scale harvests of lichens are neither ecologically sensible nor sustainable. So the new technologies in molecular biology come in light for the direct access of lichen genomes to reveal and eventually to harvest/production of novel secondary metabolites (Miao et al. 2001).

Polyketides are one of the major groups of secondary metabolites produced by lichens (Fahselt 1994). Hitherto attempts have been made to produce lichen polyketides in prokaryotic or eukaryotic host organisms but proved unsuccessful. This genetic engineering technique could be used to clone lichen polyketide pathway-participating enzymes by means of surrogate hosts to produce high yields of polyketides for industrial use. A prerequisite for producing polyketide in surrogate hosts is the heterologous expression of the functional polyketide synthase (PKS). However, genes for fungal PKS are very large DNA sequence of usually 6–8 kb (Miao et al. 2001), which often contain intron sequences. Therefore, cloning is very difficult and especially the expression of the protein (PKS) with its original function. Hence, further work is required to find out suitable surrogate host particularly fast growing fungi, to express the original lichen PKS genePKS gene (Chooi et al. 2008).

For commercial exploitation of lichen substances, huge collection of naturally grown thalli is necessary. To prevent the decline of lichen population in protected areas, the only conceivable alternative is in vitro culture of lichen (Brunauer and Stocker-Wörgötter 2005).

In vitro culture of lichen is an essential tool for investigating the biosynthetic pathways of lichen metabolites as well as the conditions that are necessary for their production (Scheme 8.1). However, to culture whole lichen thallus is very difficult and many problems are unsolved. Only a very few lichenologists have succeeded in reconstituting new lichen thalli or cell aggregates composed of both the bionts (mycobiont and photobiont). The lichen *Cladonia cristatella*, *Endocarpon pusillum* and *Staurothele clopima* were successfully cultured by Ahmadjian (1966), Ahmadjian and Heikkilä (1970). Later on, *Usnea rubescens, Ramalina yasudae* and *Cladonia vulcani* were cultured by induction of thallus fragments and reported the usnic acid production in the culture (Yamamoto et al. 1985; Yoshimura et al. 1987). They have also studied the effects of culture conditions on the growth of lichen-cultured tissue (Yamamoto et al. 1987). The cyanobacterial lichen species *Peltigera* didactyla was cultured from soredia (Stocker-Wörgötter and Turk 1988). Some *Umbilicaria* species of lichen and *Usnea flexilis* were also cultured and

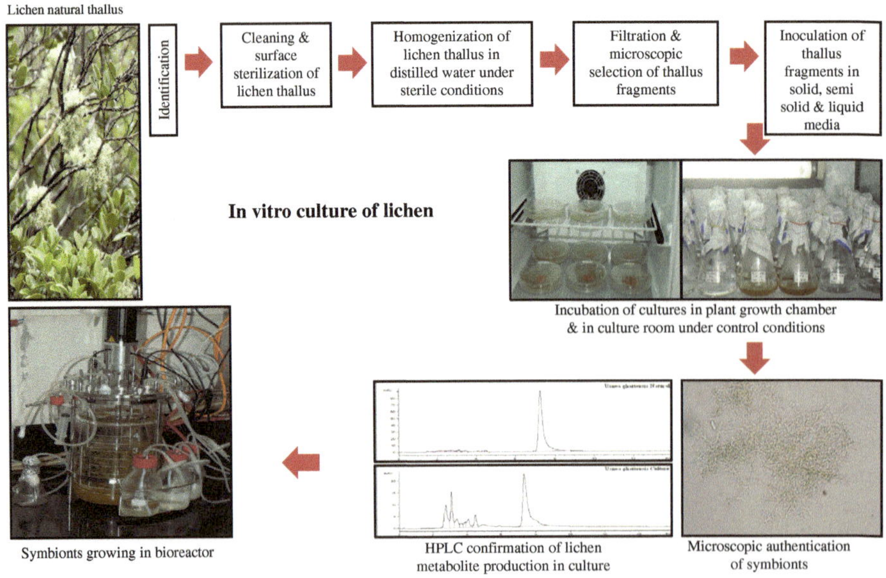

Scheme 8.1 Schematic flow chart to enhance lichen biomass through in vitro culture

produced secondary metabolites (Yoshimura et al. 1989). Further, some Antarctic lichens were also cultured up to the vegetative thallus stage by the same authors (Yoshimura et al. 1990a, b). Yamamoto et al. (1993) reported the use of lichen tissue culture in many modern biological experiments. Kinoshita et al. (1993a, b) studied the morphogenetic capacity of the isolated mycobiont from the cultured fragments of *Usnea hirta* and detected usnic acid in cell aggregates.

8.3 Influence of Physiological Conditions on Secondary Metabolite Production in Lichens

Lichens are known to produce over 1,050 secondary metabolites, which are unique to lichens only and possess various important biological activities of pharmaceutical interest (Molnar and Farkas 2010; Huneck 1999; Shukla et al. 2010). In-depth investigations on the production of these unique bioactive metabolites have suggested that secondary metabolite production in lichens is influenced by changes in culture conditions (temperature, pH, humidity, nutrient media etc.) which are prerequisite for appropriate symbiotic association of bionts and their function. Therefore, experimental lichenological study relies on the conditions and optimization of culture. Several reports also indicated that by manipulating the physiological conditions, bionts could able to produce lichen substance along with other accessory pigments which are not present in voucher specimens (Stocker-Wörgötter 2008). On the other hand, it is also been reported that most of the lichen species have diagnostic compounds that are consistently produced by genetic inheritance and species adaptation to particular conditions (Christopher et al. 2012).

8.3.1 Effects of Temperature, pH, Light and Humidity

Temperature, pH, light, and humidity are the major driving forces for the culturing of lichens. These physiological conditions are required in a balanced proportion for secondary metabolite production in lichens under culture. Culberson and Armaleo (1992) and Kinoshita et al. (1993b) reported the time period of desiccation and higher agar concentration is necessary for the production of larger amount of usnic acid in lichen culture. The production of baeomycesic and squamatic acids in *Thamnolia vermicularis* culture supported by relatively high light, low temperature, and dehydrating nutrient media (Stocker-Wörgötter 2001). High quantity of depsides and depsidones production was observed in *Ramalina siliquosa* cultures when the pH was 6.5 and incubation temperature was 15–17 °C (Hamada 1982, 1989). Stocker-Wörgötter (2001) further showed that species within the same genera produce their specific compounds only when grown on an acidic nutrient media. *Cladonia grayi* and *C. merochlorophaea* grow at lower pH than *C. chlorophaea*, and in some cases showed identical morphologies with different secondary metabolites (Culberson et al. 1988; Culberson 1986). In our observations (Behera et al. 2006d), an increased production of usnic acid was seen by the symbionts of cultured lichen *Usnea ghattensis* due to osmotic conditions created by increasing agar concentration. The observations also revealed that the excess temperature and acidic pH often dry the media, and cracking of the nutrient surface in the petriplates destroys the morphological differentiation of the bionts. This leads the cell aggregates for deprivation of required nutrient and stops the production of desirable lichen compounds. Many authors suggested that production of lichen substances have strong correlation with the light intensity for a longer period (Rundel 1969; Fahselt 1981; Hamada 1991; Bjerke et al. 2002, 2004; McEvoy et al. 2006, 2007a, b; Armaleo et al. 2008). Brodo (1973) reviewed the significance of the substratum to lichen and reported different tree bark and rock types can have different pHs, nutrients, and water holding capacity makes different lichen species to grow (Fig. 8.2). These observations suggest that culture parameters related to environmental conditions (temperature, pH, light, humidity) may trigger the production of certain compounds in certain lichen species. The adaptation of lichen cultures towards the artificial culture conditions helps them in survival and metabolite production.

8.3.2 Nutritional Effects on Secondary Metabolite Pathway

The mycobiont of lichen undergoes rapid growth, sometimes in many species when all required nutrients are available in optimal quantities and in a suitable proportion. This growth is slowed when one nutrient supply becomes exhausted from the supplied media components (macro- and micronutrients). In contrary to that workers reported that secondary metabolism is often triggered at a stage of fungal growth and development when one or more nutrients become limiting and growth slows down (Moore 1998; Behera et al. 2006d, 2009a; Verma et al. 2011). It is thought that when mycelial growth slows, carbohydrates are not used in growth processes rather metabolized and produce secondary metabolites and accumulate on the surface of the cell aggregate. Though it may not serve specific functions, rather advantageous for their survival in adverse conditions (Christopher et al. 2012).

Many lichenologists have studied the effects of various carbon and nitrogen sources on cultures. The secondary metabolism of mycobiont influenced by the carbon supplied by the photobiont. *Trebouxia* and *Trentepohlia* are the most common photobionts which produce sugar alcohol as ribitol and erythritol (Honegger 2009). Mycobiont metabolize these sugar alcohol into mannitol. Brunauer et al. (2007) demonstrated the influence of mannitol on secondary metabolite production in *Xanthoria elegans* over ribitol. The lecanoric acid production is enhanced by the catabolism of mannitol or glucose (Hamada et al. 1996). The type of carbon, nitrogen and their availability may affect metabolite production along with increased fungal growth, sporulation and sometimes production of high aflatoxin (Keller et al. 2002). The secondary metabolite yield in some cultured lichen species increased with supplementation of 10 % sucrose in nutrient media, indicates variation of culture conditions/ or algal physiology (Hamada et al. 1996).

Physiology of foliose, fruticose and crustose lichens are highly complex with differentiated layers producing different lichen secondary

Fig. 8.2 Lichen sp. *Parmotrema* growing on various substratum; tree bark, rock and on electric pole

metabolites (Honegger 2008). The lichen species *Usnea* is an epiphytic lichen which grows on the bark of trees and is attached by holdfast. As the sucrose is the transport sugar in higher plants and since most epiphytic lichens growing on trees have holdfast which tightly anchors to the tree bark, it seems reasonable to suggest that they might absorb sucrose from the tree bark (Fig. 8.3). However, such carbon transfer has yet to be demonstrated. In a study, Behera et al. (2006d) supplied the sucrose in the nutrient media and reported the relatively high growth rates of *U. ghattensis* cultures biomass and metabolite production. This suggests that the mycobiont might have a strong preference for sucrose consistent with the utilization of this sugar as a carbon source in nature. Lichenization seems to be important not only for the carbon source, but also for giving constant environmental (culture) conditions to the symbionts (Stocker-Wörgötter 2001). The experimental observation suggests that the production of lichen metabolites by the symbionts in culture may be due to the combined effects of the high osmotic pressure of the medium as well as the nutritional conditions.

8.4 Recent Advances in the Development of Lichen Culture, Bioactivity and Mass Production of Lichen Metabolites

After successful culture of lichen symbionts or isolated mycobionts (Yamamoto et al. 1985) Higuchi et al. (1993) screened 46 cultured lichen species for tyrosinase inhibitory activity and found strong activity in cultured *Hypogymnia physodes*, *Letharia vulpina*, and *Cetraria juniperina*. Yamamoto et al. (1993) reported antibacterial, antifungal, superoxide scavenging, and tyrosinase inhibitory activity in cultured lichens of family *Cladoniaceae*, *Graphidaceae*, *Parmeliaceae*, *Umbilicariaceae*, and *Usneaceae*. Kinoshita et al. (2006) reported monoamine oxidase inhibition activity in cultured mycobiont of 26 lichen species.

The foliose lichen *Bulbothrix setschwanensis* has been successfully cultured in vitro with producing the secondary metabolites (Behera et al. 2000). These authors also studied the effect of various culture conditions especially nutritional

Fig. 8.3 Lichen species growing on tree bark

modification for the growth and the production of salazinic acid in vitro, and they demonstrated the potential of biological activities like inhibition of tyrosinase and xanthine oxidase in culture and natural thallus of *B. setschwanensis* (Behera and Makhija 2001, 2002).

In a further study extracts of dried herbarium specimens of as many as 77 species belonging to the lichen family, *Graphidaceae* have been screened for the inhibition of tyrosinase, xanthine oxidase, nitro blue tetrazolium (NBT), and scavenging of superoxide activities. Several species were shown to have potential for these activities (Behera et al. 2003, 2004, 2006a). Three species of the crustose lichen genus *Graphis*, *G. guimarana*, *G. nakanishiana*, and *G. schizograpta* have been cultured in modified Bold's Basal medium, it was observed to produce the secondary metabolites norstictic acid, constictic acid, and stictic acid in 90 days as they produce in nature. The methanolic extract of these cultured lichens were found to inhibit tyrosinase, xanthin oxidase, and scavenge superoxide anions (Behera et al. 2006b). In continuation of lichen tissue culture work, further four different types of lichen species *U. ghattensis* (fruticose), *Arthothelium awasthii* (crustose), *Heterodermia podocarpa* (foliose), and *Parmotrema tinctorum* (foliose) were cultured in vitro (Table 8.1). Various culture media viz. malt–yeast extract, Bischoff and Bold, Lilly Barnet, Murashige and Skoog, Bold's Basal and Modified Bold's Basal media were tried for the culturing of lichen species (Fig. 8.4).

In case of lichen *U. ghattensis*, malt–yeast extract medium supported the growth of symbionts along with the production of lichen secondary compound usnic acid and norstictic acid. Modified Bold's basal media supported the growth of *A. awasthii*, *H. podocarpa*, and *P. tinctorum* symbionts along with the production of lichen secondary compounds atranorin, barbatic acid, lecanoric acid, and zeorin (Verma 2011). Some lichenologists (Hamada 1993; Hamada et al. 1996) reported that culture media supplemented with excess carbon or nitrogen sources, play an important role for the enhancement of the growth of mycobiont alone or the symbionts and the production of lichen substances. Keeping the view in mind, experiments were conducted by adding excess carbon sources [glucose, sucrose, polyethylglycol, in concentrations of 2, 4, 8, 16 or 32 % (w/v)] and nitrogen sources [amino acids; L- or D-asparagine, glutamine, alanine, glycine, in concentrations of 0.02, 0.04, 0.06, 0.08 or 0.1 % (w/v), vitamins; thiamin (B_1), riboflavin (B_2), ascorbic acid (C), biotin (H) in concentrations of 1, 10, 100 or 1,000 ppb] individually/in combination in the MYE and MBB media. The pH of the medium used was adjusted to 6.5 with 1 N sodium hydroxide (NaOH). Then, the inoculated petriplates were kept under 18 °C in light (400 lux) with daily cycle (8 h light/16 h dark) for 2 months.

The additional supplementation of carbon and nitrogen sources in the MYE and MBB media showed linear increment in growth rate of symbionts and the production of lichen metabolites

Table 8.1 Lichen species cultured and their biological activities

Lichen species cultured	Secondary metabolites	Biological activity studied	References
Bulbothrix setschwanesis	Atranorin Salazinic acid	Inhibition of tyrosinase activity Inhibition of xanthine oxidase activity	Behera and Makhija (2002)
Graphis guimarana	Norstictic acid	Scavenging of super oxide radical Inhibition of tyrosinase activity	Behera et al. (2006b)
Graphis nakanishiana	Constictic acid Stictic acid Norstictic acid	Inhibition of xanthine oxidase activity	
Graphis schizograpta	Not determined		
Usnea ghattensis	Usnic acid Norstictic acid	Antioxidant, Antityrosinase Antibacterial, hepatoprotective activity	Verma et al. (2008a, c) Behera et al. (2005b)
Arthothelium awasthii	Barbatic acid	Antioxidant activity Inhibition of lipid peroxidation	Verma et al. (2008b)
Heterodermia podocarpa	Atranorin Zeorin	Electron donating ability	
Parmotrema tinctorum	Atranorin Lecanoric acid	Inhibition of tyrosinase activity	
Usnea complanata	Usnic acid Psoromic acid	Cardiovascular protective activity Antioxidative potential	Behera et al. (2012)
Ramalina nervulosa	Sekikaic acid Usnic acid	Glucosidase inhibitory activity Radical scavenging potential	Verma et al. (2012a, b)
Ramalina pacifica	Salazinic acid Usnic acid		

Further readings for International scenario refer Stocker-Wörgötter (2008), Molnár and Farkas (2010)

during the growth period. In the case of lichen *U. ghattensis*, the optimum symbiont growth (6.0 g dry biomass) was found with an amount of lichen substance usnic acid (4.5 μg/g dry biomass) and norstictic acid (1.8 μg/g dry biomass) in MYE medium supplemented with 2 % sucrose and 0.06 % glycine. In *A. awasthii*, MBB medium supplemented with 4 % glucose and 1 ppb thiamin produced 5.5 g dry biomass with 3.1 μg barbatic acid/g dry biomass. Again, modified Bold's basal medium supplemented with 4 % sucrose, yielded 3.7 g dry biomass with 3.2 μg zeorin and 1.4 μg atranorin/g dry biomass in *H. podocarpa* and 3.1 g dry biomass with 2.1 μg atranorin and 2.4 μg lecanoric acid/g dry biomass in *P. tinctorum*. There was no significant increase in growth of biomass in these species was found even in addition of excess glucose or PEG or nitrogen source alone or in combination (Verma 2011). The above cultured lichen species (*U. ghattensis*, *A. awasthii*, *H. podocarpa*, *P. tinctorum*) and their secondary metabolites usnic acid, norstictic acid, atranorin, zeorin, lecanoric acid, and barbatic acid were also studied for the antioxidant, antimicrobial and antityrosinase properties and found to be as a potential candidates for biopharmaceutical research (Behera et al. 2005a, b, 2006c, d, 2009a; Verma et al. 2008a, b).

The extract of cultured lichen *U. ghattensis* showed high antioxidative potential in terms of scavenging of 1,1-diphenyl-2-picrylhydrazyl (DPPH), superoxide, nitric oxide, and hydroxyl radicals along with lipid peroxidative inhibitory activities in in vitro assay system. Further, the authors have examined the hepatoprotective effects of the same lichen extract. As the quantity of lichen extracts was the limiting factor the ethanol-induced toxicity liver slice culture model was selected (Invittox Protocol No. 42 1992). Liver slice culture is an in vitro technique that offers the advantages of an in vivo situation and hence is a

Fig. 8.4 Cultured lichen sp. **a** *Usnea ghattensis*, **b** *Parmotrema tinctorum*, **c** *Heterodermia podocarpa* and **d** *Ramalina pacifica*

more suitable model for the analysis of hepatotoxic events at microlevel experimentations. The cultured lichen extract showed hepatoprotection against ethanol-induced toxicity in the mice liver slice culture by a significant decrease in the antioxidant enzymes, glutathione peroxidase, catalase, and superoxide dismutase, along with a decrease in lipid peroxidation and lactate dehydrogenase release (Verma et al. 2008c).

In general, lichen tissue cultures grow much faster than natural thalli, but very slow than other microorganisms. Growth rates must be improved, if lichen tissue cultures are to be used for industrial production of important bioactive lichen substances (Yamamoto et al. 1993). With this view, some workers have studied to know, why the lichen mycobiont grow slowly even in enriched medium. Therefore, the study of isolated cultured fungi *C. vulcani* cytochrome profile was carried out by Fujiwara et al. (1995) and reported that the cytochrome level is lower than the fast-growing *Candida rugosa*. Cytochromes are membrane bound hemoproteins and as a fundamental component in the cell, it is responsible for the generation of ATP via the electron transport system (Yamanaka and Okunuki 1964). In a study *U. ghattensis* symbionts grew and produced lichen substances much faster (Behera et al. 2009a) and showed significantly high concentration of cytochromes a, b, and c type (Verma et al. 2011). The significant enhancement was achieved with malt–yeast extract medium supplemented with additional carbon sources 0.01 mol/l sucrose and polyethylglycol. Probably providing additional carbon source might activated the cytochrome respiratory system. Another molecular approach was also adopted as fusion of protoplast of lichen mycobiont with the protoplast of fast-growing fungi *A. nidulans* and the fusant regeneration to improve the biomass growth rate with synthesis of natural lichen substance. The result showed successful fusant regeneration with the production of lichen secondary metabolite usnic acid (Behera et al. 2009b).

However, major effort has been made to improve the growth and the metabolite production at flask level, but they are still not sufficient for commercial exploitation of lichen secondary metabolites. For industrial scale production of compounds such as salazinic acid, sekikaic acid, and usnic acid in bioreactor, culture biomass of *Ramalina nervulosa* and *Ramalina pacifica* produced biomass 10–18 g with the production of lichen secondary metabolites in higher quantity after $4^{1/2}$ day (Verma et al. 2012a). The lichen metabolites such as salazinic acid, sekikaic acid, and usnic acid detected by the lichen species *R. nervulosa* and *R. pacifica* had glucosidase inhibitory and radical scavenging properties (Verma et al. 2012b).

Although advances in experimental lichenology has been increased but still far behind due to the difficulties in culturing of lichens in vitro. Since lichen is a nutritionally specialized phenotypic organism, and their interrelationship is not properly understood. Hence, studies must have to carry out for the nutritional signaling between the bionts for their growth and triggering the production of secondary metabolites. Based on that technique, various lichens and their isolated mycobiont/photobiont were cultured in laboratory and their culture conditions were also studied for the production of lichen secondary metabolites. The growth of the cultured tissue of lichens under laboratory conditions is much faster than the growth of those lichens in natural habitat. If lichen tissue cultures are to have industrial uses for their wide range of potentially useful activities, we must have to improve the growth rate and in vitro synthesis of their natural secondary metabolites. In most of the countries including India, the lichen diversity and abundance considerably decreased due to change in ecological conditions, loss of forest cover and habitat, and growth of urban and industrial areas (Upreti et al. 2005). Therefore, it is an urgent need to improvise the lichen tissue culture technique for better understanding of their biotechnological/biomedical applications. Further, the use of bioreactor techniques complements the conventional culture method for mass production of lichen secondary metabolites under in vitro conditions within less span of time. This could be helpful in obtaining good amounts of cultured biomass and may also considerably enlarge the access to lichen metabolites towards possible applications in pharmaceutical/nutritional supplements.

8.5 Conclusion and Future Perspective

As of now lichen flora represents about 13,500 species worldwide (Rai et al. 2014). Most of the lichens from tropical, temperate, and alpine zones are poorly bioprospected for their potential bioactive compounds and genes due to remarkably little information is available about the critical stages in the life cycle of lichens. This is partly caused by the complex symbiotic partnership between fungi and photobionts and also because of the inherent difficulties in growing lichens in vitro (Crittenden and Porter 1991; Behera et al. 2006d). Mating system, generation times, and even less is known about the genetic variation within and among lichen populations. Although, the recent advances in molecular genetics have sparked the interest. But it is prerequisite to develop suitable methodology to culture lichen thallus and biosynthesis its secondary compounds in desired quantities for bioprospecting screening and downstream processing towards possible drug development. Lichen cultures are also important for molecular studies specifically to mine key genes involved in secondary compound synthesis, symbiosis, drought resistance, and many more in lichens. Cell aggregates derived from the natural thallus fragment culture were composed of algal and fungal cells, but morphological differentiation is lacking. It is not easy to identify these (algae and fungi) positively as the symbionts of the lichen. Final confirmations of the identity of the symbionts/mycobionts are possible when a true lichen thallus can be established in vitro (Yoshimura et al. 1993). Furthermore, the production of lichen substances under artificial culture conditions is similar to the natural thallus as they produce in nature.

Acknowledgments We are gratefully acknowledge the various financial supports by the Department of Biotechnology (DBT), Govt. of India, New Delhi [Grant no. BT/PR3133/BCE/08/237/2002; BT/PR8551/NDB/52/15/2006], Council of Scientific and Industrial Research (CSIR), Govt. of India, New Delhi [Grant no. 09/670(034)/2006-EMR-I; 09/670(0046)2010/EMR-I] and Science and Engineering Research Board (SERB), Govt. of India, New Delhi [Grant No. SR/FT/LS-170/2009]. We are also thankful to Director, Agharkar Research Institute, Pune for research facilities provided.

References

Ahmadjian V (1966) Lichens. In: Henry SM (ed) Symbiosis, vol I. Academic Press, New York, pp 36–97

Ahmadjian V (1993) The lichen symbiosis. Wiley, New York 250

Ahmadjian V, Heikkilä H (1970) The culture and synthesis of *Endocarpon pusillum* and *Staurothele clopima*. Lichenologist 4:256–267

Armaleo D, Zhang Y, Cheung S (2008) Light might regulate divergently depside and depsidone accumulation in the lichen *Parmotrema hypotropum* by affecting thallus temperature and water potential. Mycologia 100:565–576

Behera BC, Makhija U (2001) Effect of various culture conditions on growth and production of salazinic acid in *Bulbothrix setschwanensis* (lichenized ascomycetes) in vitro. Curr Sci 80:1424–1427

Behera BC, Makhija U (2002) Inhibition of tyrosinase and xanthine oxidase by lichen species *Bulbothrix setschwanesis*. Curr Sci 82:61–66

Behera BC, Makhija U, Adawadkar B (2000) Tissue culture of *Bulbothrix setschwanensis* (lichenized ascomycetes) in vitro. Curr Sci 78:781–783

Behera BC, Adawadkar B, Makhija U (2003) Inhibitory activity of xanthine oxidase and superoxide-scavenging activity in some taxa of the lichen family *Graphidaceae*. Phytomedicine 10:536–543

Behera BC, Adawadkar B, Makhija U (2004) Capacity of some *Graphidaceous* lichens to scavenge superoxide and inhibition of tyrosinase and xanthine oxidase activities. Curr Sci 87:83–87

Behera BC, Verma N, Sonone A, Makhija U (2005a) Evaluation of antioxidant potential of the cultured mycobiont of a lichen *Usnea ghattensis*. Phytother Res 19:58–64

Behera BC, Verma N, Sonone A, Makhija U (2005b) Antioxidant and antibacterial activities of lichen *Usnea ghattensis* in vitro. Biotechnol Lett 27:991–995

Behera BC, Adawadkar B, Makhija U (2006a) Tyrosinase-inhibitory activity in some species of the lichen family *Graphidaceae*. J Herbal Pharmacother 6:55–69

Behera BC, Adawadkar B, Makhija U (2006b) Tissue-culture of selected species of the *Graphis* lichen and their biological activities. Fitoterapia 77:208–215

Behera BC, Verma N, Sonone A, Makhija U (2006c) Determination of antioxidative potential of lichen *Usnea ghattensis* in vitro. LWT-Food Sci Technol 39:80–85

Behera BC, Verma N, Sonone A, Makhija U (2006d) Experimental studies on the growth and usnic acid production in lichen *Usnea ghattensis* in vitro. Microbiol Res 161:232–237

Behera BC, Verma N, Sonone A, Makhija U (2009a) Optimization of culture conditions for lichen *Usnea ghattensis* G. Awasthi to increase biomass and antioxidant metabolite production. Food Technol Biotechnol 47:7–12

Behera BC, Sonone A, Makhija U (2009b) Protoplast isolation from cultured lichen *Usnea ghattensis*, their fusion with protoplasts of *Aspergillus nidulans*, fusant regeneration and production of usnic acid. Folia Microbiol 54:415–418

Behera BC, Mahadik N, Morey M (2012) Antioxidative and cardiovascular-protective activities of metabolite usnic acid and psoromic acid produced by lichen species *Usnea complanata* under submerged fermentation. Pharm Biol 50:968–979

Bjerke JW, Lerfall K, Elvebakk A (2002) Effects of ultraviolet radiation and PAR on the content of usnic and divaricatic acids in two arctic-alpine lichens. Photochem Photobiol Sci 1:678–685

Bjerke JW, Joly D, Nilsen L, Brossard T (2004) Spatial trends in usnic acid concentrations of the lichen *Flavocetraria nivalis* along local climatic gradients in the Arctic (Kongsfjorden, Svalbard). Polar Biol 27:409–417

Boustie J, Grube M (2005) Lichens—a promising source of bioactive secondary metabolites. Plant Geneti Resour Charact Utiliz 3:273–283

Brodo IM (1973) Substrate ecology. In: Ahmadjian V, Hale ME (eds) The lichens. Academic Press, New York and London, pp 401–441

Brunauer G, Stocker-Wörgötter E (2005) Culture of lichen fungi for future production of biologically active compounds. Symbiosis 38:187–201

Brunauer G, Hager A, Grube M, Turk R, Stocker-Wörgötter E (2007) Alterations in secondary metabolism of aposymbiotically grown mycobionts of *Xanthoria elegans* and cultured resynthesis stages. Plant Physiol Biochem 45:146–151

Chooi YH, Stalker DM, Davis MA, Fujii I, Elix JA, Louwhoff SHJJ, Lawrie AC (2008) Cloning and sequence characterization of a non-reducing polyketide synthase gene from the lichen *Xanthoparmelia semiviridis*. Mycol Res 112:147–161

Christopher D, Brinda T, Piercey-Normore MD (2012) Effect of environmental change on secondary metabolite production in lichen-forming fungi. In: Stephen Y (ed) International perspectives on global environmental change. InTech, Rijeka, pp 197–230

Cohen PA, Towers GHN (1995) The anthraquinones of *Heterodermia obscurata*. Phytochemistry 40:911–915

Crittenden PD, Porter N (1991) Lichen forming fungi: potential source of novel metabolites. Trends Biotechnol 9:409–414

Culberson WL (1986) Chemistry and sibling speciation in the lichen-forming fungi: ecological and biological considerations. Bryologist 89:123–131

Culberson CF, Armaleo D (1992) Induction of a complete secondary-product pathway in a cultured lichen fungus. Exp Mycol 16:52–63

Culberson CF, Culberson WL, Johnson A (1988) Gene flow in lichens. Am J Bot 75:1135–1139

Elix JA (1996) Biochemistry and secondary metabolites. In: Nash TH III (ed) Lichen biology. Cambridge University Press, Cambridge, pp 154–180

Fahselt D (1981) Lichen products of *Cladonia stellaris* and *C. rangiferina* maintained under artificial conditions. Lichenologist 13:87–91

Fahselt D (1994) Secondary biochemistry of lichens. Symbiosis 16:117–165

Fujiwara T, Ohashi T, Hamade R, Kinoshita Y, Yamamoto Y, Fukumori Y, Yoshimura I, Yamanaka T (1995) Cytochromes in the cultured mycobiont of a lichen, *Cladonia vulcani* sav. Plant Cell Physiol 36:183–186

Hale ME (1973) Growth. In: Ahmadjian V, Hale ME (eds) The lichens. Academic Press, London, pp 473–492

Hale ME (1983) The biology of lichens. In: Willis AJ, Sleigh MA (eds) Edward Arnold, London, pp 108–132

Hamada N (1982) The effect of temperature on the content of the medullary depsidone salazinic acid in *Ramalina siliquosa* (lichens). Can J Bot 60:383–385

Hamada N (1989) The effect of various culture conditions on depside production by an isolated lichen mycobiont. Bryologist 92:310–313

Hamada N (1991) Environmental factors affecting the content of usnic acid in the lichen mycobiont of *Ramalina siliquosa*. Bryologist 94:57–59

Hamada N (1993) Effects of osmotic culture conditions on cultured mycobionts. Bryologist 96:569–572

Hamada N, Miyagawa H, Miyagawa H, Inoue M (1996) Lichen substances in mycobionts of crustose lichens cultured on media with extra sucrose. Bryologist 99:71–74

Higuchi M, Miura Y, Boohene J, Kinoshita Y, Yamamota Y, Yoshimura I, Yamada Y (1993) Inhibition of tyrosinase activity by cultured lichen tissues and bionts. Planta Med 59:253–255

Honegger R (2008) Mycobionts. In: Nash TH (ed) III: lichen biology, 2nd edn. Cambridge University Press, Cambridge, pp 27–39

Honegger R (2009) Lichen forming fungi and their photobionts. In: Deising H (ed) The Mycota V, plant relationships, 2nd edn. Springer, Berlin

Huneck S (1999) The significance of lichens and their metabolites. Naturwissenschaften 86:559–570

Invittox Protocol No. 42 (1992) Liver slice hepatotoxicity screening system. The ERGATT/FRAME Data Bank of in vitro techniques in toxicology. INVITTOX, England

Keller NP, Calvo A, Wilson R, Bok J (2002) Relationship between secondary metabolism and fungal development. Mol Microbiol Rev 66:447–459

Kinoshita Y, Hayase S, Yamamoto Y, Yoshimura I, Kurokawa T, Ahti T, Yamada Y (1993a) Morphogenetic capacity of the mycobiont in *Usnea* (lichenized ascomycete). Proc Jpn Acad 69:18–21

Kinoshita Y, Yamamoto Y, Yoshimura I, Kurokawa T, Yamada Y (1993b) Production of usnic acid in cultured *Usnea hirta*. Bibliotheca Lichenologica 53:137–146

Kinoshita Y, Yamamoto Y, Okuyama E, Yamazaki M (2006) Screening in natural thalli and cultured mycobionts of lichens for inhibitory activity of monoamine oxidase. Lichenology 5:1–6

McEvoy M, Nybakken L, Solhaug KA, Gauslaa Y (2006) UV triggers the synthesis of the widely distributed secondary lichen compound usnic acid. Mycol Progress 5:221–229

McEvoy M, Solhaug KA, Gauslaa Y (2007a) Solar radiation screening in usnic acid containing cortices of the lichen *Nephroma arcticum*. Symbiosis 43:143–150

McEvoy M, Gauslaa Y, Solhaug KA (2007b) Changes in pools of depsidones and melanins, and their function, during growth and acclimation under contrasting natural light in the lichen *Lobaria pulmonaria*. New Phytol 175:271–282

Miao V, Coffet-LeGal MF, Brown D, Sinnemann S, Donaldson G, Davies J (2001) Genetic approaches to harvesting lichen products. Trends Biotech 19:349–355

Molnár K, Farkas E (2010) Current results on biological activities of lichen secondary metabolites: a review. Z Naturforsch C 65:157–173

Moore D (1998) Fungal morphogenesis. Cambridge University Press, Cambridge, pp 1–469

Müller K (2001) Pharmaceutically relevant metabolites from lichens. Appl Microbiol Biotechnol 56:9–16

Oksanen I (2006) Ecological and biotechnological aspects of lichens. Appl Microbiol Biotechnol 73:723–734

Paudel B, Bhattarai HD, Koh HY, Lee SG, Han SJ, Lee HK, Oh H, Shin HW, Yim JH (2011) Ramalin, a novel nontoxic antioxidant compound from the Antarctic lichen *Ramalina terebrata*. Phytomedicine 18:1285–1290

Rai H, Khare R, Upreti DK (2014) Lichenological studies in India with reference to Terricolous lichens. In: Rai H, Upreti DK (eds) Terricolous Lichens in India. Diversity patterns and distribution ecology, vol I. Springer Science+Business Media, New York, pp 1–20

Richardson DHS (1991) Lichens and man. In: Hawksworth DL (ed) Frontiers in mycology. CABI, Oxford, pp 187–210

Rundel PW (1969) Clinal variation in the production of usnic acid in *Cladonia subtenuis* along light gradients. Bryologist 72:40–44

Shukla V, Joshi GP, Rawat MSM (2010) Lichens as a potential natural source of bioactive compounds: a review. Phytochem Rev 9:303–314

Stocker-Wörgötter E (2001) Experimental studies of the lichen symbiosis: DNA-analyses, differentiation and secondary chemistry of selected mycobionts, artificial resynthesis of two- and tripartite symbioses. Symbiosis 30:207–227

Stocker-Wörgötter E (2005) Approaches to a biotechnology of lichen forming fungi: induction of polyketide pathways and the formation of chemosyndromes in axenically cultured mycobionts. Rec Res Dev Phytochem 9:115–131

Stocker-Wörgötter E (2008) Metabolic diversity of lichen-forming ascomycetous fungi: culturing, polyketide and shikimate metabolite production, and PKS genes. Nat Prod Rep 25:188–200

Stocker-Wörgötter E, Turk R (1988) Culture of the cyanobacterial lichen *Peltigera didactyla* from soredia under laboratory conditions. Lichenologist 20: 369–376

Upreti DK, Chatterjee S (2007) Significance of lichens and their secondary metabolites: a review. In: Ganguli BN, Deshmukh SK (eds) Fungi multifaceted microbes. Anamaya Publishers, New Delhi, pp 169–188

Upreti DK, Divakar PK, Nayaka S (2005) Commercial and ethnic use of lichens in India. Econ Bot 59: 269–273

Verma N (2011) Studies on antioxidant activities of some lichen metabolites developed in vitro. Shodh Ganga, Indian ETD Repository, Issue 2, Sept 2011

Verma N, Behera BC, Sonone A, Makhija U (2008a) Cell aggregates derived from natural lichen thallus fragments: antioxidant activities of lichen metabolites developed in vitro. Nat Prod Commun 3:1911–1918

Verma N, Behera BC, Sonone A, Makhija U (2008b) Lipid peroxidation and tyrosinase inhibition by lichen symbionts grown in vitro. Afr J Biochem Res 2:225–231

Verma N, Behera BC, Makhija U (2008c) Antioxidant and hepatoprotective activity of a lichen *Usnea ghattensis* in vitro. Appl Biochem Biotechnol 151:167–181

Verma N, Behera BC, Makhija U (2011) Studies on cytochromes of lichenized fungi under optimized culture conditions. Mycoscience 52:65–68

Verma N, Behera BC, Joshi A (2012a) Studies on nutritional requirement for the culture of lichen *Ramalina nervulosa* and *Ramalina pacifica* to enhance the production of antioxidant metabolites. Folia Microbiol 57:107–114

Verma N, Behera BC, Sharma BO (2012b) Glucosidase inhibitory and radical scavenging properties of lichen metabolites salazinic acid, sekikaic acid and usnic acid. Hacet J Biol Chem 40:7–21

Yamamoto Y, Mizuguchi R, Yamada Y (1985) Tissue culture of *Usnea rubescens* and *Ramalina yasudae* and production of usnic acid in their cultures. Agric Biol Chem 49:3347–3348

Yamamoto Y, Mizuguchi R, Takayama S, Yamada Y (1987) Effects of culture conditions on the growth of *Usneaceae* lichen tissue cultures. Plant Cell Physiol 28:1421–1426

Yamamoto Y, Miura Y, Higuchi M, Kinoshita Y (1993) Using lichen tissue culture in modern biology. Bryologist 96:384–393

Yamamoto Y, Kinoshita Y, Matsubara H, Kinoshita K, Koyama K, Takanashi K, Kurokawa T, Yoshimura I (1998) Screening of biological activities and isolation of biological active compounds from lichens. Rec Res Dev Phytochem 2:23–34

Yamanaka T, Okunuki K (1964) Comparative biochemistry of cytochrome C. J Biol Chem 239:1813–1817

Yoshimura I, Kurokawa T, Nakano T, Yamamoto Y (1987) A preliminary report of cultures of *Cladonia vulcani* and the effects of the hydrogen ion concentration on them. Bull Kochi Gakuen Coll 18:335–345

Yoshimura I, Kurokawa T, Nakano T, Yamamoto Y (1989) Production of secondary metabolic substances by cultured tissue of *Usnea flexilis*. Bull Kochi Gakuen Coll 20:535–540

Yoshimura I, Kurokawa T, Kanda H (1990a) Tissue culture of some Antarctic lichens preserved in the refrigerator. In: Proceedings of NIPR symposium on polar biology, vol 3, pp 224–228

Yoshimura I, Kurokawa T, Yamamoto Y, Kinosita Y (1990b) Thallus formation of *Usnea rubescens* and *Peltigera praetextata* in vitro. Bull Kochi Gakuen Coll 21:565–576

Yoshimura I, Kurokawa T, Yamamoto Y, Kinoshita Y (1993) Development of lichen thalli in vitro. Bryologist 96:412–421

Biochemical Diversity and Ecology of Lichen-Forming Fungi: Lichen Substances, Chemosyndromic Variation and Origin of Polyketide-Type Metabolites (Biosynthetic Pathways)

9

Elfie Stocker-Wörgötter

Contents

Abstract

Many organisms containing high-value compounds are difficult to culture or are becoming endangered or even extinct by over-harvesting. Lichens, in general, are slow-growing organisms and the extraction of the naturally grown, composite thalli, in many cases is economically not feasible and profitable which may also be very limited. Mycobiont cultures are an attractive alternative to the extraction of naturally grown thalli. In Europe, the laboratory at the University of Salzburg has established a worldwide recognized unique culture collection of c. 150 different mycobionts. The modulation ("regulative manipulation") of growth conditions of microorganism and fungi is a common strategy used in biotechnology and applied microbiology to improve yields and diversity of secondary metabolites of therapeutic value. Interest in polyketide-type metabolites is considerable, as many of these natural products are of medical, and industrial, and/or agricultural importance and involve polyketide synthase

E. Stocker-Wörgötter (✉)
Department of Organismic Biology, University of Salzburg, Hellbrunner Str., 34, 5020 Salzburg, Austria
e-mail: Elfriede.stocker@sbg.ac.at

D.K. Upreti et al. (eds.), *Recent Advances in Lichenology*,
DOI 10.1007/978-81-322-2235-4_9

(PKS) pathway. In case studies, growth and culture conditions have been modulated and optimized culture conditions have been adopted to obtain increased biomass production for several selected mycobionts (e.g., *Roccella decipiens*, species of the genus *Xanthoparmelia*) by adopting particular environmental conditions in one of the culture chambers. In recent investigations, by exploring further possibilities to optimize culture conditions and biomass production, it turned out that axenically cultured mycobionts can be triggered to produce single or a whole pattern of secondary metabolites. Polyketides, as has been demonstrated, are only biosynthesized under "permissive" ecological conditions. By using the knowledge from preliminary investigations and doing further extensive test series, it was possible to achieve the production of one particular polyketide and even the production of a predictable pattern of polyketides, depending upon the investigated lichen chemotypes. Such studies could also help to elucidate the often observed variation in secondary products (chemosyndromic variation) within a lichen population growing under heterogeneous environmental conditions. Variations in chemistry actually mirror physiological, ecological, and even evolutionary responses to change in the environment and climate. The repeatable and in vitro production of higher quantities of lichen metabolites in fungal cell cultures has already and could further become a milestone elucidating the architecture and function of PKS genes that are involved in polyketide and "still unknown" genes that control other metabolic pathways, e.g., shikimate and pulvinic acid compound production. In a novel and holistic approach, functional genomics could be used to understand the molecular mechanisms that are involved in the desiccation tolerance of some lichens. By constructing a cDNA library from selected species of *Xanthoparmelia* and its cultured mycobiont, it could be of high interest to perform transcriptome sampling which could be used to detect and identify further "new genes" responsible for the control of desiccation resistance in lichens.

Keywords
Lichens • Mycobionts • Secondary chemistry • Culture • Growth modulation • PKS • Desiccation tolerance

9.1 Introduction

Lichens are composite, symbiotic organisms. Referring to a modern understanding of the lichen symbiosis, e.g., the "novel" definition of the lichen apart from a two- or three partnership would notice it as a more complex multiple partnership, (fungus, +algae, +cyanobacteria, +associated fungi and bacteria). For this reason, important topics in many recent papers (Joneson et al. 2011; Grube and Berg 2009; Cardinale et al. 2006; Hodkinson and Lutzoni 2009; Wornik and Grube 2010) deal with questions of the lichen symbiosis trying to specify the roles of the lichen's fungus major partners (algae and cyanobacteria) and associated organisms and their interactions within the lichen thallus or interpreting it even as "controlled parasitism" following the suggestion of Ahmadjian (1993).

The correct interpretation of a lichen has again (and before … since the early days of lichenology) and is de novo fomenting many discussions on workshops and congresses, but new ideas and multifocal views of scientific playgrounds lead nowadays to completely new ways of experimentation using modern technologies and research hypotheses by using more and more sophisticated molecular techniques and most recently even genomic analyses.

Still, the molecular interaction between the symbiotic partners is not resolved.

Many modern findings (Grube and Muggia 2013) have contributed to noticing the "lichen lifestyle" and the lichen's complex thallus morphologies as an amazing success in the history of living organisms. Until recently, about 18,500 lichen taxa have been described and further species are on the way to be newly discovered documenting the diversification of the "lichen symbiosis" in only few major fungal lineages. The lichen's life style, during an evolution of 400 million years (Honegger et al. 2013), has been adopted by three classes of ascomycetous fungi (Sordariomycetes, Lecanoromycetes, and Eurotiomycetes); moreover, about estimated 20–30 species of basidiomycetous lichens are known and further will be probably discovered.

Taking into consideration the morphogenetic capacity of lichen fungi to form a thallus together with photo/cyanobionts, lichens could be best interpreted as novel phenotypes that have developed by convergent evolution (Grube and Hawksworth 2007). As an evolutionary innovation, lichen thalli can be recognized as structures in which photobionts (algae, cyanobacteria), mycobionts, and further lichen inhabiting bacteria and fungi are kept in a more or less layered body in which the metabolisms of different organisms can interact in a concerted manner. As a dynamic structure, the thalli and their morphologies being highly variable, having adopted more or less complex growth forms during evolution, have proven to be resistant even to extreme environmental influences. Their metabolic capacities have become outstandingly adapted to allow the survival of the joint organism under a wide range of climatic and ecological conditions.

Lichens, as a two or even multiple partnership, are able to respond more sensibly to the various environmental signals in complex ecosystems than algae and fungi growing without protection and living aposymbiotically.

In general, understanding the adaptive changes taking place in a symbiotic organism in response to variations in the environment is a key issue of lichen biology and contemporary biology. The fact that metabolisms display structural and metabolic plasticity, turning on and off some reactions as the growth conditions are altered, has been observed in all living organisms and particularly also in lichen-forming fungi (Leuckert et al. 1990; Molina et al. 2003; Stocker-Wörgötter et al. 2009; Stocker-Wörgötter and Elix 2009). Genes and gene clusters that control different metabolic pathways in lichen fungi and algae have only been started to be studied (Schmitt et al. 2005; Chooi et al. 2008; Brunauer et al. 2009; Stocker-Wörgötter 2008).

9.2 Ecology of Lichens

Most lichens grow in temperate and arctic climate zones, though there are many lichen taxa distributed in moist, tropical, and dry desert habitats. In particular, tundra is abundantly covered with lichens, mosses, and liverworts. This cover helps to insulate the soil and may provide forage for grazing animals such as reindeer and caribou. The so-called reindeer moss is a lichen (*Cetraria islandica*) and most common in arctic tundra of Europe and North America.

Lichens are pioneers on bare rock, in semiarid and arid regions on desert sand, on soil cleared from vegetation, dead wood, living bark (epiphytic lichens on trees and shrubs), animal bones, man-made surfaces such as rusty metal, cement, and even on weathered glass and medieval glass windows of ancient cathedrals. Able to shift from the metabolically active to the inactive state during periods of unfavorable conditions, lichens are able survive extremes of heat, cold, and drought.

Lichens have been classified as poikilohydric and desiccation tolerant organisms which cannot actively regulate their water contents like higher plants. Lichens have no comparable water transport systems such as plant stems and roots, and they lack cuticles which would protect them from drying out. When desiccated, their water status is frequently in the range of 10–20 % of their dry weight (Rundel 1988). During rainfall, morning and evening dew, fog or high humidity of the air, lichens become hydrated. Lichens with green photobionts can absorb water to 200–300 % of their dry weight at saturation, but some lichens hosting cyanobacteria as bionts can absorb water as much as 2,000 % of their dry weight (Kranner et al. 2008). Some lichens can survive drying out to water contents of 5 % or less, but most lichens can endure low water contents for months at low relative humidity; for this reason, they are perfectly adapted to semiarid and arid regions. Lichens suffer from oxidative stress during desiccation, in the desiccated state and also during subsequent rehydration, for this reason they have discovered a variety of physical and physiological/biochemical mechanisms that facilitate survival in the desiccated state, and a quick management to come back to normal physiological activity upon rehydration (Kranner et al. 2008; Beckett 1995).

Beside desiccation tolerance, the vast majority of lichens can withstand daily, seasonal and annual temperature fluctuations, excess light, and high levels of UV radiation. In overall, lichens are able to withstand a wide array of abiotic and biotic stresses, which make them fit to survive under harsh and extreme environmental conditions such as Arctic, Antarctic, wet tropics, and also in Desert ecosystems.

Another strategy to study desiccation tolerance are ultra-structural investigations. Honegger (2006) demonstrated that water repellence of fungal and also algal cell walls within the lichen thalli is created both by specific proteins (hydrophobins localized in a rodlet layer) and crystals of secondary metabolites covering the outer surface of the hyphae and occasionally even algal cells.

9.3 Lichen Substances

Lichens are known to produce a high number of unusual secondary metabolites which have not been discovered in other plants. The uniqueness of many lichen substances attracted the attention of early chemists in the middle of the nineteenth century; however, man had known lichens as a source of chemical products for dyeing textiles and as additives for soap and perfume manufacture, and a considerable number of lichens were used to help to cure diseases, to a certain extent to antedate the discoveries and evidence in the twentieth century that many lichen substances are actually biologically active and could play an important role as valuable natural products for modern medicine (Huneck 1999).

The majority of lichen substances are small aromatic polyketides biosynthesized by the lichen fungus during its mutualistic relationship (symbiosis) with green photobionts. Cyanobacterial lichens frequently lack the typical lichen polyketides, but do form a variety of other interesting secondary metabolites, e.g., terpenes and terpenoids.

The core chemistry of the acetyl-polymalonyl-derived family (polyketides) is relatively uniform across all known lichen species and mainly consists of two phenolic rings with various substituents, joined by ester and/or ether linkages. The most common phenolic acid units derived from the acetate-polymalonate pathway (polyketide pathway) and combined to form the typical lichen substances represent two types: (a) orcinol-type units such as orsellinic acid and related mononuclear units and (b) ß-orcinol-type units, which have a C1 substituent at the 3-position of the aromatic ring. Considering the biosynthesis of orcinol and ß-orcinol units, two types of polyketide synthases (PKSs) can be expected to be involved in their formation.

In the simplest case, the carboxylic acid of one ring is joined with the OH group in para-position of the carboxylic acid of the second ring. Such esterifications are typical for the para-depsides. If a second esterification takes place, tridepsides are formed. If the ester linkage of the first ring joins the meta-position of the carboxylic acid at the second ring, meta-depsides are the resulting chemical structures. The orcinol-type compounds, in general, form a closely related series of metabolites in which variations in the length and oxidative state of 6-alkyl substituents are of major importance. Additionally, the orcinol compounds show modifications like O-methylations, chlorination, decarboxylation, and also lactonization. The ß-orcinol compounds, however, mainly show variation in the oxidative state of the C1 substituent at the 3- and 6-positions of the phenolic units (e.g., dCH_3, CH_2OH, CHO, and COOH). A large number of lichen substances contain chlorine (e.g., atranorin and chloroatranorin); chorine substituents are found attached to aromatic rings of depsides, depsidones, and lichen pigments such as xanthones, anthraquinones, and naphthoquinones. The latter, pigments and a number of fatty acids (aliphatic compounds) which are not peculiar to lichens do occur. Furthermore, a few large-ring compounds have been found and chemically identified. Macrolide (large ring, macrocyclic lactones) compounds representing aliphatic polyketides are often formed by nine units (one unit of acetyl-CoA and eight units of malonyl-CoA). The essential steps, after the polyketide chain has been formed, are reductions, followed by aromatization and lactonization.

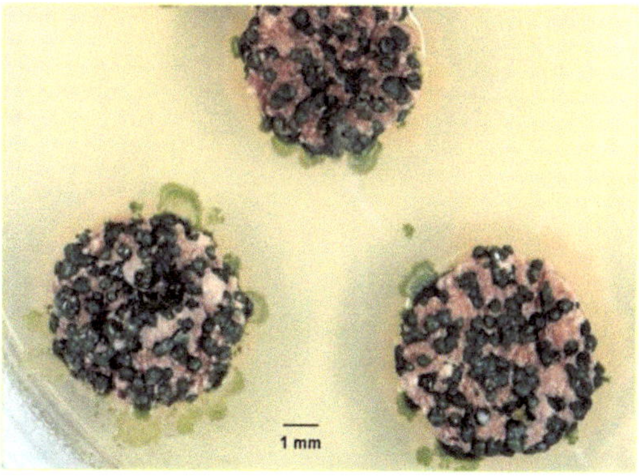

Fig. 9.1 The combination of mycobiont and photobionts in culture does not necessarily affect secondary compound production, under standard culture conditions (stable temperature of 20 °C, simulating 14/10 h day night regimes in the culture chamber)

9.3.1 Culturing of Lichen Fungi and Expression of Lichen Metabolites in Axenically Grown Mycobionts

Cultures of aposymbiotically grown lichen fungi could be triggered by change of environmental parameters (low temperature treatments, desiccation) to produce repeatedly lichen secondary metabolites (Culberson and Armaleo 1992; Hager et al. 2007; Stocker-Wörgötter 2002a, b; Stocker-Wörgötter and Elix 2006; Zocher and Stocker-Wörgötter 2005), demonstrating that lichen substances can be produced in absence of the symbiotic algal partner (Fig. 9.1).

Lichen fungi, cooperating with green algae and farming algae like a farmer growing food crop (Piercey-Normore and Deduke 2011), imply that the mycobiont itself takes algae into monoculture within a particular layer within the lichen thallus; additionally probably also further partners such as cyanobacteria and bacteria able to get further nitrogenous compounds for biomass production and survival in nutrient-deficient habitats (Fig. 9.2).

After spore germination, lichen fungi start their development like other fungi forming interconnected mycelial networks (Fig. 9.3) that scavenge for nutrient resources on favorable substrata under heterogeneous environmental conditions. The structure and architecture of the hyphal network underlies developmental transitions, which means that it is variable adapting to local nutritional cues through growth, fusion, or regression. Hyphae scanning the environment/substrate for food could have adopted algae/cyanobacteria producing sugars and sugar alcohols to complement their nutrient requirements in a habitat with scarce resources.

Regarding long-term experience with mycobiont cultures, secondary metabolites form only by mycobionts grown under unfavorable culture conditions; stress through change in the properties of substrata (low contents of sugars or sugar alcohols), changes in moisture contents (phases of desiccation over several weeks), and temperature changes (such as treatments with cold and warm temperature regimes, exposure to variable light regimes, depending upon the adaptation of the lichen in the natural environments). Such experiments can only be done in culture chambers with electronically adjustable environmental and climatic conditions.

If lichens and especially mycobionts could be cultured in higher quantities producing interesting biologically active lichen substances, culturing in large scale could open a door to the mass production of valuable lichen substances and their pharmaceutical and technical applications.

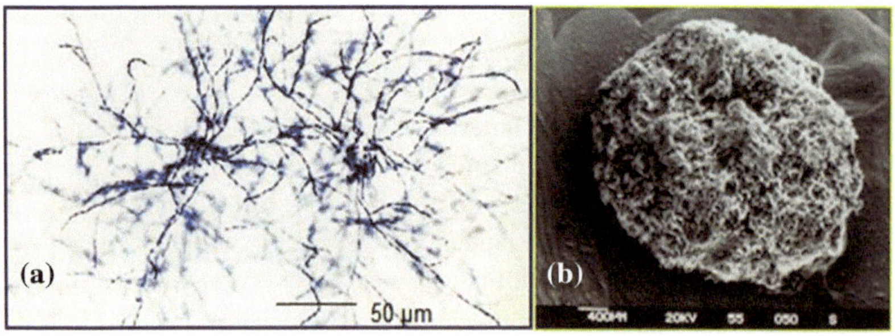

Fig. 9.2 **a** Germinated spores of *Teloschistes exilis* forming hyphal networks for scanning the agar substrate for nutrients. **b** More compact mycelium of after 370 days in culture showing strong development and increase in height (typical 3 dimensional structure of mycobionts)

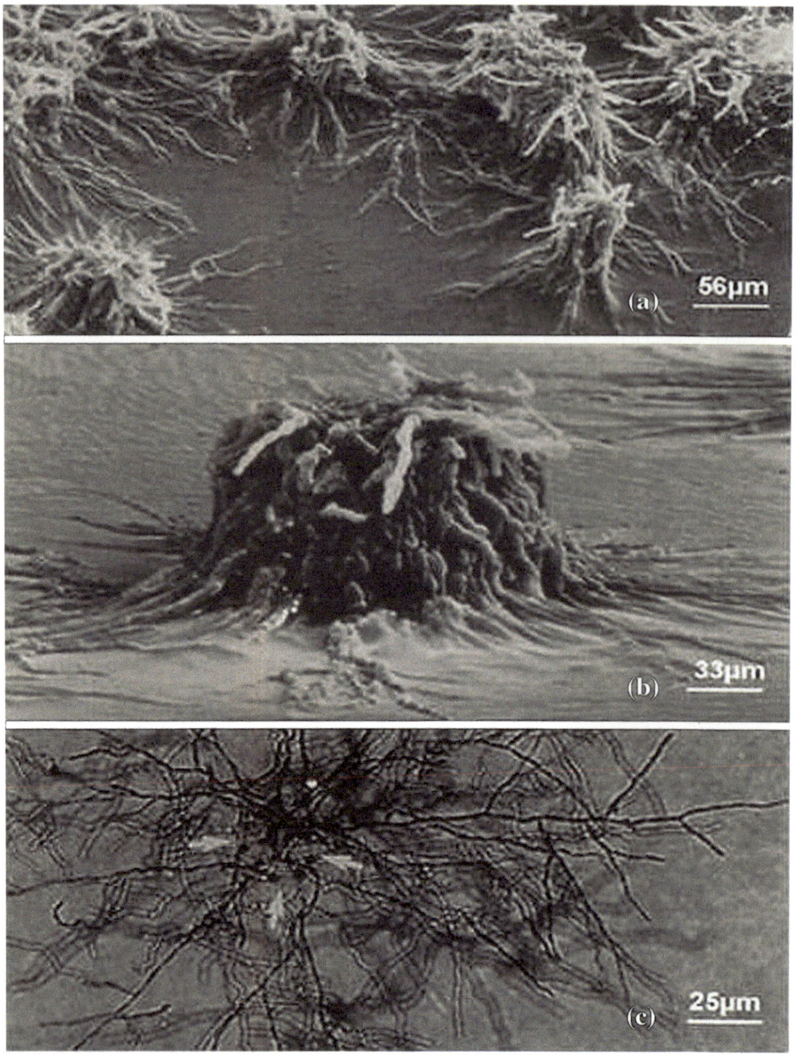

Fig. 9.3 **a** Three-dimensional growth of mycobiont as exemplified by developing mycelia of *Xanthoria parietina*, SEM. **b** Formation of aerial hyphae by a mycobiont during a period of desiccation, SEM. **c** Mycelium, 2 months in culture, investigated by the light microscope. The morphogenetic capacity of some cultured mycobionts and differentiation processes can influence the induction of diverse lichen metabolite pathways

It has to be taken into consideration that lichen fungi grow relatively slowly compared with other metabolite-producing organisms; another challenge is that lichen produce secondary metabolites only when the mycelia have developed beyond the relatively "flat" structure searching for nutrients on the substrate into a three-dimensional structure, where cell differentiation seems to be essential to initiate secondary metabolite production (Stocker-Wörgötter and Elix 2006) in many cases (even higher plants do produce secondary metabolites only in the more differentiated parts of the plant body); details of the biochemistry of differentiation and morphogenetic processes and the effects of morphogenetic stimuli on initiation of biochemical metabolic pathways are not known (at least not for lichens) and would be an exciting topic to investigate.

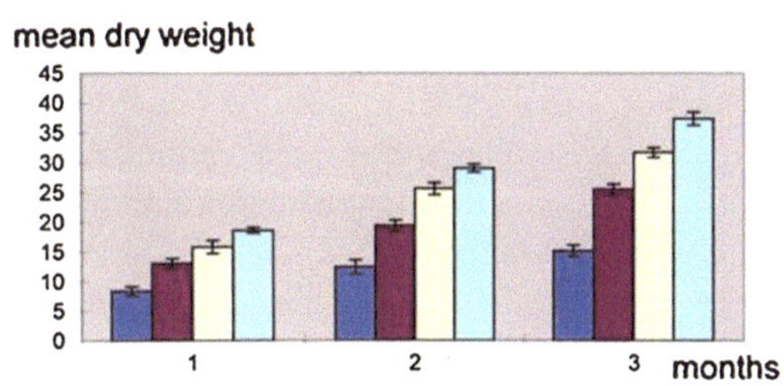

Fig. 9.4 Growth rate of the cultured *Roccella decipiens* fungus under standardized conditions (23 °C, 14: 10 h light dark cycle, 22 °C; Lilly and Barnett medium, +1–4 % mannitol): *blue* 1 % mannitol; *magenta* 2 % mannitol; *yellow* 3 % mannitol; *bright blue* 4 % mannitol; mean dry weight in mg

We have started to do case studies and have grown the mycobiont of *Roccella decipiens* under standardized and optimized culture conditions (Figs. 9.4, 9.5 and 9.6). In further experiments (Figs. 9.7, 9.8 and 9.9), the different developmental stages of 5- and 6-month-old mycobiont cultures were compared and the results quite surprising: mycobionts 7 and 8 show different combinations of metabolites. The compounds are closely related, orsellinic acid is a mononuclear metabolite (one aromatic ring) and probably a precursor of lecanoric acid (two rings) and erythrin is basically lecanoric acid esterified with the sugar alcohol erythriol, coming from the *Trentepohlia*-type algae (produce erythriol as polyol transfer metabolites in the intact lichens) (Figs. 9.10 and 9.11).

9.3.2 Chemosyndromic Variation as Exemplified by Species of the Genus *Xanthoparmelia*

Examples of high interest are species of the genus *Xanthoparmelia*, representing the largest genus of the lichen family Parmeliaceae, which live preferentially in the warm and hot semiarid zones of the Southern Hemisphere, having their main areas of speciation in Australia and South Africa (Elix 1994).

Species of *Xanthoparmelia* have been investigated mainly because of their highly diverse thallus morphologies, but also because of their complex chemistry, their high variability in types, and contents of secondary metabolites. In general, there are several possibilities of

Fig. 9.5 a, b *Roccella decipiens* and *Niebla* lichens in the natural environment; Baja California, Mexico

Fig. 9.6 Growth of *Roccella decipiens* mycobiont under optimized culture conditions, MS-medium with 6 % mannitol and 3 % Erythritol = polyol/sugar alcohol of *Trentepohlia* transferred to the lichen fungus. 1 g of cultured mycobiont has been calculated—produces 40 mg pure erythrin and 16 mg orsellinic acid and 30 mg of lecanoric acid

chemical variation (chemotypic variation) found in lichens (Scheme 9.1). In an ongoing investigation about Australian species of *Xanthoparmelia*, we studied several taxa for chemosyndromic variation (Figs. 9.12 and 9.13).

A major question was as follows: Is it possible to trigger chemical variation in cultured mycobionts by changing ecological parameters in the culture chambers? All taxa of *Xanthoparmelia* are well known to produce a high number of different secondary metabolites, an outstandingly rich diversity of lichen-specific fungal products such as depsides and depsidones, pigments, nowadays compiled as polyketides that are biosynthesized by the acetate-polymalonate pathway. More than 40 chemosyndromes have been found and structurally elucidated in species of *Xanthoparmelia* worldwide. Chemosyndromic variation has been correlated with morphology, geographic variation, and also with ecology. Our investigations focussed on selected species of *Xanthoparmelia* from semiarid locations. We screened several ecological factors that could be responsible for chemosyndromic variations among the selected taxa.

All taxa of *Xanthoparmelia* tested in this study contained usnic acid in the cortex (Fig. 9.14).

Thalli of *Xanthoparmelia flavecentireagens* (in the natural environment) (Fig. 9.15) produce two chemically related depsidones, norlobaridone, loxodin, and the depside divaricatic acid (+usnic acid) (Figs. 9.16 and 9.17a).

The next step in this investigation was the culture of one selected mycobiont (*X. flavecentireagens*, Australian species) on various nutrient media and under different culture conditions (Fig. 9.17b) to find out if changed ecological parameters would trigger the biosynthesis of another set of depsidones and cycloaliphatic acids.

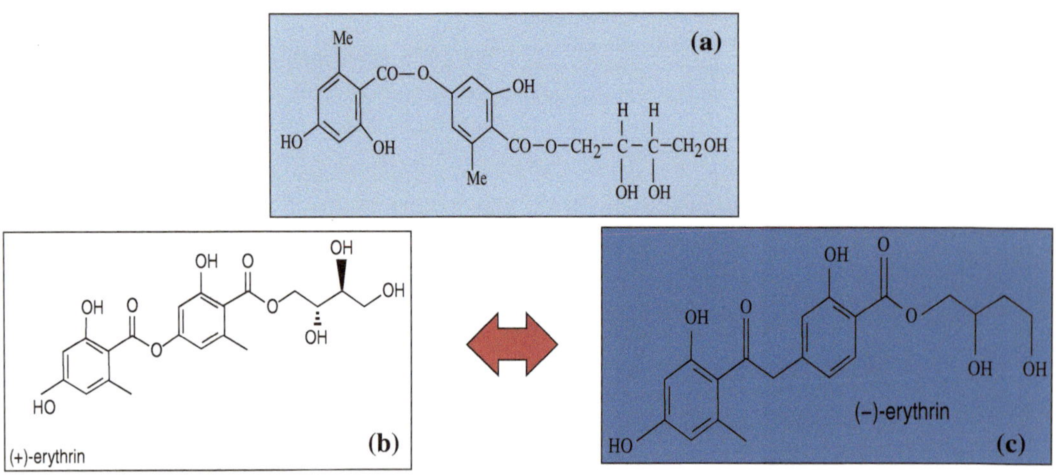

Fig. 9.7 a–c Erythrin: is a depside which is further esterified with the sugar alcohol (polyol) erythritol. In the lichen and also in the mycobiont (+) and (−) enantiomers of erythrin are found (Basset et al. 2010), which can be separated by chiral HPLC

Fig. 9.8 Mycobiont on Murashige Skoog medium +6 % mannitol (5-month-old culture, showing cell differentiation and morphogenetic capacities), producing only erythrin

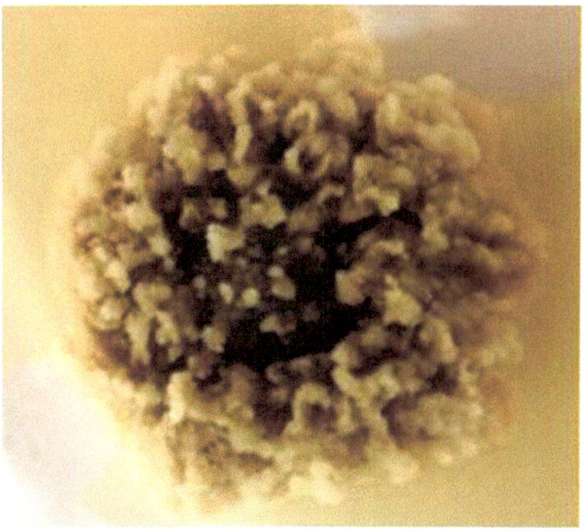

Fig. 9.9 Mycobiont of *Roccella decipiens* (6 months in culture, more differentiated, "lobe-like" mycelia): producing orsellinic acid, lecanoric acid, and erythrin

This specimen (and others) showed that changes of nutrient medium (LBM → SAB 2 %) could have an influence which metabolites are produced; in this case which lichen substances do occur as major or minor metabolites; further satellite substances that do not form in the natural thalli are found, like connorlobaridone; moreover mycobiont 1 additionally produced an unknown reddish brown pigment.

Fig. 9.10 Erythrin crystals, recrystallized from cultured mycobiont of *Roccella decipiens*

Comparison of mycobionts grown under stable and variable culture conditions/production of lichen substances of the thallus (natural environment, Canberra, Australia): Winter season (June–August): day 15–18 °C, Night: −5 to 10 °C; to, Summer (December–March): day 15/40 °C, night: 3/15 °C. Winter: rainy and foggy; summer: hot, very dry with occasional strong rain falls with chemical substances Norlobaridone as major while loxodin, divaricatic acid as minor contents (Fig. 9.18). Mycobiont under stable conditions were as on MY Medium: Day/night 20 °C; produces cyclo-aliphatic acids such as constipatic/protoconstipatic acid (Fig. 9.19).

A. ***Xanthoparmelia flavecentireagens-mycobiont*** 1:

Nutrient Medium: LBM + soil extract; Temperature shifts: day: 20/27 °C, Night: 4/10 °C; Desiccation treatment: 6 weeks/3 months/6 weeks.

Chemistry: Major: Unknown pigment

Minor: Norlobaridone, connorlobaridone, Divaricatic, loxodin acid.

B. ***Xanthoparmelia flavecentireagens-mycobiont*** 2:

Nutrient Medium: F: SAB 2 %; Temperature shifts: day: 12/20, Night: 1/8 °C

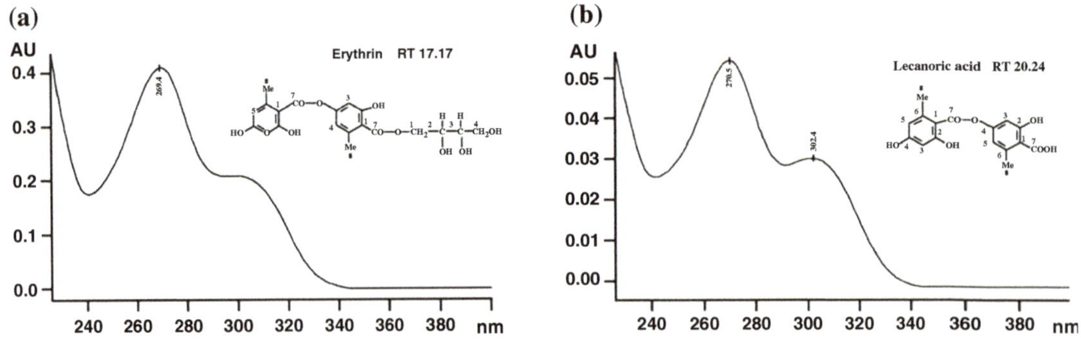

Fig. 9.11 a, b UV spectrum of erythrin and lecanoric acid (HPLC analysis), from mycobiont of *Roccella decipiens* (from Figs. 9.7 and 9.8)

Xanthoparmelia antleriformis: Australia
CHEMISTRY: containing usnic acid, salazinic acid (major), consalazinic acid and ±traces of norstictic acid.

***Xanthoparmelia cheelii*:**
CHEMISTRY: containing usnic acid, salazinic acid and consalazinic acid.

Xanthoparmelia filarskyana:
CHEMISTRY: containing usnic acid, ±loxodin, norlobaridone, ±protoconstipatic acid and ±constipatic acid; rarely containing conorlobaridone (trace) and conloxodin (trace).

Xanthoparmelia flavecentireagens:
CHEMISTRY: containing usnic acid, loxodin, norlobaridone, ±constipatic acid, ±protoconstipatic acid, ±conloxodin and ±connorlobaridone, +4 scabrosin esters.

Fig. 9.12 Investigated species of Xanthoparmelia from Australia

Desiccation treatment: 8 weeks/3 months/ 8 weeks; Chemistry: Major: Norlobaridone Minor: Connorlobaridone, divaricatic acid, loxodin (Fig. 9.20).

At least for this example it was shown that chemical variation can be influenced by selections of nutrient medium and variation of ecological conditions in the culture chambers.

A	A	A
1	2	3

Replacements series: one substance is replaced by a chemically closely related substance.

A	A	A
2 + 3	4+(-)5	None

Accessory substances: additive, accessory and none (acid deficient chemotype); substance may or may not occur.

B	B	B	B
1	2	3	4
	1		1
2	4	2	2
4	5	5	5

Scheme 9.1 Chemosyndromic variation. Each species has one major substance and several minor substances, one of the minor substances can become a major substance in another species. Several substances regularly occur together, e.g., in some taxa of the genus *Xanthoparmelia* the stictic acid chemosyndrome

9.3.3 Modulating Growth and Culture Conditions of Lichen Fungi and Effects on the Biosynthesis of Secondary Metabolites

The modulation ("regulative manipulation") of growth conditions of microorganism and fungi is a common strategy used in biotechnology and applied microbiology to improve yields and diversity of secondary metabolites of therapeutic interest. As mentioned above, interest in polyketide-type metabolites is considerable, as many of these natural products are of medical, industrial, and/or agricultural importance (Calvo et al. 2002).

Lichen fungi are remarkable organisms that readily produce a wide range of secondary metabolites, polyketides, and shikimate derivatives with potential pharmaceutical applications (Boustie and Grube 2005; Stocker-Wörgötter 2008; Boustie et al. 2011) and that still remain unexploited, because of the lack of optimized culture conditions for these fungi.

In many "case" studies and test series performed in our laboratory, it was shown that "permissive" culture conditions (Fig. 9.21) can influence the expression of secondary lichen metabolites or even can induce the formation of chemosyndromes in the cultured mycelia (Stocker-Wörgötter and Elix 2004, 2009; Stocker-Wörgötter et al. 2004, 2009; Stocker-Wörgötter and Hager 2008).

Relatively fast growing mycobionts (*Bunodophoron* sp., *R. decipiens*, species of *Pseudocyphellaria*, *Xanthoparmelia*) were selected and optimized culture conditions for obtaining a particular depside or depsidone in high yield. Further modulation of growth conditions could give more details about expression patterns of lichen polyketides and shikimic acid-derived compounds, for the first time.

Xanthoparmelia lineola:
CHEMISTRY: containing usnic acid, salazinic, ±constipatic acid, ±protoconstipatic acid.

Xanthoparmelia metaclystoides:
CHEMISTRY: containing usnic acid, norstictic acid, connorstictic acid, ±salazinic acid, ±constipatic acid and ± protoconstipatic acid.

Xanthoparmelia substrigosa:
CHEMISTRY: containing usnic acid, norstictic acid (major), connorstictic acid, ±salazinic acid and ±consalazinic acid.

Xanthoparmelia tasmanica:
CHEMISTRY: containing usnic acid, salazinic acid, consalazinic acid, rarely protocetraric acid (trace) and rarely norstictic acid (trace).

Fig. 9.13 Further species of Xanthoparmelia from Australia, studied in our laboratory

In general, the use of cell cultures when compared to plant and mushroom cultivations is of great advantage, because of their faster growth and lower requirements of space and cost.

The production of fungal secondary metabolites, influenced by culture conditions, usually occurs during the stationary phase (idiophase) and after fungal growth has decreased (tropophase).

Parameters such as specific nutrients in the media, incubation periods, pH, temperature, and light (for photosynthetic active organisms) can be easily modified or regulated ("modulated") in electronic adjustable culture chambers and finally also in small-scale bioreactors, which would allow the production of water-soluble pigments of lichens such as anthraquinones and naphthoquinones in high quantity.

9.3.4 Origin of Polyketide-Type Metabolites and Biosynthetic Pathways

Polyketides represent a large "family" of structurally and biochemically highly diverse secondary metabolites. They are produced by PKSs (large enzyme complexes) which condense and use activated acetate units (e.g., malonyl-CoA) in

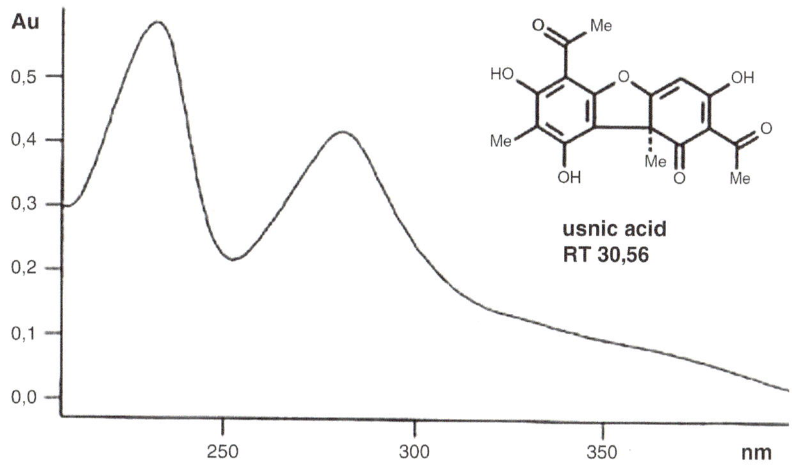

Fig. 9.14 Spectrum of usnic acid, obtained by HPLC-analyses

successive cycles of catalysis (Keller 2005). In ascomycetes, multi-domain enzymes (iterative Type I) start to produce first a polyketide chain, then the ketosynthase domain (KS) which is responsible for the actual condensation step, the acyl-transferase (AT) loads malonyl-CoA onto the enzyme complex, and the acyl carrier protein (ACP, could be interpreted as an anchor or arm), to which the growing polyketide chain is bound. In a further step, the terminated polyketide chain is off-loaded and/or cyclized by the thioesterase domain (TE).

Lichen fungi produce a diversity of polyketide-type metabolites. As the polyketides are diverse in form and function, they have a common biosynthetic origin, which means they are formed by repeated condensation of carboxylic acids, using coenzyme A-activated malonate in a mechanism that resembles fatty acid biosynthesis (Hopwood and Sherman 1990).

In lichens, as mentioned above, polyketide biosynthesis is catalyzed by iterative type I PKSs.

Fungal PKSs and lichen PKSs consist of a succession of enzymatic domains, a ketosynthase (KS), an acyl-transferase (AT), dehydratase (DH), enoyl reductase (ER), keto reductase (KR), ACP, and a thioesterase (TE). Better: surprisingly, the PKSs of *Xanthoria elegans* (Brunauer et al. 2009) and one of the PKSs of *Xanthoparmelia*, we have deciphered and investigated in our laboratory have two ACPs, two PP-binding positions (phospho-pantetheine binding sites, Hametner).

Fig. 9.15 Usnic acid crystals

The simplest fungal PKS has a KS, an AT, and ACP domains (required for carboxylic acid condensations), further domains are optional depending upon the metabolites (final metabolites) that are produced (Fig. 9.22).

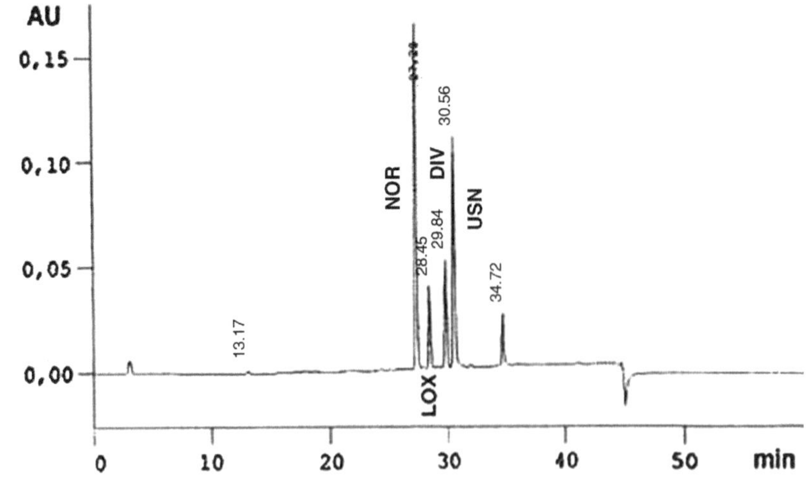

Fig. 9.16 Chromatogram of *X. flavecentireagens*, HPLC-analysis

Most polyketide-type lichen substances are oxidized metabolites, which are biosynthesized by non-reducing polyketide synthases (NR-PKSs). Instead of reductions, aldol and Claisen-type cyclizations are performed, which are responsible for the formation of aromatic ring structures; many lichen substances, e.g., depsides, consist of 2 or even three aromatic rings.

Sequence comparisons of the ketosynthase domains also support the existence of the two major groups of PKSs, the reducing PKSs and the non-reducing PKSs (Fig. 9.23). Kroken et al. (2003) conducted a thorough phylogenetic analysis of putative PKS genes from the genomes of *Neurospora crassa*, *Cochliobolus heterostrophus*, *Gibberella moniliformis*, *Botryotinia fuckeliana*, *Saccharomyces cerevisiae*, *Eremothecium gossypii*, *Schizosaccharomyces pombe*, *Gibberella zeae*, and previously characterized fungal and bacterial PKS genes. This analysis suggests that the filamentous fungi rival the actinomycetes (actinobacteria) in polyketide number and diversity. Lichen fungi do also have reducing PKSs-producing secondary compounds such as bourgeanic acid; however, most of the PKSs are non-reducing, producing oxidized polyketides such as depsides, depsidones, and dibenzofuranes.

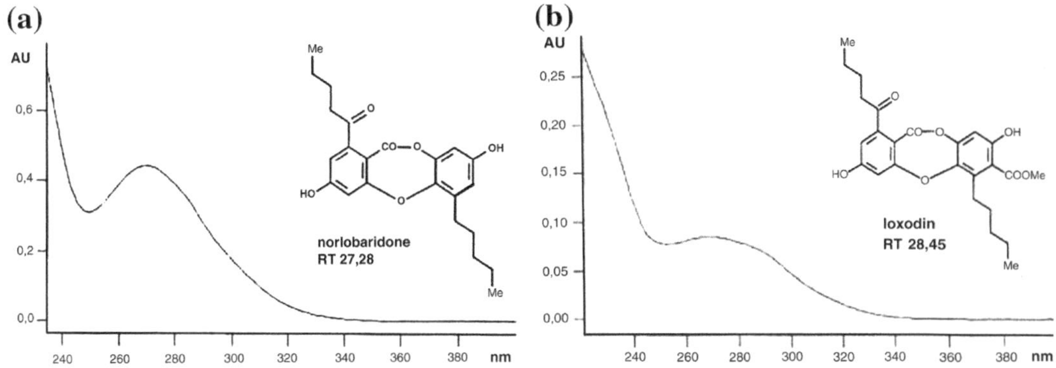

Fig. 9.17 UV-spectra of norlobaridone and loxodin

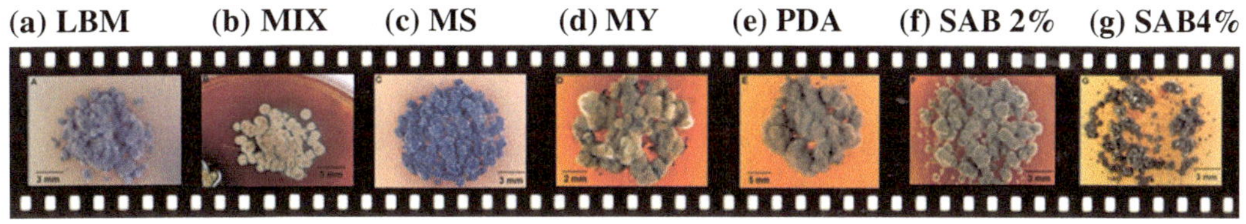

Nutrient Media: A: Lilly & Barnett medium, + 20 ml soil extract; B: Mix medium; C: Murashige Skoog-Medium;D: Malt Yeast Medium; E: Potato-Detrose Agar; F: Sabouraud 2% Glucose Agar; G: Sabouraud 4% Glucose Agar

Fig. 9.18 **a–g** Mycobiont of *X. flavecentireagens* on 7 different media; 4 months in subculture

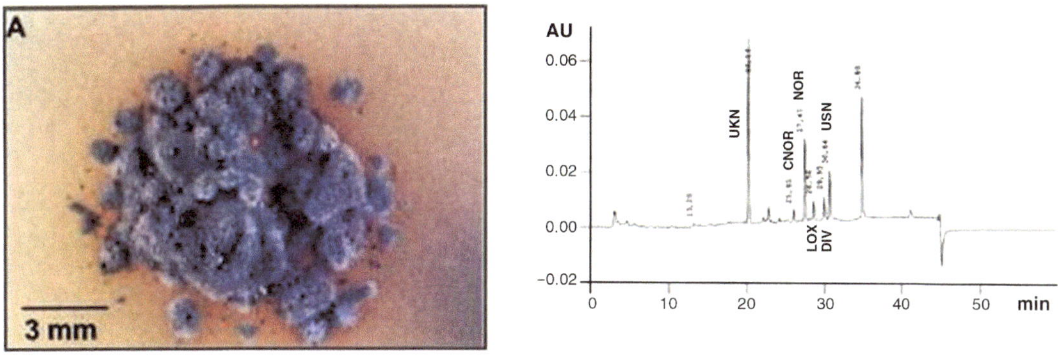

Fig. 9.19 Mycobiont 1: on LBM with soil extract: Major: unknown pigment; Minor: norlobaridone, Minor: connorlobaridone, divaricatic, loxodin

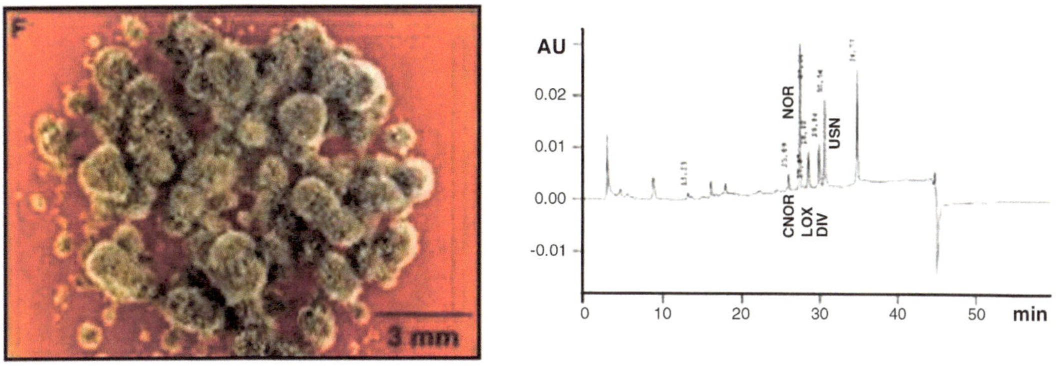

Fig. 9.20 Mycobiont 2: Major: Norlobaridone; Minor: Connorlobaridone, divaricatic acid, loxodin

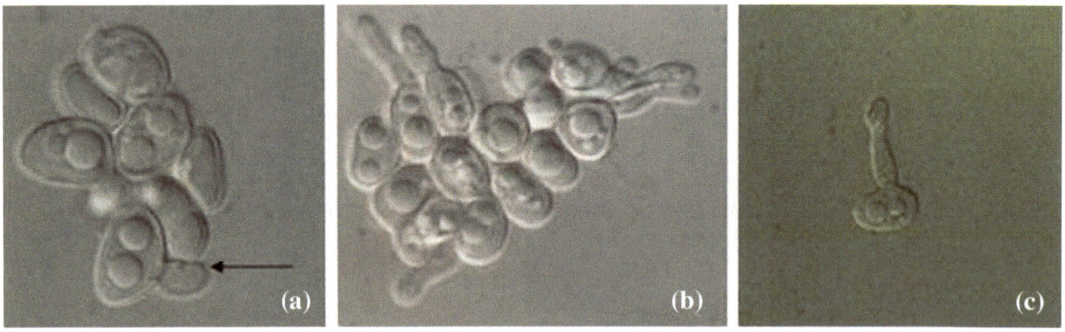

Fig. 9.21 **a**, **b** Germinated spores of *X. conspersa*, germinate and grow only, when they are exposed to day light. **c** Single spore isolate with germination tube

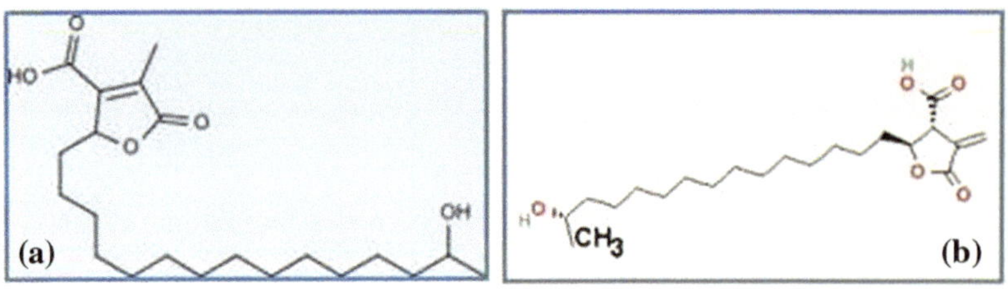

Fig. 9.22 Constipatic and protoconstipatic acids

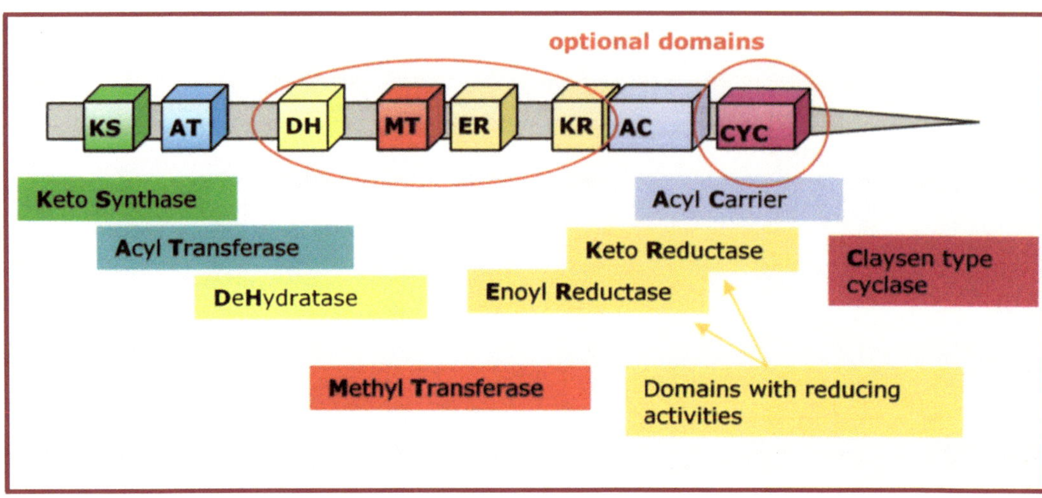

Fig. 9.23 Architecture of PKS I multienzyme complex with essential and optional domains that can be used reiteratively

Investigations on PKSs, in general, can help to understand the genetic and chemical potential —how polyketides are synthesized and how their biosynthesis is controlled by PKS genes and can also contribute to elucidate the evolution of the metabolic diversity (evolution of chemical compounds) known from lichen-forming fungi.

9.4 Outlook and Future

Construction of cDNA libraries representing the cultured fungal and mixed (fungal and algal) genomes of lichens under standard/control and stress conditions are the top priority approaches to be achieved in future. In a novel and holistic approach, functional genomics could be used to understand the molecular mechanisms that are involved or "behind" the desiccation tolerance of lichens.

To construct a cDNA library for a selected species of *Xanthoparmelia conspersa*, we will adopt a method (Junttila et al. 2009). A protocol is presented that allows inexpensive RNA extractions combined with commercial cleanup kits for lichens and cultured mycobionts-containing secondary metabolites such as polyketides. Main constraints in the methodology are the high contents of secondary metabolites and polysaccharides in lichen fungi that have to be removed by cleaning procedures before cDNA construction. An alternative method could be to use axenically cultured mycobionts grown under stable conditions which do not produce or express secondary metabolites. As soon as the cDNA library is successfully constructed, by use of sufficient quantity of high-quality total RNA from the model lichen, there are several interesting applications. The presence of many novel sequences would allow transcriptome sampling which could be used to detect and identify still unknown and "new" genes.

Such information could also be useful to characterize the molecular mechanisms that are responsible for the desiccation tolerance of lichens, a long-lasting unresolved basic question of lichen ecophysiology that would add basic knowledge to recent findings about biochemical and physical adaptations to tolerate desiccation stress in lichens and lichen fungi (Kranner et al. 2003). The knowledge of genes and proteins involved in desiccation resistance could be of high interest for the development of desiccation tolerant crops to save water supplies in semiarid and arid regions, some of the expected findings could be applicable to plants, in general.

Acknowledgments Research presented in this book chapter was generously supported by the Austrian Science Foundation (FWF), by grant P-20887 and P23570 to ESTW. I am very grateful to Prof. J.A. Elix for his long-term cooperation and support with the chemistry of Australian Xanthoparmelias.

References

Ahmadjian V (1993) The Lichen Symbiosis. Wiley, New York

Beckett RP (1995) Some aspects of the water relations of lichens from habitats of contrasting water status studied using thermocouple psychrometry. Ann Bot 76:211–217

Boustie J, Grube M (2005) Lichens—a promising source of bioactive secondary metabolites. Plant Genet Resour 3:273–287

Boustie J, Tomasi S, Grube M (2011) Bioactive lichen metabolites: alpine habitats as an untapped source. Phytochem Rev 10(3):287–307

Brunauer G, Muggia L, Stocker-Wörgötter E, Grube M (2009) A transcribed polyketide synthase gene from *Xanthoria elegans*. Mycol Res 113(1):82–92

Calvo AM, Wilson AR, Bok JW, Keller NP (2002) Relationship between secondary metabolism and fungal development. Microbiol Mol Biol Rev 66(3):447–459

Cardinale M, Puglia AM, Grube M (2006) Molecular analysis of lichen-associated bacterial communities. FEMS Microbiol Ecol 57:484–495

Chooi YH, Stalker D, Davis MA, Fujii I, Louwhoff SJJ, Lawrie AC (2008) Cloning and sequence characterization of a non-reducing polyketide synthase gene from the lichen *Xanthoparmelia semiviridis*. Mycol Res 112(2):147–161

Culberson CF, Armaleo D (1992) Induction of a complete secondary-product pathway in a cultured lichen fungus. Exp Mycol 16:52–63

Elix JA (1994) Flora of Australia: lichens, vol 55. Lecanorales 2, Parmeliaceae

Grube M, Berg G (2009) Microbial consortia of bacteria and fungi with focus on the lichen symbiosis. Fungal Biol Rev 23:72–85

Grube M, Hawksworth DL (2007) Trouble with lichen: the re-evaluation and re-interpretation of thallus form and fruit body types in the molecular era. Mycol Res 111:1116–1132

Grube M, Muggia L (2013) Success by flexible management of algal partners. Biochemistry (Biochemistry Society) August 2013:10–13

Hager A, Brunauer G, Türk R, Stocker-Wörgötter E (2007) Investigations on production, biological functions and bioactivity of common lichen metabolites as exemplified by the Australian lichen *Heterodea mülleri*. J Chem Ecol, 1561–1573

Hodkinson B, Lutzoni F (2009) A microbiotic survey of lichen-associated bacteria reveals a new lineage from the *Rhizobiales*. Symbiosis 49:163–218

Honegger R (2006) In: Gadd GM, Watkinson SC, Dyer P (eds) Fungi in the environment .Cambridge University Press, Cambridge. 2007 Brit Myc Soc Symposia Ser No. 25, pp 185–200

Honegger R, Edwards D, Axe L (2013) The earliest records of internally stratified cyanobacterial and algal lichens from the Lower Devonian of the Welsh Borderland. New Phytologist 197:264–275

Hopwood DA, Sherman DH (1990) Molecular genetics of polyketides and its comparison to fatty acid biosynthesis. Annu Rev Genet 24:37–62

Huneck S (1999) The significance of lichens and their metabolites. Naturwissenschaften 86:559–570

Joneson S, Armaleo D, Lutzoni F (2011) Fungal and algal gene expression in early developmental stages of lichen-symbiosis. Mycologia 103(2):291–306

Junttila S, Lim K-J, Rudd S (2009) Optimization and comparison of different methods for RNA isolation for cDNA library construction from the reindeer lichen *Cladonia rangiferina*. BMC Res Notes 2:204

Keller NP, Turner G, Bennett JW (2005) Fungal secondary metabolism—from biochemistry to genomics. Nat Rev Microbiol 3:937–947

Kranner I, Zorn M, Turk B, Wornik S, Beckett RP, Batic F (2003) Biochemical traits of lichens differing in relative desiccation tolerance. New Phytol 160:167–176

Kranner I, Beckett R, Hochman A, Nash TH III (2008) Desiccation-tolerance in lichens: a review. Bryologist 111(4):576–593

Kroken S, Glass NL, Taylor JW, Yoder OC, Turgeon BG (2003) Phylogenomic analysis of type I polyketide synthase genes in pathogenic and saprobic ascomycetes. PNAS 100:15670–15675

Leuckert C, Ahmadjian V, Culberson CF, Johnson A (1990) Xanthones and depsidones of the lichen *Lecanora dispersa* in nature and of its mycobiont in culture. Mycologia 82:370–378

Molina MC, Crespo A, Vicente C, Elix JA (2003) Differences in the composition of phenolics and fatty acids of cultured mycobiont and thallus of *Physconia distorta*. Plant Physiol Biochem 41:175–180

Piercey-Normore M, Deduke C (2011) Fungal farmers or algal escorts: lichen adaptation from the algal perspective. Mol Ecol 20:3708–3710

Rundel PW (1988) Water relations. In: Galun M (ed) Handbook of lichenology, vol 2. CRC Press, Boca Raton, pp 17–36

Schmitt I, Martin MP, Kautz S, Lumbsch HT (2005) Diversity of non-reducing polyketide synthase genes in the *Pertusariales* (lichenized Ascomycota): a phylogenetic perspective. Phytochemistry 66:1241–1253

Stocker-Wörgötter E (2002a) Resynthesis of photosymbiodemes. In: Kranner I, Beckett RP, Varma AK (eds) Methods in lichenology. Springer, Heidelberg, pp 47–60 (Springer Lab Manual)

Stocker-Wörgötter E (2002b) Analysis of secondary compounds in cultured mycobionts. In: Kranner I, Beckett RP, Varma AK (eds) Methods in lichenology. Springer, Heidelberg, pp 296–306 (Springer Lab Manual)

Stocker-Wörgötter E (2008) Metabolic diversity of lichen forming, ascomycetous fungi: culturing, production of polyketides and shikimic acid derivatives, and PKS genes. Nat Prod Rep (Cambridge) 25:188–200

Stocker-Wörgötter E, Elix JA (2006) Morphogenetic strategies and induction of secondary metabolite biosynthesis in cultured lichen-forming Ascomycota, as exemplified by *Cladia retipora* (Labill.) Nyl. and *Dactylina arctica* (Richards.) Nyl. Symbiosis 41(1):9–20

Stocker-Wörgötter E, Elix JA (2009) Experimental studies of lichen-forming fungi: formation of depsidones and shikimic-acid derivatives by the cultured mycobionts of three selected species of *Rhizocarpon* (Lecideaceae, lichenized Ascomycota). Bibl Lichenologica 100:495–512

Stocker-Wörgötter E, Hager A (2008) Culture methods for lichens and lichen symbionts. In: Nash TH (ed) Lichen biology. Cambridge University Press, Cambridge, pp 353–363

Stocker-Worgötter E, Elix JA, Grube M (2004) Secondary chemistry of lichen-forming fungi: and DNA-analyses of cultures and chemotypes in the *Ramalina farinacea* complex. Bryologist 107:152–162

Stocker-Wörgötter E, Hager A, Elix JA (2009) Intraspecific chemical variation within the crustose lichen genus Haematomma: anthraquinone production in selected cultured mycobionts as a response to stress and nutrient supply. Phytochem Rev 8:561–569

Wornik S, Grube M (2010) Joint dispersal does not imply maintenance of partnership in lichen symbioses. Microb Ecol 59:150–157

Zocher B, Stocker-Wörgötter E (2005) Secondary chemistry of lichen-forming ascomycetes: culture experiments with isolates of Xanthoparmelia species from Australia. Folia Cryptogamia Estonica 41:123–134

Lichen Substances and Their Biological Activities 10

Yoshikazu Yamamoto, Kojiro Hara, Hiroko Kawakami, and Masashi Komine

Contents

Y. Yamamoto (✉) · K. Hara · H. Kawakami · M. Komine
Faculty of Bioresource Sciences, Akita Prefectural University, 241-438, Kaidobata-Nishi, Shimoshinjo-Nakano, Akita 010-0195, Japan
e-mail: yyamamoto@akita-pu.ac.jp

D.K. Upreti et al. (eds.), *Recent Advances in Lichenology*,
DOI 10.1007/978-81-322-2235-4_10

Abstract

Lichens are known to synthesize variety of secondary metabolites which show wide range of biological activities. Many substances have been successfully isolated and identified from natural thalli and mycobiont cultures. Various biological activities have been screened for chemicals produced by mycobionts in culture medium and natural thalli of lichens. In a study conducted two decades back, lichen substances and their biological activities were reviewed, in which were thirty-nine lichen products, mainly novel phenolic compounds, synthesized by mycobiont cultures. Here, we review studies regarding synthesis of lichen substances by liquid cultures of lichen mycobionts, new sophisticated analytical methods for isolation and characterization of lichen substances, and various sensitive bioassays to assess the bioactivity of the isolated compounds. Biological properties of mycobiont cultures and natural thalli of lichens for range of potential bioactivity as anti-oxidation, inhibitions of monoamine oxidase, acetylcholinesterase, and photosynthesis, and growth inhibitions of animal-diseased bacteria, wood-decaying fungi and tumor cells have been discussed in detail.

Keywords

Secondary metabolites · Natural thalli · Lichen culture · Anti-oxidation · Anti-bacterial · Anti-fungal · Growth inhibition of tumor cells · Biological activities

10.1 Introduction

Lichens are symbiotic associations composed of fungal (mycobiont) and algal (photobiont) partners and have been used as medicines, dyes, perfumes, foods, and drink stuffs since ancient times all over the world. They produce characteristic secondary metabolites known as "lichen substance" (Asahina and Shibata 1954). Lichen substances are synthesized by fungal partners; therefore, mycobionts are separated from thalli and cultured in vitro by two methods (Yoshimura et al. 2001; Yamamoto et al. 2001).

10.2 Lichen Substances

Lichens produce many characteristic phenols, such as depsides, depsidones, dibenzofurans, pulvinates, chromones, and quinones (Culberson 1969) which are structurally diverse and potentially bioactive toward number of bioassays. Huneck and Yoshimura (1996) showed spectral data and other information of more than 500 lichen substances derived from natural thalli and cultured mycobionts. This book affords additional information on analyses of lichen substances. In the book entitled "Protocols in Lichenology," Lumbsch (2001) described analytical techniques of phenolic lichen substances for identification and taxonomy.

In recent two decades, many new lichen substances have been isolated from lichen cultures, production of lichen substances using liquid cultures has been reported, and analytical methods have been developed. Here, these three topics have been discussed.

10.2.1 New Products from Mycobiont Cultures

Twenty percent of about 100 lichen-derived cultures produced the same metabolites as those of their natural thalli, 20 % had extraordinary products that were not synthesized in their natural thalli, and 60 % did not have produced as detected by HPLC (Yoshimura et al. 1994b). Mycobiont cultures without the algal partner often synthesize novel and extraordinary products as shown in Table 10.1, e.g., Aliphatic [1] (Miyagawa et al. 1994), aza-anthraquinones [2–4] (Kawakatsu et al. 2006; Moriyasu et al. 2001; '="Q4' Text="Kindly note that references et al. (2001), Tanahashi et al. (2000), Lawrey (1984) are cited in the text but not provided in the reference list. Please provide the respective references in the list or delete these citations." –>Yamamoto et al. 2002a), azaphilones [5–6] (Arai et al. 2011), dibenzopyranones [7–9] (Tanahashi et al. 1997), furanoquinone [10] (Miyagawa et al. 1994), isochromenes [11–13] (Takenaka et al. 2011), isocoumarins [14–18] (Tanahashi et al. 2000; Takenaka et al. 2011), isoquinolines [19–26] (Kinoshita et al. 2003, 2005), macrolide [27] (Yamamoto et al. 2002b), naphthopyran [28] (Takenaka et al. 2010), and naphthoquinones [29–38] (Amano et al. 2000, Ernst-Russell et al. 1999, Kawakatsu et al. 2006, Kinoshita et al. 2009, Yamamoto et al. 1996, 2002a), and phenanthrenequinone [39] (Arai et al. 2012; Luo et al. 2013). These products were similar substances known as mycotoxins.

10.2.2 Production of Lichen Substances by Liquid Cultures of Lichen Mycobionts

The production of lichen substances was influenced by various culture conditions. The first study on the factors that affect cell growth and the production of secondary metabolites of a lichen mycobiont in liquid culture was reported by Yamamoto et al. (1995). An ascospore-derived strain of *Cladonia cristatella* mycobiont accumulated and excreted red-pigmented naphthoquinones (cristazarin [33] and 6-methylcristazarin [36]) into a liquid medium. Tested factors were sugars and sugar alcohols, sucrose concentration, amino acids, L-asparagine concentration, additional nutrients, medium pH, culture temperature, and inoculum weight.

Growth of the mycobiont was increased by using liquid Lilly–Barnett medium containing 16 % (w/v) sucrose, 0.2 % (w/v) L-glutamine, and 0.2 % (w/v) polypeptone, adjusting pH to 5.0

Table 10.1 Unique chemicals produced by mycobiont cultures

Product	Species	Reference
Aliphatic		
Graphenone [1]	*Graphis scripta*	Miyagawa et al. (1994)
Aza-anthraquinone		
Bostrycoidin [2]	*Arthonia cinnabarina*	Yamamoto et al. (2002a)
5-Deoxy-7-methylbostrycoidin [3]	*Haematomma* sp.	Moriyasu et al. (2001)
	Haematomma sp.	Kawakatsu et al. (2006)
8-*O*-Methylbostrycoidin [4]	*Arthonia cinnabarina*	Yamamoto et al. (2002a)
Azaphilone		
BE-1 [5], 3 [6]	*Cladonia bellidiflora*	Arai et al. (2011)
Dibenzopyranones		
Graphislactone A [7] - C [9]	*Graphis scripta*	Tanahashi et al. (1997)
Furanoquinone		
Graphisquinone [10]	*Graphis desquamescens*	Miyagawa et al. (1994)
Isochromene		
Proserin A [11]–C [13]	*Graphis proserpens*	Takenaka et al. (2011)
Isocoumarins		
4,6-Dihydroxy-3,9-dehydromellein [14]	*Graphis proserpens*	Takenaka et al. (2011)
8-*O*-methyldichorodiaportin [15]	*Graphis* sp.	Tanahashi et al. (2000)
6,8-Di-*O*-methylcitreoisocoumarin [16]	*Graphis* sp.	Tanahashi et al. (2000)
5-Hydroxy-7-methoxy-3-(1-hydroxyethyl) phthalide [17]	*Graphis proserpens*	Takenaka et al. (2011)
6-Hydroxy-3-hydroxymethyl-8-methoxyisocoumarin [18]	*Graphis proserpens*	Takenaka et al. (2011)
Isoquinoline		
Panaefluorolines A [19] - C [21]	*Amygdalaria panaeola*	Kinoshita et al. (2003)
Panaefluorolines D [22] - H [26]	*Amygdalaria panaeola*	Kinoshita et al. (2005)
Macrolide		
Baeomycenone [27]	*Baeomyces placophyllus*	Yamamoto et al. (2002b)
Naphthopyran		
Lcanopyrone [28]	*Lecanora leprosa*	Takenaka et al. (2010)
Naphthoquinone		
Arthoniafurone A [29], B [30]	*Arthonia cinnabarina*	Yamamoto et al. (2002a)
Boryquinone [31]	*Sphaerophorus fragilis*	Kinoshita et al. (2009)
Cinnabarinal [32]	*Arthonia cinnabarina*	Takahashi et al. (2005)
Cristazarin [33]	*Cladonia cristatella*	Yamamoto et al. (1996)
7-Demthylcristazarin [34]	*Sphaerophorus fragilis*	Kinoshita et al. (2009)
Hybocarpone [35]	*Lecanora hybocarpa*	Ernst-Russell et al. (1999)
6-Methylcristazarin [36]	*Cladonia cristatella*	Yamamoto et al. (1996)
5,7-Dihydroxy-6-hydroxymethyl-2-methoxy-1,4-naphthoquinone [37]	*Opegrapha* sp.	Amano et al. (2000)
5-deoxy-7-methylanhydrofusarubin lactol [38]	*Haematomma* sp.	Kawakatsu et al. (2006)
Phenanthrenequinone		
Biruloquinone [39]	*Cladonia fruticulosa*	Arai et al. (2012)
	Cladonia macilenta	Luo et al. (2013)

before autoclaving and incubating cultures at 20 °C. Pigment production by the mycobiont was increased by liquid Lilly–Barnett medium containing 4 % (w/v) sucrose, 0.2 % (w/v) L-asparagine, and 0.2 % (w/v) malt extract, adjusting pH to 5.0 before autoclaving and incubating cultures at 20 °C. Kinoshita et al. (2001) also reported effects of the nitrogen sources in the liquid medium for the production of usnic acid in a mycobiont of the lichen *Usnea hirta*. The production was higher in the liquid medium containing ammonium and nitrate ions than in those containing amino acids.

There are few reports on the mass production of cultured lichen-forming fungi and their products by liquid cultures. Red-pigmented naphthoquinones (cristazarin [33] and 6-methylcristazarin [36]) were produced by liquid cultures of *Cladonia cristatella* (Yamamoto et al. 1995, 1996); they were effective to dye silk (Nagashima et al. 2002). In order to make the dyes useful, they will need to be produced in large quantities for the dyeing industry. Komine et al. (2014) showed the result of naphthoquinone pigment production by the 3-l jar fermentation cultures of the *C. cristatella*. The growth and production were influenced by culture medium, culture temperature, rotation speed, and inoculum weight. Agitating Lilly–Barnett medium (pH 5.5) containing the mycobiont at 20 °C at 120 rpm for 28 days resulted in pigment production up to 25.3 mg/l day.

10.2.3 New Analytical Methods for Lichen Substances

Since 1970s, lichen substances have been analyzed by the thin-layer chromatography (TLC) (Culberson 1972a) or high-performance liquid chromatography (HPLC) (Culberson 1972b). TLC provided Rf values and colors of spots as important data for the identification of lichen substances; HPLC did only retention time data; therefore, HPLC was not a satisfied method. Recently, a photodiode array spectrometer was attached to HPLC and provided the complete UV–VIS spectra of the corresponding peaks, and this was a great help for the identification of substances. Yoshimura et al. (1994a) reported on the UV spectral analysis and identification of lichen substances using HPLC with a photodiode array detector (HPLC-PDA). In our laboratory, about 4,000 freezed specimens were analyzed and HPLC-PDA data were accumulated in a database, where data of UV spectra and retention times of about 360 lichen substances were registered. This database is a useful tool for the taxonomical identification. A new analytical method using HPLC with two mass spectrometries (LC-MS/MS) becomes to be used in identification of lichen substances. Kawakami and Yamamoto (2014) showed that constituents of two species, *Parmotrema praesorediosum* and *Canoparmelia aptata*, were identified by LC-MS/MS. The combination of HPLC-PDA and LC-MS/MS is a strong weapon in the field of chemical lichenology. Sato et al. (2011) established a new method to analyses of lichen triterpenoids using HPLC with a differential refractive index detector (HPLC-RID). The method enabled quantitative analysis of the compounds that do not absorb ultraviolet rays as triterpenoids. Kinoshita et al. (1997) attempted a new method to analyze optically active lichen substances using HPLC with a chiral column; they proved concentrations of (+) and (−) usnic acid contained in about 30 species.

10.3 Screening of Biological Activities and Biologically Active Compounds in Natural Thalli and Cultures

According to Vartia (1973), *Evernia furfuracea* was apparently used for medicinal purposes in Egypt in the seventeenth and eighteenth centuries B.C., and Hippocrates recommended *Usnea barbata* for uterine trouble and reviewed the use of lichens as folklore medicines and pharmacological studies. Yamamoto et al. (1993) also reviewed the folklore, pharmacological studies on natural thalli and cultures of lichens.

Lichens are natural and potential source of biologically active novel molecules, which can play a vital role in pharmacological and cosmetic industries. Owing to their probably higher pharmaceutical potential, about 5,000 freezed natural thalli and 500 cultures of lichen symbionts are preserved in the laboratory of Akita Prefectural University, Japan, in order to further investigate even higher potential by genetic and biochemical manipulation. In this chapter, an attempt has been made to review the studies on the biological activities of natural thalli as well as cultured mycobiont carried out in the last two decades.

10.3.1 Anti-oxidation

Reactive oxygen species (ROS) such as superoxide anion are produced in human cells by extracellular processes such as UV rays and cause lipid peroxidation to damage cell membranes, leading to promote aging and tumor. Normally, ROS are scavenging by enzymes such as superoxide dismutase (SOD) and catalase. Plants have secondary metabolites having anti-oxidation function to protect themselves. Lichens can grow in places received high energy levels of UV rays such as high mountains and polar regions; therefore, recent studies on lichen metabolites showing a potential property for anti-oxidation have been investigated by using various anti-oxidation systems. Yamamoto et al. (1993) first reported the anti-oxidation activity in lichens measuring SOD. Bhattarai et al. (2008), Paude et al. (2008), and Luo et al. (2009) have further carried out researches on antioxidant properties of lichens. Bhattarai et al. (2008) prepared methanol–water (90:10 v/v) extracts of five polar lichen species: *Stereocaulon alpinum*, *Ramalina terebrata*, *Caloplaca* sp., *Lecanora* sp., and *Caloplaca regalis,* collected at King George Island (Antarctica) and analyzed the major anti-oxidative constituents of lichen extracts using TLC, followed by a 2,2-diphenyl-l-picrylhydrazyl (DPPH) spray technique. Paude et al. (2008) assayed antioxidant activities of extracts of same species from King George Island by DPPH and $ABTS^{\cdot+}$ [2,2′-azinobis-(3-ethyl-benzothiazoline-6-sulfonate)] radical scavenging capacities and compared with those of commercial standards BHA (butylated hydroxyanisole) and Trolox [(±) - 6- hydroxyl -2,5,7,8- tetramethyl chromane -2- carboxylic acid]. The results indicated that *Caloplaca* species and *Lecanora* species exhibited comparatively higher antioxidant activities than the remaining other three species.

Luo et al. (2009) collected *Cetraria aculeata*, *Cladonia furcata*, *Pseudephebe pubescens*, *Sphaerophorus globosus*, *Stereocaulon alpinum*, *Umbilicaria antarctica*, *Usnea antarctica*, and *U. aurantiacoatra* in King George Island and evaluated the antioxidant activities of their methanol and acetone extracts by anti-linoleic acid peroxidation activity. The extract of *U. antarctica* was found to have the strongest antioxidant property among test species, and lecanoric acid was the main antioxidant compound in the acetone extract. The antioxidant properties of methanol extract of 46 lichen species, collected from the highly UV-exposed alpine areas of southwestern China, were evaluated by the anti-linoleic acid peroxidation and DPPH methods (Luo et al. 2010b). Extracts of *Peltigera praetextata* and *Sticta nylanderiana* were found to exhibit the most potent activity in all of the antioxidant tests. *S. nylanderiana* possessed the strongest free radical scavenging activity among all the tested species, with an inhibition rate of 90.4 % at concentration of 330 μg/ml. TLC and HPLC analyses revealed lecanoric acid [40] to be primarily responsible for the effective antioxidant properties of *S. nylanderiana*. Evaluation of the antioxidant properties of an edible lichen *Ramalina conduplicans* by anti-linoleic acid peroxidation method has been carried out (Luo et al. 2010a). Sekikaic acid [41] (IC_{50} 0.082 mg/m) and homosekikaic acid [42] (IC_{50} 0.276 mg/ml) were identified as the main free radical scavenging compounds in *R. conduplicans* extract.

Hara et al. (2011) carried out screening test on anti-oxidation activity using DPPH method for 99 ethanol extracts of 86 species of natural thalli of lichens in order to find novel anti-oxidation compounds. The 21 extracts of natural thalli showed high anti-oxidation activity, among which the activities of *Hypogymnia vittata*, *Pseudevernia furfuracea*, *Nephromopsis ornata*, *Peltigera elizabethae*, *P. aphthosa*, and *P. rufescens* were higher (Table 10.2). Extracts of *Peltigera* spp. showed higher activity than those of other genera. Ethanol extract of *P. aphthosa*

Table 10.2 Anti-oxidant activity of ethanol extracts of natural thalli of lichens expressed as Trolox equivalent (Hara et al. 2011)

Species	Trolox equivalent	Species	Trolox equivalent
Alectoria ochroleuca	107	*Letharia vulpina*	7
Alectoria sarmentosa	5	*Lobaria fuscotomentosa*	196
Anzia opuntiella-1	37	*Lobaria isidiophora*	0
Anzia opuntiella-2	50	*Lobaria linita*	0
Anzia opuntiella-3	140	*Lobaria pulmonaria*	0
Arctoparmelia centrifuga	0	*Lobaria scrobiculata*	87
Baeomyces placophyllus	72	*Lobaria spathulata*	14
Bryoria fremontii	128	*Menegazzia terebrata*	83
Bryoria nadvornikiana	51	*Nephroma laevigatum*	31
Canoparmelia aptata	64	*Nephromopsis nephromoides*	79
Cetraria laeviganda	56	*Nephromopsis ornata*	256
Cetrelia braunsiana-1	0	*Ochrolechia yasudae*	92
Cetrelia braunsiana-2	17	*Pannoparmelia angustata*	115
Cladia aggregata	0	*Parmelia adaugescens*	105
Cladina arbuscula	20	*Parmelia laevior*	109
Cladina rangiferina-1	12	*Parmelia praesquarrosa*	97
Cladina rangiferina-2	29	*Parmelia sinanoana*	56
Cladina subrangiferina	38	*Parmelia squarrosa*	38
Cladonia amaurocraea	75	*Parmelia submontana*	80
Cladonia arbuscula var. *mitis*	64	*Parmelia sulcata*	100
Cladonia convolute	49	*Parmotrema chinense*-1	55
Cladonia crispata	0	*Parmotrema chinense*-2	121
Cladonia furcata	103	*Peltigera aphthosa*	207
Cladonia gracilis	57	*Peltigera elizabethae*	214
Cladonia krempelhuberi-1	6	*Peltigera neopolydactyla*	184
Cladonia krempelhuberi-2	24	*Peltigera praetextata*	139
Cladonia macilenta	90	*Peltigera pruinosa*	96
Cladonia rangiferina-1	4	*Peltigera rufescens*	201
Cladonia rangiferina-2	7	*Peltigera aphthosa*	182
Cladonia rangiferina-3	11	*Pseudevernia furfuracea*	257
Cladonia scabriuscula	1	*Pseudocyphellaria crocata*	83
Cladonia sp.	26	*Punctelia rudecta*	5
Cladonia vulcani-1	184	*Pyxine limbulata*	0
Cladonia vulcani-2	31	*Ramalina capitata*	47
Coenogonium luteum	10	*Rimelia clavulifera*	0
Collema subflaccidum	0	*Stereocaulon intermedium*-1	26
Evernia divaricata	0	*Stereocaulon intermedium*-2	31
Evernia esorediosa	34	*Stereocaulon intermedium*-3	119
Evernia prunastri	45	*Stereocaulon japonicum*	84
Flavoparmelia caperata	0	*Stereocaulon sorediiferum*	3
Heterodermia diademata	8	*Sticta nylanderiana*-1	57
Heterodermia hypoleuca	37	*Sticta nylanderiana*-2	197
Heterodermia isidiophora	0	*Sulcaria sulcata*	76

(continued)

Table 10.2 (continued)

Species	Trolox equivalent	Species	Trolox equivalent
Heterodermia japonica-1	8	*Umbilicaria* sp.	0
Heterodermia japonica-2	11	*Usnea bismolliuscula*	31
Hypogymnia hypotrypella	43	*Usnea filipendula*	120
Hypogymnia vittata	313	*Usnea rubrotincta*	38
Hypotrachyna pseudosinuosa	9	*Usnea trichodeoides*	36
Leptogium pedicellatum	0	*Xanthoparmelia tuberculiformis*	19
Letharia columbiana	69		

-1,-2,-3; different localities

had separated ethyl acetate (EtOAc)-soluble and water-soluble fractions. Water extract had two anti-oxidative spots, and the lower spot had the same Rf value and color of authentic solorinine that was previously found as a unique quaternary ammonium compound from *Peltigera* spp (Matsubara et al. 1994). Water extract had two peaks, and the earlier peak had the same Rt and UV spectrum compared with authentic solorinine [43]. Solorinine showed similar anti-oxidation activity (EC50 120 μM) as Trolox (150 μM) and better than lecanoric acid (43 mM).

10.3.2 Inhibition of Tyrosinase and Melanin Biosynthesis

Tyrosinase, a kind of phenol oxidases, acts as a catalyst from tyrosine to dopaquinone via dopa in the biosynthetic pathway of melanin; therefore, tyrosinase inhibitors such as ascorbate, arbutin, and kojic acid have been used as whitening agents in cosmetics.

Yamamoto et al. (1998) reviewed tyrosinase inhibitory activity of tissue cultures of lichens and tyrosinase inhibitors by screening the inhibitory activity of tyrosinase obtained from mushroom in ethanol extracts of 108 natural thalli of 91 species. Extracts of *Baeomyces placophyllus* and *Stereocaulon intermedium* showed the strong activity among them. Higuchi et al. (1993) studied the inhibitory activity obtained from mushroom in tissue cultures of 45 species. Besides, lichen mycobionts were cultured in malt–yeast extract liquid medium; 4 weeks later, mycelia were separated from liquid medium by filtration. Acetone extracts from mycelia were obtained from mycobiont cultured in multi-yeast extract liquid medium after 4 weeks, and EtOAc extracts showed that inhibition of melanin biosynthesis in B16 melanoma cells was screened in 34 acetone extracts and 28 EtOAc extracts. Acetone extracts of *Chaenotheca brunneola* and *Myelochroa aurulenta* and EtOAc extract of *Pyrenula* sp. showed the strong activity among them.

10.3.3 Inhibition of Monoamine Oxidase

Monoamine oxidase (MAO) is a key enzyme which plays an essential role in the turnover of biogenic amines. MAO inhibitors have been used for the treatment of depression, hypertension, etc. The norsolorinic acid, solorinic acid, and averantin 6-*O*-methyl ether from *Solorina crocea* also have MAO inhibition properties (Okuyama et al. 1991). Endo et al. (1994) reported that confluentic acid [44] and 2′-*O*-methylperlatolic acid [45] isolated from the Brazilian plant showed inhibitory effect of MAO. On the grounds that the synthetic analogues, 5-alkylresorcinol (including 5-acylresorcinol) and 4-alkylresorcinol (including 4-acylresorcinol) derivatives, were considered to be the common part of the structure of MAO inhibitory compounds, these synthetic analogues were examined for MAO inhibitory by Kinoshita et al. (2002), among them 4-acylresorcinol which has $R=C_6H_{13}$ as the side chain showed potent activity (IC_{50} 4.27 μM), which showed that

Table 10.3 Monoamine oxidase inhibition of the methanol extracts of natural thalli and cultured mycobionts of lichens, excluding ineffective samples (Kinoshita et al. 2006)

Species	Inhibitory ratio (%)		
	1.0×10^{-4} (g/ml)	2.5×10^{-5} (g/ml)	1.0×10^{-5} (g/ml)
Natural thallus			
Bunodophoron melanocarpum	11.9	0.1	0
Erioderma pulchrum	3.6	0	
Evernia prunastri	6.1	4.6	1.6
Leptogium tremelloides	3.2	0	
Lobaria crenulata	15.0	0.2	1.6
Lobaria isidiophora	3.7	4.7	2.7
Lobaria orientalis	7.0	4.5	2.7
Lobaria quercizans	4.1	4.3	3.3
Lobaria subexornata	35.1	13.6	5.9
Nephroma arcticum	2.9	0	
Nephroma helveticum	8.8	1.2	0
Ochrolechia sp.	2.3	3.7	4.3
Parmeliella mariana	83.0	33.5	14.6
Peltigera aphthosa	8.0	2.2	0.9
Peltigera horizontalis	29.6	4.3	1.5
Peltigera leucophlebia	6.2	0.3	0
Peltigera nigripunctata	10.6	2.1	0.1
Pseudocyphellaria crocata	6.7	0	
Stereocaulon intermedium	0.6	3.6	0
Sticta weigelii	0.5	0	
Vulpicida juniperinus	0.9	0	
Cultured mycobiont			
Caloplaca scopularis	3.6	4.1	1.3
Cetraria islandica	0.5	0	
Evernia prunastri	9.1	5.7	1.0
Graphis scripta		10.5	8.4
Gymnoderma cococarpum	7.3	7.0	3.6
Hypogymnia enteromorpha	0.9	0	
Lecidea sp.	0.9	2.5	0
Letharia vulpina	1.0	0.8	0
Mycoblastus sanguinarius	0.2	0	
Pseudevernia furfuracea	16.7	8.3	2.7
Ramalina farinacea	4.6	0	
Stereocaulon intermedium	2.0	0.9	0.5
Sulcaria sulcata	0		
Umbilicaria kisovana	4.5	0.4	1.4
Usnea diffracta	1.9	0.2	3.4
Usnea trichodeoides	2.0	1.4	2.8
Vulpicida juniperinus	1.5	0	
Xanthoria mandschurica	0.8	3.5	4.0

0 no effect, *blank* not tested

compounds having longer aliphatic side chain showed higher MAO inhibitory effects.

Screening test on MAO inhibition was performed for the methanol extracts of 30 species of natural thalli and 26 species of cultured mycobionts of lichens by Kinoshita et al. (2006). The extracts of natural thalli of *Parmeliella mariana*, *Lobaria subexornata*, and *Peltigera horizontalis* showed strong MAO inhibitory activity, while the extract of a cultured mycobiont of *Graphis scripta* showed the strong inhibition (Table 10.3).

10.3.4 Inhibition of Acetylcholinesterase (AChE)

In order to develop new and effective agents for anti-AD (Alzheimer's disease) from lichen products, both the AChE inhibitory and the neuroprotective effects were evaluated by Luo et al. (2013). The AChE inhibitory assay was performed based on Ellman's reaction, and the neuroprotective effect was evaluated by using the MTT method on injured PC12 cells. Biruloquinone [39] as an AChE inhibitor (IC_{50} 27.1 μg/ml) was isolated from the extract of lichen-forming fungus *Cladonia macilenta*, which showed the most potent AChE inhibitory activity.

10.3.5 Growth Inhibition of Animal-Diseased Bacteria

Lichens have been used as medicines from ancient times all over the world. In particular, anti-bacterial activity had been studied since 1940s and was reviewed by Vartia (1973). Yamamoto et al. in the year 1993 first reported that an anti-bacterial activity was screened among extracts from lichen cultured mycobionts and culture metabolites of lichen mycobionts had anti-bacterial activity (Yamamoto et al. 1998). Furthermore, Yamamoto et al. (2010) screened on growth inhibition against 15 animal-diseased bacteria among acetone extracts of 70 species of natural thalli and 34 species of cultured mycobionts of lichens. The extracts of natural thalli of *Anaptychia palmulata* that do not contain usnic acid, depsides, and depsidones previously reported as antibiotics showed moderate anti-bacterial activity against three bacterial strains. On the other hand, the extracts of three cultured mycobionts of *Cladonia boryi*, *C. cristatella*, and *Haematomma* sp. remarkably inhibited the growth of several bacteria. Two pigments 5-deoxy-7-methylanhydrofusarubin lactol [38] and 5-deoxy-7-methylbostrycoidin [3] isolated from the cultured mycobiont of *Haematomma* species also showed high anti-bacterial activity (Table 10.4).

10.3.6 Growth Inhibition of Wood-Decaying Fungi

It is well known that lichens have allelopathic effects inhibiting the growth of plants and fungi. Lundström and Henningsson (1973) reported that lichens prevent to decay wood by fungi. Vartia (1973) reported that characteristic secondary metabolites of lichens, such as usnic, divaricatic, and lichesterinic acids, inhibit the growth of some filamentous fungi, and Yamamoto et al. (1993) proved that lichen tissue cultures showed growth inhibition of plant-diseased fungi. The screening result on growth inhibition of two wood-decaying fungi, *Trametes versicolor* and *Fomitopsis palustris*, by 46 strains of cultured lichen mycobionts was shown in Table 10.5 Yamamoto et al. (2002c).

Cell aggregates of each mycobiont strain were placed on agar plates of malt–yeast extract medium, glucose peptone medium, and potato dextrose medium and preincubated. After a month, wood-decaying fungi were inoculated onto the agar plate on which different mycobionts grew, and they were cultured together. Mycobionts of *Acarospora fuscata*, *Arthonia cinnabarina*, and *Ramalina exilis* inhibited the growth of both fungi. Out of the 54 natural thalli powder of lichen species, *Bunodophoron melanocarpum*, *Evernia esorediosa*, *Heterodermia isidiophora*, *H. japonica*, *Myelochroa aurulenta*, *M. leucotyliza*, *Parmotrema tinctorum*, and *Pseudevernia furfuracea*

Table 10.4 Anti-bacterial activity of the acetone extracts of natural thalli and cultured mycobionts of lichens, excluding ineffective samples (Yamamoto et al. 2010)

Species	Animal-diseased bacteria														
	BS	ML	SA	SA2	SE	PM	EF	SM	SP	**LA**	**ER**	**AP**	**BP**	**CP**	**PA**
Natural thallus															
Anaptychia palmulata	B	–	–	–	B	–	–	–	–	–	–	–	–	B	–
Anzia opuntiella	B	B	B	B	A	–	–	–	–	–	–	–	C	B	–
Bryoria trychodes	–	–	–	B	–	–	–	–	–	–	–	–	–	–	–
Bunodophoron melanocarpum	B	B	B	B	A	–	–	–	–		–	B	B	B	–
Cetraria laevigata	–	B	B	B	B	–	–	–	–	–	B	–	–	B	–
Cetrelia braunsiana	B	B	B	B	A	–	–	B	–	–	–	–	C	B	–
Cladia aggregata	C	B	B	–	B	–	–	–	–	–	–	C	C	A	–
Cladonia arbuscula subsp. *mitis*	A	B	B	B	A	B	B	–	B	A	B	A	B	A	B
Cladonia furcata	–	–	B	–	–	–	–	–	–	–	–	–	–	–	–
Cladonia macilenta	–	–	–	–	B	–	–	–	–	–	–	–	–	–	–
Cladonia rangiferina	C	–	–	–	B	–	–	–	–		–	–	C	A	–
Cladonia vulcani	B	B	B	B	A	–	–	B	–	B	B	B	C	B	C
Coelocaulon steppae	B	B	B	B	B	B	B	B	–	–	–	–	–	B	–
Collema subflaccidum	–	–	–	–	–	–	–	–	–	–	–	–	–	A	–
Evernia esorediosa	B	B	B	B	A	–	–	B	–	–	B	B	C	B	–
Evernia prunastri	C	B	B	B	B	–	–	–	–	–	–	–	C	B	–
Flavoparmelia caperata	B	B	–	–	A	–	–	–	–		–	B	C	B	C
Heterodermia diademata	–	–	B	–	–	–	–	–	–	–	–	–	–	C	–
Heterodermia japonica	–	–	B	–	B	–	–	–	–	–	–	–	–	–	–
Heterodermia isidiophora	–	–	B	–	–	–	–	–	–	–	–	–	–	–	–
Heterodermia pseudospeciosa	B	B	B	B	A	–	–	B	–	–	B	B	C	B	–
Hypotrachyna pseudosinuosa	B	–	–	–	–	–	–	–	–	–	–	–	–	–	–
Lobaria isidiophora	–	–	B	–	–	–	–	–	–	–	–	–	–	B	–
Lobaria scrobiculata	B	–	B	–	A	–	–	B	–	B	B	B	C	B	C
Lobaria spathulata	–	–	–	–	–	–	–	–	–	–	–	–	–	B	–
Myelochroa leucotyliza	–	–	B	–	–	–	–	–	–	–	–	–	–	C	–
Nephroma helveticum	B	–	–	–	–	–	–	–	–	–	–	–	–	–	–

(continued)

Table 10.4 (continued)

Species	Animal-diseased bacteria														
	BS	ML	SA	SA2	SE	PM	EF	SM	SP	**LA**	**ER**	**AP**	**BP**	**CP**	**PA**
Nephroma laevigatum	B	–	–	–	–	–	–	–	–		–	–	–	B	–
Nephromopsis asahinae	B	B	B	B	A	–	–	B	–	B	–	B	C	B	–
Nephromopsis nephromoides	B	B	B	B	B	–	–	–	–	–	–	–	–	B	C
Nephromopsis ornata	B	B	B	B	B	–	–	–	–	–	–	–	–	B	C
Pannoparmelia angustata	B	B	B	B	A	B	B	B	–	–	–	–	B	B	–
Parmelia praesquarrosa	–	–	B	–	–	–	–	–	–	–	–	–	–	–	–
Parmelia shinanoana	B	–	B	–	–	–	–	–	–	–	–	–	–	B	–
Parmelia squarrosa	–	–	B	–	–	–	–	–	–	–	–	–	–	B	–
Parmelia submontana	B	–	–	–	B	–	–	–	–	–	–	–	–	–	–
Parmotrema chinense	–	–	–	–	B	–	–	–	–	–	–	–	–	–	–
Peltigera pruinosa	–	–	B	–	–	–	–	–	–	–	–	–	–	–	–
Pyxine limbulata	B	–	–	–	B	–	–	–	–	–	–	–	–	–	C
Ramalina conduplicans	C	B	B	B	A	–	–	–	–	B	–	C	C	B	–
Ramalina exilis	C	–	–	–	A	B	–	B	–	–	B	B	C	A	C
Ramalina fraxinea	B	B	B	B	B	–	–	–	–	–	–	–	C	B	–
Ramalina peruviana	B	B	B	B	A	–	–	–	–		–	B	C	B	–
Ramalina roeslei	A	B	B	B	A	–	–	–	–	–	B	B	C	A	C
Ramalina siliquosa	B	B	B	B	A	–	–	–	–	–	–	B	C	B	C
Rimelia clavulifera	–	–	–	–	B	–	–	–	–	–	–	–	–	–	–
Stereocaulon exutum	B	B	B	B	A	–	–	–	–	–	–	B	–	B	–
Stereocaulon intermedium	B	–	B	B	B	–	–	–	–	–		–	C	B	–
Stereocaulon sorediiferum	B	–	B	–	B	–	–	–	–	–	–	–	–	B	–
Sticta nylanderiana	–	–	B	–	–	–	–	–	–	–	–	–	–	–	–
Sulcaria sulcata	–	C	B	C	B	B	B	B	–	–	–	–	–	–	–
Tuckermannopsis americana	–	B	A	B	B	–	–	–	–	–	–	–	C	–	–
Tuckneraria pseudocomplicata	B	B	B	B	B	–	–	–	–	–	–	–	–	B	–
Usnea diffracta	B	B	B	B	A	B	–	B	–	B	B	B	C	A	C
Usnea trichodeoides	B	B	A	B	A	B	B	B	–	A	A	B	B	A	B

(continued)

Table 10.4 (continued)

Species	Animal-diseased bacteria														
	BS	ML	SA	SA2	SE	PM	EF	SM	SP	**LA**	**ER**	**AP**	**BP**	**CP**	**PA**
Xanthoparmelia tuberculiformis	A	B	B	B	A	B	B	–	B	A	B	B	B	A	B
Cultured mycobiont															
Arthonia cinnabarina	C	–	–	–	–	–	–	–	–	–	–	–	–	–	–
Arthonia spadicea	B	C	–	C	B	–	–	–	–	–	–	–	–	–	–
Cetraria aculeata	–	–	–	–	A	–	–	–	–	–	–	–	–	–	–
Cladonia boryi	–	B	B	B	B	–	–	–	B	–	–	–	–	B	B
Cladonia cristatella	B	B	C	B	B	–	–	–	B	–	–	–	–	–	B
Cladonia pleurota	–	–	C	C	C	–	–	–	–	–	–	–	–	–	–
Cladonia vulcani	–	–	–	–	–	–	–	–	–	–	–	–	–	–	C
Dibaeis absoluta	–	–	–	–	–	–	–	–	B	–	–	–	–	–	C
Haematomma sp.	B	B	B	B	B	–	–	–	–	–	–	–	–	B	B
Opegrapha ochrocheila	–	–	–	–	C	–	–	–	–	–	–	–	–	–	–
Porpidia macrocarpa	–	–	–	C	–	–	–	–	–	–	–	–	–	–	–
Thelotrema subtile	–	–	–	C	–	–	–	–	–	–	–	–	–	–	B
Vermilacinia combeoides	–	–	–	–	–	–	–	–	–	–	–	–	–	–	C

BS *Bacillus subtilis*; ML *Micrococcus luteus*; SA & SA2 *Staphylococcus aureus*; SE *S. epidermidis*; PM *Pasteurella multocida*; EF *Enterococcus faecalis*; SM *Streptococcus mutans*; SP *S. pyogenes*; LA *Lactobacillus acidophilus*; ER *Erysipelothrix rhusiopathiae*; AP *Actinomyces pyrogenes*; BP *Bifidobacterium pseudolongum*; CP *Clostridium perfringens*; PA *Propionibacterium acnes*. No effect *Escherichia coli*

A clear zone ≥20 mm; *B* clear zone <20 mm; *C* turbid zone; – no inhibition

Table 10.5 Growth inhibition of two wood-decaying fungi (TV = *Trametes versicolor* and FP = *Fomitopsis palustris*) by cultured lichen mycobionts on agar plates of malt–yeast extract (MY), glucose peptone (GP), and potato dextrose (PD) media for 1 week at 20 °C in the dark (Yamamoto et al. 2002c)

Species	FP			TV		
	MY	GP	PD	MY	GP	PD
Acarospora fuscata	+	±	±	±	+	+
Amygdalaria panaeola	−	−	−	−	−	−
Arthonia cinnabarina	±	±	+	+	+	±
Arthothelium sp.	−	−	−	−	−	−
Asahinea sholanderi	−	−	−	−	−	−
Brigantiaea ferruginea	−	−	−	−	−	−
Cetraria aculeata	−	−	−	−	−	−
Cetraria islandica var. *orientalis*	−	−	−	−	−	−
Cetrariella delisei	−	−	−	−	−	−
Cladia aggregata	+	+	+	−	−	−
Cladonia boryi	−	−	−	−	−	−
Cladonia vulcani	−	−	−	−	−	−
Dermatocarpon miniatum	±	±	±	−	−	−
Dermatocarpon reticulatum	−	−	−	−	−	−
Dibaeis absoluta	±	−	±	+	−	+
Diploschistes scoposum	−	−	−	−	−	−
Flavoparmelia caperata	−	−	−	−	−	−
Haematomma sp.	+	−	−	−	−	±
Heterodermia obscurata	−	−	−	−	−	−
Hypogymnia physodes	−	−	−	−	−	−
Icmadophila ericetorum	−	−	±	−	−	−
Lasallia papulosa	−	−	−	−	−	−
Lecanora pulverulenta	−	−	−	−	−	−
Myelochroa irrugans	−	−	−	−	−	−
Nephromopsis ornata	−	−	−	−	−	−
Pertusaria corallina	−	−	−	−	−	−
Physcia adscendens	−	−	−	−	−	−
Platismatia interrupta	−	−	−	−	−	−
Pyrenula japonica	−	−	−	−	−	−
Ramalina exilis	±	+	−	−	±	+
Ramalina litoralis	−	−	−	−	−	−
Ramalina subbreviuscula	−	−	−	−	−	−
Rhizocarpon geographicum	−	−	−	−	−	−
Sphaerophorus meiophorus	−	−	−	−	−	−
Stereocaulon alpinum	−	−	−	−	−	−
Stereocaulon paschale	−	−	−	−	−	−
Stereocaulon sorediiferum	−	−	−	+	+	+
Thamnolia vermicularis	−	−	−	−	−	−
Umbilicaria caroliniana	−	−	−	−	−	−
Umbilicaria vellea	±	−	−	−	−	−
Usnea arizona	−	−	−	−	−	−
Vermilacinia combeoides	−	−	−	−	−	−
Verrucaria sp.	−	−	±	−	−	−
Xanthoria elegans	−	−	+	−	−	−

were remarkably inhibited the growth of both mushroom among them.

10.3.7 Growth Inhibition of Tumor Cells

The first report on the anti-tumor activity of lichens was published by Fukuoka et al. (1968) utilizing crude polysaccharide fractions by adding ethanol to the aqueous extracts of 9 species lichens. Out of the 9 species, inhibition ratio against subcutaneously implanted sarcoma 180 was calculated. In *Gyrophora esculenta*, the active principle, a partially acylated β-1,6-glucan, was isolated in a pure state which had inhibition ratio of 99.1 %.

Lichenin and isolichenin were roughly isolated from *Cetraria islandica* var. *orientalis* and showed inhibition ratios of 100 and 99.6 %, respectively. Kupchan and Kopperman (1975) reported that lichens have been used to treat cancers from the time of Abu Mansur (circa A.D. 970), and in the screening for tumor inhibitors of plant origins, an alcohol–water (1:1) extract of the lichen *Cladonia lepioclada* from New Zealand showed significant inhibitory activity against the Lewis lung carcinoma. The principal tumor inhibitory constituent obtained by the fractionation of an active extract was proved to be identified as l-usnic acid.

Hur et al. (2008) studied the cytotoxicity of acetone extracts of about 100 natural lichens from Yunnan Province, China, against six human cancer cell lines (AGS, human stomach; A549, human lung; HL-60, human blood; SK-OV3, human ovary; HT-29, human colon). *Allocetraria stracheyi* exhibited strong cytotoxic activity against AGS (LD50 0.66 μg/ml) and HL-60 (0.58 μg/ml). The cytotoxic values were quite similar to those of the anticancer drug doxorubicin hydrochloride (0.50 and 0.53 μg/ml against AGS and HL-60) used as reference compound. *Tuckneraria laxa* also showed the strong activity against HL-60 (0.58 μg/ml).

Sato et al. (2015) reported that the acetone extracts of natural thalli of 114 species of lichens screened showed the extracts of natural thalli of *Myelochroa aurulenta*, *M. leucotyliza*, and *Usnea bismolliuscula* had high HL-60 anti-proliferative activity. A lichen triterpenoid, 16-*O*-acetylleucotylic acid (46, ALA), was isolated from the acetone extract of *M. aurulenta* that exhibited the anti-proliferative activity (IC_{50} 21 μM) against HL-60 (Tokiwano et al. 2009). It was further observed that *Myelochroa* extracts containing no or trace ALA showed also higher activities, which indicated the presence of other bioactive components in the lichen species showing an anti-proliferative activity. Sato et al. (2015) thus isolated an active and new compound 12β-acetoxy-16β-*O*-acetylleucotylic acid (47, AAL) from natural thalli of *M. entotheiochroa*. AAL has about half activity (IC_{50} 44 μM) of ALA and about twice of leucotylic acid (72 μM).

The mechanisms of biological activity of lichen secondary metabolites on cancer cells are still almost entirely unknown. Bačkorová et al. (2012) investigated the mechanisms of cytotoxicity of four lichen secondary metabolites (parietin, atranorin, usnic acid, and gyrophoric acid) on A2780 and HT-29 cancer cell lines and found that usnic acid [48] and atranorin [49] were more effective anticancer compounds when compared to parietin and gyrophoric acid and usnic acid and atranorin were capable of inducing a massive loss in the mitochondrial membrane potential, along with caspase-3 activation (only in HT-29 cells) and phosphatidylserine externalization in both tested cell lines, and concluded that usnic acid and atranorin are activators of programmed cell death in A2780 and HT-29, probably through the mitochondrial pathway.

Yamamoto et al. in the year 1998 first reported that methanol extracts from 169 lichen cultures showed the growth inhibition of mouse

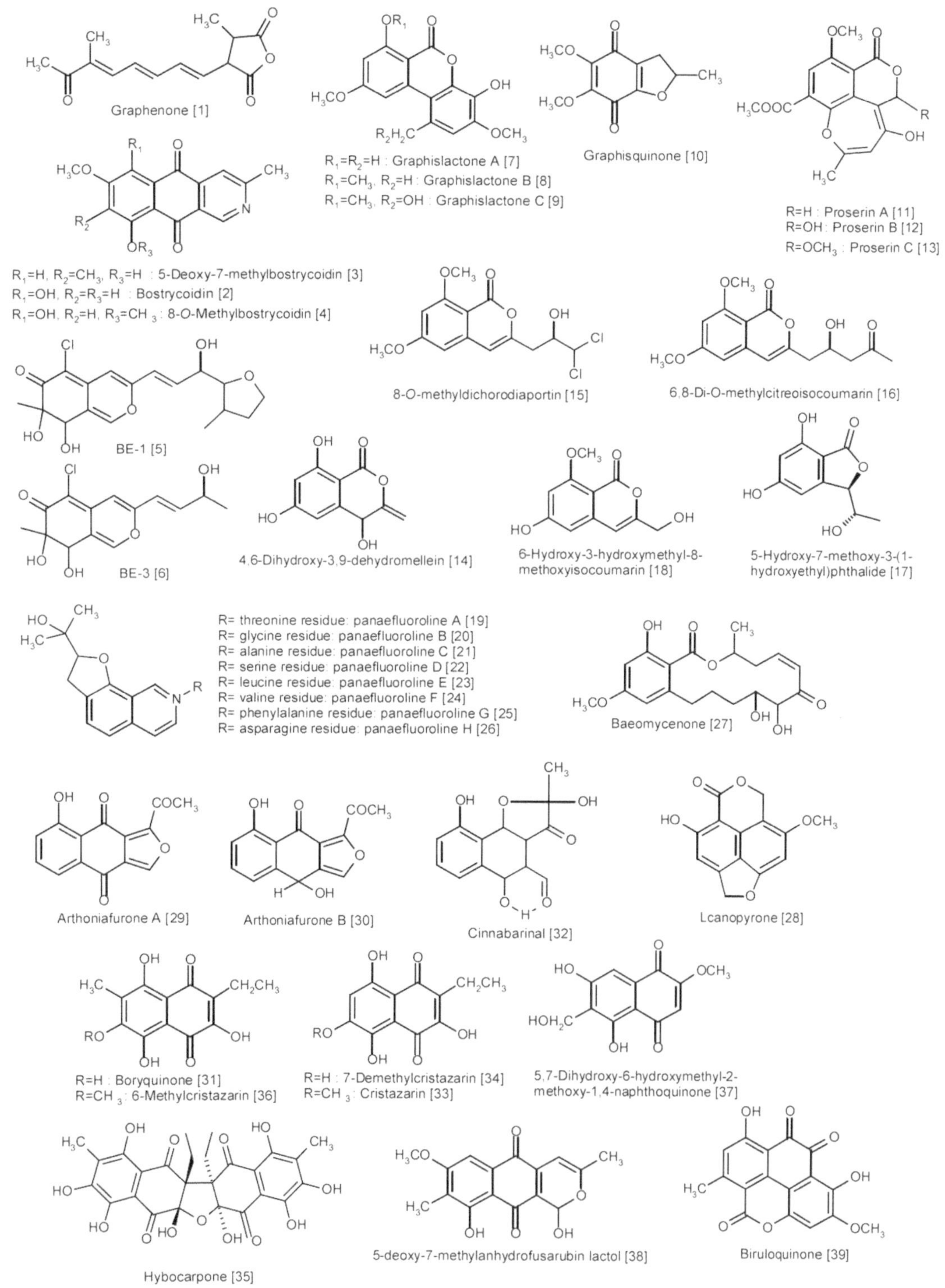

Fig. 10.1 Chemical structures of extraordinary products from mycobiont cultures

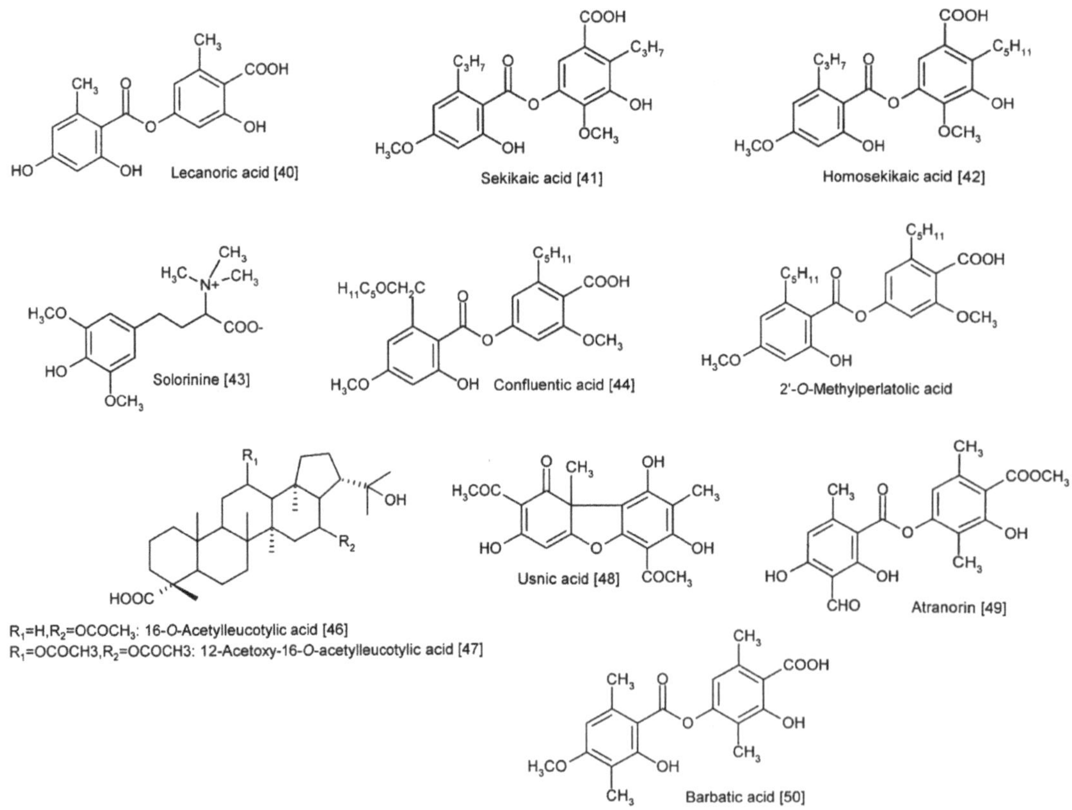

Fig. 10.2 Chemical structures of biological active substances

B-16 melanoma and human endothelial cells. Many cultures showed no effect on growth of two distinct cells, but some cultures ceased the growth of either endothelial or tumor cells. Hereafter, we cultured 32 lichen mycobionts of 31 species in malt–yeast extract liquid medium; 4 weeks later, mycelia were separated from liquid medium by filtration. Acetone extracts from mycelia and EtOAc extracts from liquid medium were obtained. Growth inhibition of HL-60 was screened in 16 acetone extracts and 28 EtOAc extracts. Acetone extracts of *Myelochroa aurulenta* (IC_{50} 23 ppm) and EtOAc extract of *Arthonia cinnabarina* (IC_{50} 2.5 ppm) and *Graphis connectans* (IC_{50} 3.5 ppm) showed the strong activity against various cancer cell lines.

Takahashi et al. (2005) revealed cytotoxicity of arthoniafurone A [29] ($EC50 < 30$ μM) and B [30] (<30 μM), and "Please check whether the usage of the term 'cinnabarinal' is OK." –>cinnabarinal (25 μM) was obtained from a cultured mycobiont of *Arthonia cinnabarina* against U937 human leukemia cells; therefore, these substances may also inhibit the growth of HL-60.

10.3.8 Inhibition of Plant Cell Growth and Photosynthesis

Nishitoba et al. (1987) identified eight depsides as growth inhibitors of lettuce seedlings from *Usnea longissima*, but the inhibition mechanism was not clarified. To identify the sites of inhibition by lichen-derived depsides in photosynthesis of higher plants, effects of 8 lichen acids on chlorophyll fluorescence and oxygen evolution in the thylakoid membranes of spinach were examined (Endo et al. 1998), and among them, barbatic acid [50], a lichen-derived depside, was the most potent inhibitor for both the reducing and oxidizing sites in the PS II complex.

10.3.9 Enhancement for Activity of the Hippocampal Long-Term Potentiation

Smriga et al. (1996) investigated central effectiveness of PC-2, a glucan from lichen *Parmelia caperata* (=*Flavoparmelia caperata*) with regard to the long-term potentiation (LTP) of evoked potential. The studies reported that the extent of LTP, induced by high-frequency stimulation of medial perforant pathway, was evaluated as fractional increase in population spike amplitude in dentate gyms in anesthetized rats and oral and intravenous application of PC-2 resulted in significant enhancement of LTP elicited by a weak, but not by a strong, tetanic stimulation.

10.4 Conclusion

Thus, studies related to screening of biological activities and bioactive compounds in natural thalli and cultures provide vital information about the structural and conformational diversity of lichen compounds. The advent of sophisticated analytical instruments and equipments has resulted in yielding high biomass which facilitates isolation and characterization of bioactive component much easier in the recent years. Lichens are treasure-house of structurally diverse bioactive components, which can be utilized in various pharmacopoeial and other industries as such or structurally altered to increase its potency (Figs. 10.1 and 10.2).

Acknowledgment Authors are thankful to Emeritus Prof. I. Yoshimura of Kochi Gakuen College for identification of lichen specimens, and Prof. K. Takahashi and Dr. K. Kinoshita of Meiji College of Pharmacy and Prof. H. Miyagawa of Kyoto University for identification of lichen substances. Prof. T. Inamoto, Prof. Doi, and Prof. Yoshizawa of Akita Prefectural University are acknowledged for assays of anti-bacterial and antifungal activities and growth inhibition of HL-60.

References

Amano T, Miyagawa H, Ueno T, Hamada N (2000) Production of 5,7-dihydroxy-6-hydroxymethyl-2-methoxy-1,4-naphthoquinone by the cultured lichen mycobiont of *Opegrapha* sp. no. 9771836. Z. Naturforsch 55b:667–669

Arai Y, Kinoshita K, Usuniwa Y, Yamamoto Y, Koyama K, Takahashi K (2011) Structures of yellow pigments from cultured lichen mycobiont, *Cladonia bellidiflora*. The Lichenology 9:77 (in Japanese)

Arai Y, Kinoshita K, Usuniwa Y, Yamamoto Y, Koyama K, Takahashi K (2012) Search for the substances of cultured lichen mycobiont—focusing on *Cladonia fruticlosa*. The Lichenology 10:199 (in Japanese)

Asahina Y, Shibata S (1954) Chemistry of lichen substances, Tokyo

Bačkorová M, Jendželovsky R, Kello M, Bačkor M, Mikeš J, Fedoročko JMP (2012) Lichen secondary metabolites are responsible for induction of apoptosis in HT-29 and A2780 human cancer cell lines. Toxicol In Vitro 26:462–468

Bhattarai HD, Paude B, Hong SG, Lee JS, Yim JH (2008) Thin layer chromatography analysis of antioxidant constituents of lichens from Antarctica. J Nat Med 62:481–484

Culberson CF (1969) Chemical and botanical guide to lichen products. University of North Carolina Press, Chapel Hill

Culberson CF (1972a) Improved conditions and new data for the identification of lichen products by a standardized thin-layer chromatographic method. J Chromatogr 72:113–125

Culberson CF (1972b) High-speed liquid chromatography of lichen extracts. The Bryologist 75:54–62

Endo Y, Hayashi H, Sato T, Maruno M (M. Chin), Ohia X, Nozoe S (1994) Confluentic acid and 2′-*O*-methylperlatolic acid, monoamine oxidase B inhibitors in a Brazilian plant, *Himatanthus sucuuba*. Chem Pharm Bull 42:1193–1201

Endo T, Takahagi T, Kinoshita Y, Yamamoto Y, Sato F (1998) Inhibition of photosystem II of spinach by lichen-depsides. Biosci Biotech Biochem 62:2023–2027

Ernst-Russell MA, Elix JA, Chal CLL, Willis AC, Hamada N, Nash III TH (1999) Hybocarpone, a novel cytotoxic naphthazarin derivative from mycobiont cultures of the lichen, *Lecanora hybocarpa*. Tetrahedron Lett 1999:6321–6324

Fukuoka F, Nakanishi M, Shibata S, Nishikawa Y, Takeda T, Tanaka M (1968) Polysaccharides in lichens and fungi. II. Antitumor activities on sarcoma-180 of the polysaccharide preparations from *Gyrophora esculenta*

Miyoshi, *Cetraria islandica* (L.) Ach. var. orientalis Asahina, and some other lichens. Gann 59:421–432

Hara H, Endo M, Kawakami H, Komine M, Yamamoto Y (2011) Anti-oxidation activity of ethanol extracts from natural thalli of lichens. Mycosystema 30:950–954

Higuchi M, Miura Y, Boohene J, Kinoshita Y, Yamamoto Y, Yoshimura I, Yamada Y (1993) Inhibition of tyrosinase activity by cultured lichen tissues and bionts. Plant Med 59:253–255

Huneck S, Yoshimura I (1996) Identification of lichen substances. Springer, Heidelberg

Hur JS, Yee ST, Koh YJ (2008) Cytotoxic activity of some Chinese lichen extracts on human cancer cell lines. The Lichenology 7:179–182 (in Japanese)

Kawakami H, Yamamoto Y (2014) Application of HPLC for studies on lichen substances. The Lichenology 12:81–83 (in Japanese)

Kawakatsu M, Yamamoto Y, Miyagawa H (2006) Pyranonaphthoquinone pigment from cultured lichen mycobiont of *Haematomma* sp. The Lichenology 5:31–36 (in Japanese)

Kinoshita Y, Yamamoto Y, Yoshimura I, Kurokawa T, Huneck S (1997) Distribution of optical isomers of usnic and isousnic acids analyzed by high performance liquid chromatography. J Hattori Bot Lab 83:173–178

Kinoshita Y, Yamamoto Y, Kurokawa T, Yoshimura I (2001) Influences of nitrogen sources on usnic acid production in a cultured mycobiont of lichen *Usnea hirta* (L.) Wigg. Biosci Biotechnol Biochem 61:1900–1902

Kinoshita K, Saito D, Koyama K, Takahashi K, Sato Y, Okuyama E, Fujimoto H, Yamazaki M (2002) Monoamine oxidase inhibitory effects of some lichen compounds and their synthetic analogues. J Hattori Bot Lab 92:277–284

Kinoshita K, Yamamoto Y, Koyama K, Takahashi K, Yoshiinura I (2003) Novel fluorescent isoquinoline pigments, panaefluorolines A–C from cultured mycobiont of a lichen, *Amygdalaria panaeola*. Tetrahedron Lett 3:8009–8011

Kinoshlta K, Yamamoto Y, Takatori K, Koyama K, Takahashi K, Kawai K, Yoshimura I (2005) Fluorescent compounds from the cultured mycobiont of *Amygdalaria panaeola*. J Nat Prod 68:1723–1727

Kinoshita Y, Yamamoto Y, Okuyama E, Yoshida E, Yamazaki M (2006) Screening for inhibitory activity of monoamine oxidase in natural thalli and cultured mycobionts of lichens. The Lichenology 5:1–6

Kinoshita K, Usuniwa Y, Yamamoto Y, Koyama K, Takahashi K (2009) Red pigments from the cultured mycobiont of a lichen, *Sphaerophoius fragilis* (L.) Pers. The Lichenology 8:1–4

Komine M, Tsuchiya S, Hara K, Yamamoto Y (2014) Preliminary study of pigment production by *Cladonia cristatella* mycobiont in a jar-fermenter. The Lichenology 12:61–65

Kupchan SM, Kopperman HL (1975) l-Usnic Acid: tumor inhibitor isolated from lichens. Experientia 31:625–626

Lumbsch HT (2001) Analysis of phenolic products in lichens for Identification and taxonomy. In: Kranner I, Becknett R, Varma A (eds) Protocols in lichenology. Springer, Heidelberg, pp 281–295

Lundström H, Henningsson B (1973) The effect of ten lichens on the growth of wood-destroying fungi. Mater Organismen 5:233–246

Luo H, Yamamoto Y, Kim JA, Jung JS, Koh YJ, Hur JS (2009) Lecanoric acid, a secondary lichen substance with antioxidant properties from *Umbilicaria antarctica* in maritime Antarctica (King George Island). Polar Biol 32:1033–1040

Luo H, Wei X, Yamamoto Y, Liu Y, Wang L, Jung JS, Koh YJ, Hur JS (2010a) Antioxidant activities of edible lichen *Ramalina conduplicans* and its free radical-scavenging constituents. Mycoscience 51:391–395

Luo H, Yamamoto Y, Liu Y, Jung JS, Kahng HY, Koh YG, Hur JS (2010b) The *in vitro* antioxidant properties of Chinese highland lichens. J Microbiol Biotechnol 20:1524–1528

Luo H, Li C, Kim JC, Liu Y, Jung JS, Koh YJ, Hur JS (2013) Biruloquinone, an acetylcholinesterase inhibitor produced by lichen-forming fungus *Cladonia macilenta*. J Microbiol Biotechnol 23:161–166

Matsubara H, Kinoshita K, Koyama K, Ye Y, Takahashi K, Yoshimura I, Yamamoto Y, Kawai K (1994) An amino acid from *Solorina crocea*. Phytochemistry 37:1209–1210

Miyagawa H, Hamada N, Sato M, Ueno T (1994) Pigments from the cultured lichen mycobionts of *Graphis scripta* and *G. desquamescens*. Phytochemistry 36:1319–1332

Moriyasu Y, Miyagawa H, Hamada N, Miyawaki H, Ueno T (2001) 5-Deoxy-7-methylbostrycoidinfrom cultured mycobionts from Heamatomma sp. Phytochemistry 58:239–241

Nagashima N, Yamamoto Y, Kinoshita Y, Ogasawara S, Katayama A (2002) Dyeing of silk with a new naphthazarin pigment. J Seric Sci Jpn 71:141–146 (in Japanese)

Nlshitoba Y, Nishimura H, Nishiyama T, Mizutani J (1987) Lichen acids, plant growth inhibitors from *Usnea longissima*. Photochemistry 26:3181–3185

Okuyama E, Hossain CFE, Yamazaki M (1991) Monoamine oxidase inhibitors from a lichen, *Solorina crocea* (L.) Ach. Shoyakugaku Zasshi 45:159–162

Paude B, Bhattarai HD, Lee JS, Hong SG, Shin HW, Yim JH (2008) Antioxidant activity of polar lichens from King George Island (Antarctica). Polar Biol 31:605–608

Sato H, Hara K, Komine M, Yamamoto Y (2011) Analysis of lichen substances including triterpenoids by high performance liquid chromatography with a differential refractive index detector and a photodiode array detector. Mycosystema 30:944–949

Sato H, Obara T, Hara K, Komine M, Yoshizawa Y, Yamamoto Y (2015) Proliferation inhibitory activity of natural thalli of lichens against HL-60 cells. The Lichenology 13:1–8

Smriga M, Saito H, Shibata S, Narui T, Okuyama T, Nishiyama N (1996) PC-2, linear homoglucan with α-linkages, peripherally enhances the hippocampal long-term potentiation. Pharm Res 13:1322–1326

Takahashi K, Kinoshita K, Yamamoto Y, Koyama K, Yoshimura I (2005) Chemical constituents from lichens for pharmaceutical and industrial uses. Folia Cryptog. Estonia Fasc 41:109–114

Takenaka Y, Hamada N, Tanahashi T (2010) Structure and biosynthesis of lecanopyrone, a Naphtho[1,8-cd] pyran-3-one Derivative from cultured lichen mycobionts of *Lecanora leprosa*. Z. Naturforsch 65c:637–641

Takenaka Y, Morimoto N, Hamada H, Tanahashi T (2011) Phenolic compounds from the cultured mycobionts of *Graphis proserpens*. Phytochemistry 72:1431–1435

Tanahashi T, Kuroishi M, Kuwahara A, Nagakura N, Hamada N (1997) Four phenols from the cultured lichen mycobiont of *Graphis scripta* var. *pulverulenta*. Chem Pharm Bull 45:1183–1185

Tanahashi T, Takenaka Y, Nagakura N, Hamada N, Miyawaki H (2000) Two isocoumarins from the cultured lichen mycobiont of Graphis sp. Heterocycles 53:723–728

Tokiwano T, Sato H, Obara T, Hirota H, Yoshizawa Y, Yamamoto Y (2009) A lichen substance as an antiproliferative compound against HL-60 human leukemia cells: 16-O-acetylleucotylic acid isolated from *Myelochroa aurulenta*. Biosci Biotechnol Biochem 73:2525–2527

Vartia KO (1973) Lichen antibiotics. In: Ahmadjian V, Hale ME (eds) The lichens. Academic Press, New York, pp 547–561

Yamamoto Y, Miura Y, Higuchi M, Kinoshita Y, Yoshimura I (1993) Using lichen tissue cultures in modern biology. The Bryologist 96:384–393

Yamamoto Y, Kinoshita Y, Yoshimura I, Ahmadjian V, Yamada Y (1995) Cell growth and pigment production in suspension culture of a mycobiont isolated from the lichen *Cladonia cristatella*. Can J Bot 73 (Suppl 1):S590–S594

Yamamoto Y, Matsubara H, Kinoshita Y, Kinoshita K, Koyama K, Takahashi K, Ahmadjian V, Kurokawa T, Yoshimura I (1996) Naphthazarin derivatives from cultures of the lichen *Cladonia cristatella*. Phytochemistry 43:1239–1242

Yamamoto Y, Kinoshita Y, Matsubara H, Kinoshita K, Koyama K, Takahashi K, Kurokawa T, Yoshimura I (1998) Screening of biological activities and isolation of biological-active compounds from lichens. Recent Res Develop Phytochem 2:23–34

Yamamoto Y, Kinoshita Y, Yoshimura I (2001) Culture of thallus fragments and redifferentiation of lichens. In: Kranner I, Becknett R, Varma A (eds) Methods in lichenology. Springer, Heidelberg, pp 34–46

Yamamoto Y, Kinoshita Y, Thor G, Hasumi M, Kinoshita K, Koyama K, Takahashi K, Yoshimura I (2002a) Isofuranonaphthoquinone derivatives from cultures of the lichen *Arthonia cinnabarina* (DC.) Wallr. Phytochemistry 60:741–745

Yamamoto Y, Kinoshita Y, Kinoshita K, Koyama K, Takahashi K (2002b) A zearalenone derivative from the liquid culture of the lichen, *Baeomyces placophyllus*. J Hattori Bot Lab 92:285–289

Yamamoto Y, Hara K, Komine M, Doi S, Takahashi K (2002c) Growth inhibition for wood decaying fungi by lichen mycobionts. The Lichenology 1:45–49

Yamamoto Y, Takeda M, Sato Y, Hara K, Komine M, Inamoto T (2010) Screening for growth inhibition of animal-diseased bacteria in natural thalli and cultured mycobionts of lichens. The Lichenology 9:11–17

Yoshimura I, Kinoshita Y, Yamamoto Y, Huneck S, Yamada Y (1994a) Analysis of secondary metabolites from lichen by high performance liquid chromatography with a photodiode array detector. Phytochem Anal 5:197–205

Yoshimura I, Kurokawa T, Kinoshita Y, Yamamoto Y, Miyawaki H (1994b) Lichen substances in cultured lichens. J Hattori Bot Lab 76:249–261

Yoshimura I, Yamamoto Y, Nakano T, Finnie J (2001) Isolation and culture of lichen photobionts and mycobionts. In: Kranner I, Becknett R, Varma A (eds) Methods in lichenology. Springer, Heidelberg, pp 3–33

Anticancer Activity of Lichen Metabolites and Their Mechanisms at the Molecular Level

11

Hangun Kim, Kyung Keun Kim and Jae-Seoun Hur

Contents

Abstract

As a result of symbiotic associations between mycobiont and photobiont, lichens can produce distinct and unique metabolites. Moreover, lichens are occasionally subjected to protect themselves from harsh environmental conditions during the growth. Therefore, lichen metabolites are expected to have various biological activities including antineoplastic action, for which it should suppress more than one of characteristics of the tumor progression. Here, we reviewed anticancer activity of lichen metabolites with molecular mechanisms by exerting cytotoxicity through regulation of cell cycle or induction of cell death, and by modulation of immune activity, angiogenesis, or energy metabolism. Alongside with these points, research will be expanded to yet unexplored areas by targeting cancer-specific signaling pathways.

Keywords

Lichens · Secondary metabolites · Anticancer activity · Molecular mechanism

11.1 Introduction

Cancer is a disease characterized by uncontrolled cell growth accompanied by avoidance of immune destruction, induction of angiogenesis, and deregulation of cellular energy metabolism with malignant behavior via invasion and metastasis (Hanahan and Weinberg 2011). Cancer can be treated by eradicating cancer cells via one or more methods such as surgery, radiation therapy, chemotherapy, and/or targeted therapy. Among these, chemotherapy uses chemical compounds that exert cytotoxicity to the neoplastic cells. However, as these cytotoxic chemicals can also non-specifically affect to rapid growing cells in the body, their administration

H. Kim
College of Pharmacy and Research Institute of Life and Pharmaceutical Sciences, Sunchon National University, Sunchon, Republic of Korea

K.K. Kim
Medical Research Center for Gene Regulation Chonnam National University Medical School, Gwangju, Republic of Korea

J.-S. Hur (✉)
Korean Lichen Research Institute, Sunchon National University, Sunchon, Republic of Korea
e-mail: jshur1@sunchon.ac.kr

D.K. Upreti et al. (eds.), *Recent Advances in Lichenology*,
DOI 10.1007/978-81-322-2235-4_11

commonly evokes several unwanted side effects, especially on the hematopoiesis, the immune system, and the gastrointestinal tract. To overcome these issues, targeted therapeutic approaches have been developed. These are designed to retrieve cellular signal transduction pathways misregulated within cancer cells or surrounding stromal cells. By targeting distinct cancer signaling pathways in each cancer type or in different patients, targeted therapy can selectively intervene against cancer with less severe side effects on other tissues. Currently, small molecules and antibodies are generally applied for targeted therapy, and lichen products have been found to be a promising source for such small molecules.

11.2 Lichens as Cytotoxic Chemotherapeutic Agents

To date, most studies of the anticancer effects of lichen secondary metabolites have focused on revealing cytotoxic activity (Molnar and Farkas 2010; Shrestha and St. Clair 2013; Shukla et al. 2010; Zambare and Christopher 2012). During the 1970s, anticancer activity of lichen secondary products was explored by several groups (Cain 1966; Fukuoka et al. 1968; Kupchan and Kopperman 1975; Shibata et al. 1968). Since these early studies, many other lichen compounds have been screened for their cytotoxicity against various cancer cell lines (Shrestha and St. Clair 2013). Studies using crude extracts or purified single compounds including usnic acid, cristazarin, protolichesterinic acid, polyporic acid, depsidone, and lichenin on various cancer cells such as melanoma, leukemia, breast cancer, prostate cancer, and pancreatic cancer are the examples (Zambare and Christopher 2012). For these cytotoxic anticancer effects, inhibition of cell proliferation by regulating cell cycle and increment of cell death by inducing apoptosis or necrosis have been suggested for the underlying mechanisms.

As expected, the mechanisms of cytotoxicity by lichen metabolites vary among cancer cell lines. In a study of cell cycle alteration and subsequent cell death, Singh et al. (2013) recently reported that usnic acid inhibits lung cancer cell (A549) growth by arresting the cell cycle at the G0/G1 phase through alteration of cyclin D1, cyclin-dependent kinases (CDKs), and cyclin-dependent kinase inhibitor (CDKI) protein expression levels and induces apoptotic cell death via mitochondrial membrane depolarization. Einarsdottir et al. (2010) also reported that both (+) and (−) usnic acid inhibited entry of the cell cycle into the S phase and led to a reduction in cell size of the breast cancer cell line T47D and pancreatic cancer cell line Capan-2; however, apoptosis was not observed in these cells and necrosis was only seen in Capan-2 cells. When considering dosage-dependent induction of apoptosis or necrosis by pannarin from *Psoroma* spp. (Russo et al. 2006), one might argue whether it is worth distinguishing between apoptosis and necrosis in terms of molecular mechanistic aspects. It has been reported that lichen acids including usnic acid, atranorin, and gyrophoric acid led to accumulation of cell cycle in the S phase in nine different human cancer cell lines (Backorova et al. 2011). Similarly, retigeric acid B from the lichen *Lobaria kurokawae* Yoshim enforced S-phase arrest in androgen-independent prostate cancer cells (PC-3) with increased $p21^{Cip1}$, cyclin E, cyclin A and phosphorylated retinoblastoma protein (pRb) levels, and decreased cyclin B levels (Liu et al. 2010). Retigeric acid B also induced caspase-dependent and–independent apoptotic cell death of PC-3 cells and, interestingly, reduced expression of androgen receptors (AR) in androgen-sensitive LNCaP cells, which led to decreased AR activity. Ren et al. (2009) reported that the treatment of acetone extract of *Lethariella zahlbruckneri* to human colon cancer cells (HT-29) increased the sub-G1 population and induced cell death accompanying apoptosis characteristics including apoptotic bodies, nuclear condensation, caspase activation, Bid cleavage, increased Bax expression, decreased Bcl-2 expression, and increased AIF expression.

During apoptotic cell death, the death receptor-mediated (extrinsic) pathway and mitochondria-mediated (intrinsic) pathway activate the caspase cascade, and activated caspase in turn cleaves various substrate proteins such as poly (ADP-ribose) polymerase (PARP) and Lamins. As a death receptor-mediated apoptosis, Lin et al. (2003) reported that lichenin *Cladonia furcata* polysaccharide-2 (CFP-2) induced up-regulation of Fas and FasL expression in HL-60 promyelocytic leukemia cell lines. In these cells, CFP-2 also decreased telomerase activity, suggesting its possible cancer therapeutic potential. In addition, Russo et al. (2012) reported that vicanicin and protolichesterinic acid significantly increased the expression of TNF-related apoptosis-inducing ligand (TRAIL) in LNCaP prostate cancer cells. The possibility of Hsp70 involvement in induction of intrinsic pathway-mediated apoptosis by vicanicin and protolichesterinic acid was also raised in this report since Hsp70 can inhibit key effectors of the apoptotic machinery at the mitochondria (Rerole et al. 2011). As an intrinsic apoptosis pathway, Backorova et al. (2012) reported that usnic acid and atranorin induced massive loss of mitochondrial membrane potential and activated programmed cell death in an ovarian cancer cell line (A2780) and colon cancer cell line (HT-29). Liu et al. (2010) and Ren et al. (2009) also identified a mitochondria-mediated apoptosis pathway by showing changes in the ratio of Bax/Bcl-2 proteins after treatment with retigeric acid B or acetone extract of *L. zahlbruckneri*, respectively. At least for usnic acid, alterations in the formation and/or stabilization of microtubules are not responsible for the antimitotic and antiproliferative activities (O'Neill et al. 2010).

11.3 Lichens with Immune-modulatory Activity

Inflammation and immunity governs tumor development at different stages (Grivennikov et al. 2010). Lichen products can exert anticancer effects by regulating tumor-promoting inflammation and/or antitumor immunity, although most of the mechanisms underlying these effects have not yet been elucidated. There are no studies demonstrating cytotoxic T cell-mediated antitumor immunity by lichen metabolites. However, Omarsdottir et al. (2006) reported that chromatographically purified lichen-derived polysaccharides including lichenan, pustulan, Ths-2 (categorized into β-glucan), and thamnolan (categorized into heteroglycan) exert anti-inflammatory effects via stimulation of dendritic cell maturation into Th2-like response. As a mechanistic explanation for dendritic cell maturation and anti-inflammatory effects, the authors suggested that lichen-derived polysaccharides may act as presented antigens to T cells after phagocytized by dendritic cells or as direct ligands that can transmit intracellular signals within dendritic cells. With this regard, Freysdottir et al. (2008) found that the aqueous extract of *Cetraria islandica* exerted anti-inflammatory effects in an antigen-induced arthritis model in rats and polysaccharide lichenan only showed an in vitro immune-modulatory effect among the quantified pure compounds. Induction of dendritic cell maturation by β-glucans isolated from different species was reported by another group (Kim et al. 2010a, b), and immune-pharmacological activities of β-glucans in immune cells were reviewed (Kim et al. 2011). β-Glucans are known to be able to raise invariant natural killer T cell responses against fungi (Cohen et al. 2011).

The lichen secondary metabolites, protolichesterinic acid and lobaric acid, were also shown to exert anti-inflammatory effects through inhibition of proliferative response in mitogen-stimulated lymphocytes, possibly via inhibition of 5-Lipoxygenase activity (Ogmundsdottir et al. 1998). Later, more lichen metabolites were found to have 5- and 12-lipoxygenase inhibitory activity and in vitro anti-proliferative effects against various human cancer cell lines and human platelets (Bucar et al. 2004; Haraldsdottir et al. 2004). Jin et al. (2008) reported that usnic acid exerts anti-inflammatory effects via inhibition of tumor necrosis factor-α (TNF-α) and inducible nitric oxide synthase (iNOS) expression through regulation of nuclear factor-κB (NF-κB) signaling activity in lipopolysaccharide (LPS)-stimulated macrophage RAW 264.7 cells. In these cells, usnic acid treatment suppressed degradation of

I-κBα and induced NF-κB nuclear translocation, which modulates transcription of immune and inflammatory genes including TNF-α and iNOS. In striking contrast to this, Choi et al. (2009) reported that methanol extract of *Caloplaca regalis* increased production of nitric oxide (NO) and TNF-α in inflammatory peritoneal macrophages. Through activation of macrophages, the extract induced tumoricidal activity against co-incubated B16 melanoma cells. Since p38 mitogen-activated protein kinase (MAPK) activity was increased by the extract and macrophage-mediated tumoricidal activity was decreased by p38 MAPK inhibitor, it seems that the tumoricidal activity of macrophages treated with the extract was mediated through p38 MAPK signaling. Direct evidence showing whether increased TNF-α and NO production is correlated with p38 MAPK activity was not presented, but the authors claimed that NK-κB activity was increased by the extract treatment. Since the role of macrophages in cancer development is complex and multifaceted (Bingle et al. 2002; Klimp et al. 2002; Sica et al. 2006), elaboration is required to delineate the anticancer effects of lichen metabolites by which maturation of macrophages is regulated. Conversely, lichen acids can give rise to toxicity in immunity. Pavlovic et al. (2013) reported that some lichen acids from *Hypogymnia physodes* exert thymocytes toxicity through induction of oxidative stress.

11.4 Lichens with Novel Strategies for Anticancer Therapy

Angiogenesis is essential to the development of cancer, especially for the proliferation and metastatic spread of cancer cells (Ellis and Fidler 1996). It has been reported that olivetoric acid isolated from acetone extract of the lichen *Pseudevernia furfuracea* (var. *ceratea*) displayed potent anti-angiogenic activities (Koparal et al. 2010). Olivetoric acid inhibited proliferation of rat adipose tissue endothelial cells (RATECs) and disrupted formation of endothelial tubes in these cells. Given the dose-dependent depolymerization of F-actin stress fibers, disorganization of the actin cytoskeleton seems to be involved in inhibition of tube formation in RATECs. More recently, Song et al. (2012) reported that usnic acid has anti-angiogenic activity that occurs via inhibition of endothelial cell proliferation, migration, tube formation, and induction of apoptotic changes in the morphology of endothelial cells. In these endothelial cells, usnic acid blocked vascular endothelial growth factor receptor (VEGFR) 2-mediated extracellular signal-regulated protein kinases 1 and 2 (ERK1/2) and the AKT/P706S6K signaling pathways.

Since most cancer cells use the glycolytic metabolic pathway to generate ATP for their main energy supply source, which is known as Warburg effect (Pelicano et al. 2006), targeting

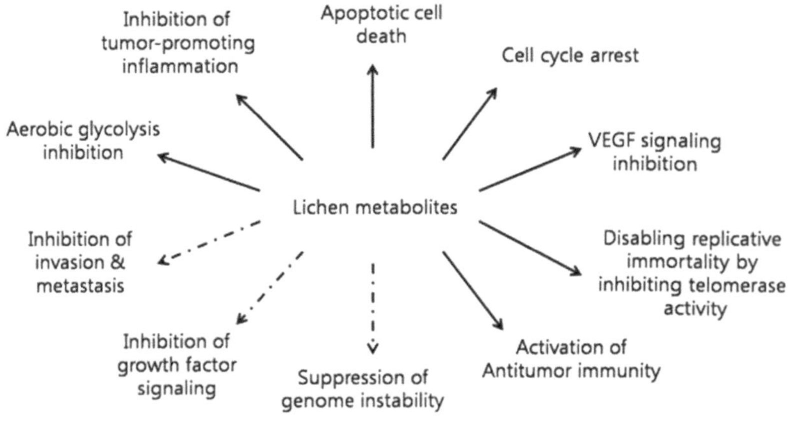

Fig. 11.1 Schematic illustration of possible anticancer mechanisms of lichen metabolites. *Solid lines* indicate studies reporting the mechanism were found. *Dotted lines* indicate that no studies reporting the mechanism were found

Table 11.1 Summary of anticancer lichen products with molecular mechanisms

Lichen products	Mechanisms	References
Cytotoxic		
• Cell cycle arrest		
Usnic acid	G0/G1 arrest through alteration of cyclin D1, cyclin-dependent kinases (CDKs) and cyclin-dependent kinase inhibitor (CDKI) protein expression levels	Singh et al. (2013)
(+) and (−) Usnic acid	Inhibition of entry of cell cycle into the S phase	Einarsdottir et al. (2010)
Usnic acid Atranorin Gyrophoric acid	Accumulation of cell cycle distribution in the S phase	Backorova et al. (2011)
Retigeric acid B from lichen *Lobaria kurokawae*	S-phase arrest with increased $p21^{Cip1}$, cyclin E, cyclin A and phosphorylated retinoblastoma protein (pRb) levels and decreased cyclin B levels	Liu et al. (2010)
Acetone extract of *Lethariella zahlbruckneri*	Increase insub-G1 population	Ren et al. (2009)
• Apoptotic Cell death		
Extrinsic pathway-mediated		
CFP-2	Up-regulation of Fas and FasL expression Decreased telomerase activity	Lin et al. (2003)
Vicanicin Protolichesterinic acid	Increased expression of TNF-related apoptosis-inducing ligand (TRAIL) Involvement of Hsp70 in inducing intrinsic pathway-mediated apoptosis	Russo et al. (2012)
Intrinsic pathway-mediated		
Usnic acid	Induction of apoptotic cell death via mitochondrial membrane depolarization	Singh et al. (2013)
Usnic acid Atranorin	Induction of massive loss of mitochondrial membrane potential and activation of programmed cell death	Backorova et al. (2012)
Retigeric acid B	Activation of mitochondria-mediated apoptosis pathway via changes in ratio of Bax/Bcl-2 proteins Induction of caspase-dependent and—independent apoptotic cell death	Liu et al. (2010)
Acetone extract of *Lethariella zahlbruckneri*	Induction of cell death with apoptotic characteristics including apoptotic bodies, nuclear condensation, caspase activation, Bid cleavage, increased Bax expression, decreased Bcl-2 expression, and increased AIF expression	Ren et al. (2009)
Immune-modulatory		
• Anti-inflammatory		
Lichenan Pustulan Ths-2 Thamnolan	Anti-inflammatory effects via stimulation of dendritic cell maturation into Th2-like response	Omarsdottir et al. (2006)
Protolichesterinic acid Lobaric acid	Anti-inflammatory effects through inhibition of proliferative response in mitogen-stimulated lymphocytes, possibly through inhibition of 5-Lipoxygenase activity	Ogmundsdottir et al. (1998)

(continued)

Table 11.1 (continued)

Lichen products	Mechanisms	References
Usnic acid	Anti-inflammatory effects via inhibition of tumor necrosis factor-α (TNF-α and inducible nitric oxide synthase (iNOS) expression through regulation of nuclear factor-κB (NF-κB) signaling activity by suppressing degradation of I-κB	Jin et al. (2008)
• Anti-tumor immunity		
Methanol extract of *Caloplaca regalis*	Activation of macrophages possibly mediated by p38 MAPK signaling	Choi et al. (2009)
Angiogenesis blocker		
Olivetoric acid	Inhibition of proliferation of rat adipose tissue endothelial cells (RATECs) and disruption of endothelial tube formation in this cell, possibly through disorganization of the actin cytoskeleton	Koparal et al. (2010)
Usnic acid	Inhibition of endothelial cell proliferation, migration, tube formation, and induction of apoptotic changes in morphology of the endothelial cells. Usnic acid blocked vascular endothelial growth factor receptor (VEGFR) 2-mediated extracellular signal-regulated protein kinases 1 and 2 (ERK1/2) and AKT/P706S6 K signaling pathways in the cells	Song et al. (2012)
Energy metabolism modulator		
Lichen secondary metabolites in the Antarctic lichen, *Lecidella carpathica*	Inhibitory activity against protein tyrosine phosphatase 1B (PTP1B)	Seo et al. (2011)

this metabolic process can be a therapeutic strategy for anticancer drugs (Porporato et al. 2011). Seo et al. (2011) found inhibitory lichen secondary metabolites in the Antarctic lichen, *Lecidella carpathica*, against protein tyrosine phosphatase 1B (PTP1B), which plays a major negative role in insulin receptor signaling.

The most life-threatening malignant behaviors of cancer are invasion and metastasis. However, no studies showing inhibitory activity against cancer cell motility from lichen products have been reported to date. Accordingly, research should be expanded to these unexplored areas by targeting cancer-specific signaling pathways (Fig. 11.1; Table 11.1).

References

Backorova M, Backor M, Mikes J, Jendzelovsky R, Fedorocko P (2011) Variable responses of different human cancer cells to the lichen compounds parietin, atranorin, usnic acid and gyrophoric acid. Toxicol In Vitro 25:37–44

Backorova M, Jendzelovsky R, Kello M, Backor M, Mikes J, Fedorocko P (2012) Lichen secondary metabolites are responsible for induction of apoptosis in HT-29 and A2780 human cancer cell lines. Toxicol In Vitro 26:462–468

Bingle L, Brown NJ, Lewis CE (2002) The role of tumour-associated macrophages in tumour progression: implications for new anticancer therapies. J Pathol 196:254–265

Bucar F, Schneider I, Ogmundsdottir H, Ingolfsdottir K (2004) Anti-proliferative lichen compounds with inhibitory activity on 12(S)-HETE production in human platelets. Phytomedicine 11:602–606

Cain BF (1966) Potential anti-tumour agents. IV. Polyporic acid series. J Chem Soc Perkin Trans 1 11:1041–1045

Choi HS, Yim JH, Lee HK, Pyo S (2009) Immunomodulatory effects of polar lichens on the function of macrophages in vitro. Mar Biotechnol 11:90–98

Cohen NR, Tatituri RV, Rivera A, Watts GF, Kim EY, Chiba A, Fuchs BB, Mylonakis E, Besra GS, Levitz SM et al (2011) Innate recognition of cell wall beta-glucans drives invariant natural killer T cell responses against fungi. Cell Host Microbe 10:437–450

Einarsdottir E, Groeneweg J, Bjornsdottir GG, Harethardottir G, Omarsdottir S, Ingolfsdottir K, Ogmundsdottir HM (2010) Cellular mechanisms of the anticancer effects of the lichen compound usnic acid. Planta Med 76:969–974

Ellis LM, Fidler IJ (1996) Angiogenesis and metastasis. Eur J Cancer 32A:2451–2460

Freysdottir J, Omarsdottir S, Ingolfsdottir K, Vikingsson A, Olafsdottir ES (2008) In vitro and in vivo immunomodulating effects of traditionally prepared extract and purified compounds from Cetraria islandica. Int Immunopharmacol 8:423–430

Fukuoka F, Nakanishi M, Shibata S, Nishikawa Y, Takeda T, Tanaka M (1968) Polysaccharides in lichens and fungi. II. Anti-tumor activities on sarcoma-180 of the polysaccharide preparations from Gyrophora esculenta Miyoshi, Cetraria islandica (L.) Ach. var. orientalis Asahina, and some other lichens. Gann = Gan 59:421–432

Grivennikov SI, Greten FR, Karin M (2010) Immunity, inflammation, and cancer. Cell 140:883–899

Hanahan D, Weinberg RA (2011) Hallmarks of cancer: the next generation. Cell 144:646–674

Haraldsdottir S, Guolaugsdottir E, Ingolfsdottir K, Ogmundsdottir HM (2004) Anti-proliferative effects of lichen-derived lipoxygenase inhibitors on twelve human cancer cell lines of different tissue origin in vitro. Planta Med 70:1098–1100

Jin JQ, Li CQ, He LC (2008) Down-regulatory effect of usnic acid on nuclear factor-kappa B-dependent tumor necrosis factor-alpha and inducible nitric oxide synthase expression in lipopolysaccharide-stimulated macrophages RAW 264.7. Phytother Res 22:1605–1609

Kim HS, Hong JT, Kim Y, Han SB (2011) Stimulatory effect of beta-glucans on immune cells. Immune Netw 11:191–195

Kim HS, Kim JY, Lee HK, Kim MS, Lee SR, Kang JS, Kim HM, Lee KA, Hong JT, Kim Y et al (2010a) Dendritic cell activation by glucan isolated from umbilicaria esculenta. Immune Netw 10:188–197

Kim HS, Kim JY, Ryu HS, Park HG, Kim YO, Kang JS, Kim HM, Hong JT, Kim Y, Han SB (2010b) Induction of dendritic cell maturation by beta-glucan isolated from Sparassis crispa. Int Immunopharmacol 10:1284–1294

Klimp AH, de Vries EG, Scherphof GL, Daemen T (2002) A potential role of macrophage activation in the treatment of cancer. Crit Rev Oncol/Hematol 44:143–161

Koparal AT, Ulus G, Zeytinoglu M, Tay T, Turk AO (2010) Angiogenesis inhibition by a lichen compound olivetoric acid. Phytother Res 24:754–758

Kupchan SM, Kopperman HL (1975) l-usnic acid: tumor inhibitor isolated from lichens. Experientia 31:625

Lin X, Cai YJ, Li ZX, Chen Q, Liu ZL, Wang R (2003) Structure determination, apoptosis induction, and telomerase inhibition of CFP-2, a novel lichenin from *Cladonia furcata*. Biochim Biophys Acta 1622:99–108

Liu H, Liu YQ, Xu AH, Young CY, Yuan HQ, Lou HX (2010) A novel anticancer agent, retigeric acid B, displays proliferation inhibition, S phase arrest and apoptosis activation in human prostate cancer cells. Chem Biol Interact 188:598–606

Molnar K, Farkas E (2010) Current results on biological activities of lichen secondary metabolites: a review. Zeitschrift fur Naturforschung C 65:157–173

O'Neill MA, Mayer M, Murray KE, Rolim-Santos HM, Santos-Magalhaes NS, Thompson AM, Appleyard VC (2010) Does usnic acid affect microtubules in human cancer cells? Braz J Biol 70:659–664

Ogmundsdottir HM, Zoega GM, Gissurarson SR, Ingolfsdottir K (1998) Anti-proliferative effects of lichen-derived inhibitors of 5-lipoxygenase on malignant cell-lines and mitogen-stimulated lymphocytes. J Pharm Pharmacol 50:107–115

Omarsdottir S, Olafsdottir ES, Freysdottir J (2006) Immunomodulating effects of lichen-derived polysaccharides on monocyte-derived dendritic cells. Int Immunopharmacol 6:1642–1650

Pavlovic V, Stojanovic I, Jadranin M, Vajs V, Djordjevic I, Smelcerovic A, Stojanovic G (2013) Effect of four lichen acids isolated from *Hypogymnia physodes* on viability of rat thymocytes. Food Chem Toxicol 51:160–164

Pelicano H, Martin DS, Xu RH, Huang P (2006) Glycolysis inhibition for anticancer treatment. Oncogene 25:4633–4646

Porporato PE, Dhup S, Dadhich RK, Copetti T, Sonveaux P (2011) Anticancer targets in the glycolytic metabolism of tumors: a comprehensive review. Front Pharmacol 2:49

Ren MR, Hur JS, Kim JY, Park KW, Park SC, Seong CN, Jeong IY, Byun MW, Lee MK, Seo KI (2009) Anti-proliferative effects of *Lethariella zahlbruckneri* extracts in human HT-29 human colon cancer cells. Food Chem Toxicol 47:2157–2162

Rerole AL, Jego G, Garrido C (2011) Hsp70: anti-apoptotic and tumorigenic protein. Methods Mol Biol 787:205–230

Russo A, Caggia S, Piovano M, Garbarino J, Cardile V (2012) Effect of vicanicin and protolichesterinic acid on human prostate cancer cells: role of Hsp70 protein. Chem Biol Interact 195:1–10

Russo A, Piovano M, Lombardo L, Vanella L, Cardile V, Garbarino J (2006) Pannarin inhibits cell growth and induces cell death in human prostate carcinoma DU-145 cells. Anticancer Drugs 17:1163–1169

Seo C, Yim JH, Lee HK, Oh H (2011) PTP1B inhibitory secondary metabolites from the Antarctic lichen *Lecidella carpathica*. Mycology 2:18–23

Shibata S, Nishikawa Y, Tanaka M, Fukuoka F, Nakanishi M (1968) Antitumour activities of lichen polysaccharides. Zeitschrift fur Krebsforschung 71:102–104

Shrestha G, Clair St LL (2013) Lichens: a promising source of antibiotic and anticancer drugs. Phytochem Rev 12:229–244

Shukla V, Joshi GP, Rawat MSM (2010) Lichens as a potential natural source of bioactive compounds: a review. Phytochem Rev 9:303–314

Sica A, Schioppa T, Mantovani A, Allavena P (2006) Tumour-associated macrophages are a distinct M2 polarised population promoting tumour progression: potential targets of anti-cancer therapy. Eur J Cancer 42:717–727

Singh N, Nambiar D, Kale RK, Singh RP (2013) Usnic acid inhibits growth and induces cell cycle arrest and apoptosis in human lung carcinoma A549 cells. Nutr Cancer 65(Suppl 1):36–43

Song Y, Dai F, Zhai D, Dong Y, Zhang J, Lu B, Luo J, Liu M, Yi Z (2012) Usnic acid inhibits breast tumor angiogenesis and growth by suppressing VEGFR2-mediated AKT and ERK1/2 signaling pathways. Angiogenesis 15:421–432

Zambare VP, Christopher LP (2012) Biopharmaceutical potential of lichens. Pharm Biol 50:778–798

Lichen Dyes: Current Scenario and Future Prospects

12

Preeti Shukla and D.K. Upreti

Contents

Abstract

Lichens are well-known organism used as colouring agent since ancient times. Earlier traditional cow urine method (CUM) was employed for extraction of dyes from lichens which was replaced with ammonia fermentation method (AFM) and later boiling water method (BWM) was introduced. In addition to the traditional methods, DMSO extraction method (DEM) used for extraction of lichen dyes is described with the underlying chemical reactions. Lichens are extremely slow-growing organism, and reports reveal that optimally grown mycobiont (fungal part of lichen) can also produce lichen compounds which are responsible for production of coloured pigments. Thus, cultured mycobiont can be used for extraction of dyes. The growth of mycobiont can be enhanced after manipulating the culture media to obtain greater biomass. Lichen mycobiont culture or lichen culture in laboratory will not only provide useful information about the potential dye-yielding lichens but also help to conserve these unique and peculiar slow-growing organisms in nature. Information on extraction of lichen dyes from natural thallus and from cultured mycobiont of lichens and dye yield of different lichen taxa are provided in the present study.

Keywords

Dyes · Fermentation · Mycobiont · Secondary metabolites · Extraction

P. Shukla · D.K. Upreti (✉)
Lichenology Laboratory, CSIR-National Botanical Research Institute, Rana Pratap Marg, Lucknow 226001, Uttar Pradesh, India
e-mail: upretidk@rediffmail.com

D.K. Upreti et al. (eds.), *Recent Advances in Lichenology*,
DOI 10.1007/978-81-322-2235-4_12

12.1 Introduction

Lichens are self-supporting organization of a fungus with one or more green or blue-green algae. As a result of symbiosis, the lichen's photobiont and mycobiont have expanded into many habitats where separately they would be rare or non-existent. Lichens can grow under rather diverse and sometimes extreme ecological conditions, in very cold and dry environments, either at polar latitudes or at extreme altitudes (up to 7400 m).

The lichen symbiosis represents a valuable source for commercially interesting compounds including antimicrobial agents, dyeing agents, ingredient of spices and perfumes. Owing to the presence of secondary metabolites, lichens are considered as economically important herb. The characteristic secondary metabolites known 'lichen acids' are the main source in production of dyes which can colour fibres (Richardson 1988). Lichen dyes have a particular affinity for natural fibres. Although the other commodities which can be dyed with lichens include wood, marble, leather, wine, and food materials, the use of lichens as source of dyeing agent has a long history, and the indigenous knowledge system particularly associated with extraction and processing of natural dyes from lichens is ancient. The purple orchils have been the most important groups of lichen dyes in historical trade and during the classical Greek and Roman period. Casselman (2001) detailed a list of over 100 common species of lichen worldwide that produce reliable colour dyes. Cow urine method (CUM), ammonia fermentation method (AFM), boiling water method (BWM) and dimethyl sulphoxide extraction method (DEM) are some popular lichen dye extraction methods which yield beautiful purple, pink, yellow, brown, orange, and green colours. Together with the colour, lichen dyes also give characteristic odour to the dyed fibres.

With the discovery of first synthetic dye in 1856 (Margareta 1981), the use of natural dyes was replaced completely by synthetic compounds due to their easy extraction methods and cost-effectiveness. As the synthetic dyes have tremendous environmental impact due to their toxic, carcinogenic and non-biodegradable nature, the demand of dyes and colours for textile, food and cosmetics from natural sources has increased in the recent years. Several attempts are now being made for the development of user-friendly pigments from the natural sources (Table 12.1).

The lichens are extremely slow-growing organism and not available in greater amount for their commercial exploitation. Recent researches have shown that the mycobiont cultured on artificial media is able to produce some of the secondary metabolites that can produce coloured compounds through chemical reactions. Because majority of organic compounds found in lichens are secondary metabolites of fungal origin, the mycobiont culture is the only way to exploit the secondary metabolites produced by lichens. In recent years, methods for cultivating lichen mycobiont and lichen tissues have been developed, and these have given rise to hopes for the production of metabolites which are otherwise difficult to obtain. Research reports on extraction methods, other than traditional ones, are available, and extraction of dyes from natural thalli and cultured mycobiont is possible.

Here, some approaches have been discussed in context to the extraction of lichen dyes from natural thallus as well as from the mycobiont (cultured fungal part). Further information on lichen dyes extracted from Indian lichen species has also been presented.

12.2 History of Lichen Dyes

Lichens have been used for making dyes since time immemorial. Orchil (purple dye) was the first documented dye produced from *Roccella* spp. and was a trade commodity developed by Phoenicians. Orchil (purple dye) was historically important as 'royal purple' throughout the Europe as purples symbolize royalty and power and were very expensive. Orchil was often used as base dye for dyeing wool with more expensive Tyrian purple, produced by a small gland in shellfish (mollusc) native to the Mediterranean

Table 12.1 Lichen species used for producing colour dyes

S. No.	Species name	Dye colour produced	References
1	*Acarospora chlorophana* (Wahlenb.) Zopf	Yellow to grey	Brough (1988), Upreti et al. (2010)
2	*Bryoria lactinea* (Nyl.) Brodo and Hawksw	Brown to yellow	Upreti et al. (2010), Shukla and Upreti (2014b)
3	*Caloplaca trachyphylla* (Tuck.) Zahlbr.	Yellow to red-purple	Brough (1988), Upreti et al. (2010)
4	*Candelariella vitellina* (Mull. Arg.) Zahlbr.	Yellow, brown to grey	Brough (1988), Upreti et al. (2010)
5	*Cladonia pyxidata*	Purple	Kok (1966)
6	*Dermatocarpon miniatum* (L.) Mann	Brown to grey	Brough (1988), Upreti et al. (2010)
7	*Dermatocarpon vellerum* Zachacke.	Green to yellow	Shukla and Upreti (2014b)
8	*Evernia mesomorpha* Nyl.	Pink to yellow	Shukla and Upreti (2014a)
9	*Evernia prunastri* (L.) Ach.	Purple to yellow	Kok (1966), Hodge (2006), Upreti et al. (2010)
10	*Lecanora alphoplaca* (Wahlenb. Ex Ach.) Ach.	Brown to yellow	Brough (1988)
11	*Lecanora frustulosa* (Dickson) Ach.	Pink, yellow to brown	Brough (1988), Upreti et al. (2010)
12	*Lobaria pulmonaria*	Green to brown	Kok (1966)
13	*Nephromopsis nephromoides* (Nyl.) Ahti and Randl	Green, brown to yellow	Shukla and Upreti (2014a)
14	*Ochrolechia* spp.	Purple	Kok (1966), Casselman (1994)
15	*Parmelia* spp.	Brown to yellow	Kok (1966), Brough (1988), Casselman (1994), Hodge (2006)
16	*Parmotrema reticulatum* (Taylor) Choisy	Brown to yellow	Shukla and Upreti (2014a)
17	*Parmotrema tinctorum*	Purple to golden brown	Upreti et al. (2010), Shukla and Upreti (2014a)
18	*Peltigera rufescens* (Weiss) Humb.	Green to yellow	Upreti et al. (2010), Shukla and Upreti (2014a, b)
19	*Pseudoevernia intense*	Yellow to brown	Dean et al. (2012)
20	*Punctelia reducta* (Nyl.) Hale	Pink to yellow	Shukla and Upreti (2014a)
21	*Ramalina* spp.	Yellow	Kok (1966), Upreti et al. (2010)
22	*Rhizoplaca melanophthalma* Leuck and Poelt	Orange to brown	Brough (1988), Upreti et al. (2010)
23	*Roccella montagneii* Bel. em. D. D. Awasthi,	Purple to red	Kok (1966), Casselman (1994), Upreti et al. (2010), Shukla and Upreti (2014a, b)
24	*Teloschistes exilis*	Yellow, grey to blue	Dean et al. (2012)
25	*Umbilicaria pustulata*	Purple, red	Kok (1966), Casselman (1994)
26	*Umbilicaria phaea* Tuck.	Brown to grey	Brough (1988)

(continued)

Table 12.1 (continued)

S. No.	Species name	Dye colour produced	References
27	*Usnea ghattensis* G. Awasthi		Upreti et al. (2012), Shukla and Upreti (2014b)
28	*Usnea longissima* Ach.	Violet, brown to yellow	Shukla and Upreti (2014b)
29	*Usnea undulata* Stirt.	Orange to brown and yellow	Kok (1966), Dean et al. (2012); Shukla and Upreti (2014a, b)
30	*Xanthoparmelia chlorochroa* (Tuck.) Hale	Brown to yellow	Brough (1988), Hart and Cox (2000), Upreti et al. (2010), Dean et al. (2012)
31	*Xanthoria elegans* (Link.) Th. Fries	Red to yellow	Shukla and Upreti (2014b)
32	*Xanthoria parietina* (L.) Th. Fr.	Pink to yellow	Upreti et al. (2010), Shukla and Upreti (2014b)
33	*Xanthoria polycarpa* (Ehrh.) Oliv.	Purple	Brough (1988), Upreti et al. (2010)

and Atlantic coast. Being much more easily collected and prepared than mollusc dye, the lichen dyes became more popular. 'Crottle' was prepared from *Parmelia* spp., a brown lichen dye popular in Scotland. The wool dyed with crottle was used in well-known Harris Tweed (http://www.chriscooksey.demon.co.uk/lichen/orcein.htm).

The recipes of lichen dyeing were kept as secrets in early times. The earliest known description of the preparation of orcein (purple dye) was given by Roseto in the year 1540. The process generally consisted of obtaining the desired lichen, adding it to stale urine and slaked lime. During the seventeenth century, a growth in usage of orchils developed with the discovery of ammonia treatment process. Later, shortage of supplies developed and new sources of lichens were developed. In twentieth century, usage of lichen dyes of all types has declined due to competition from lower costs and progressively improving synthetic dyestuffs.

Along with dyeing textiles, orecin was also used as food colouring agent, microscopical stain and preparation of litmus (acid/base indicator). Traditional dyeing methods such as fermentation in urine and ammonia were followed in ancient times to dye natural fibres. Sometimes, ammonium salts were added to the fermenting solution that functions as mordant. BWM and DMSO extraction methods were later developed. The colours obtained through ammonia/urine fermentation methods were best when compared with the colours obtained with BWM and DEM.

12.3 Processing of Lichens and Mycobiont for Dye Preparation

The lichens generally grow on bark of trees, rocks, over mosses and soil as a result the surface of the lichens gets contaminated with many foreign particles. Thus, the collected samples from which dyes have to be extracted must be washed thoroughly and desired parts should be cut off like fruiting bodies for 'spore-derived cultures' and thallus fragment for 'thallus fragment-derived cultures'. It is emphasized that only naturally detached or found lichen should be used for dyeing and whole thallus must not be harvested in order to conserve them in their natural habitat

12.4 Mycobiont Culture

The culture of selected lichens should start within few days after the collection of samples as best results can be observed from freshly collected samples. The mycobiont culture can be done following the 'spore discharge method' and 'thallus fragment-derived culture method' (Yoshimura et al. 2001).

12.4.1 Spore-Derived Culture

It is the standard method of lichen culture where culture is initiated from lichen spores. The method is established by Ahmadjian (1973), and later, many amendments have been made by practitioners (Ahmadjian 1993; Yoshimura et al. 2001). In this method, the thalli collected are screened under dissecting zoom microscope, and healthy thalli bearing fruiting bodies (apothecia) are selected for sterilization. The mature apothecia are cut off from the selected part of thalli. The apothecia are re-hydrated by placing them in a sterile water-saturated atmosphere for 24 h at 18–20 °C. The rehydrated apothecia are thoroughly cleaned with tap water, under a dissecting microscope. The apothecia are washed in running tap water for 30 min to 1 h and then treated with Tween 80 (2 %) for 5 min. The surfactants are removed by washing the apothecia in double distilled water. The sterilized apothecia are dried in autoclaved petri plates lined with dry filter paper and used for mycobiont culture. The sterilized apothecia are attached to the inside of the petri plate lids with the help of petroleum jelly. Petri plates containing solidified media were then inverted over the lids, and ascospores are allowed to discharge onto the agar medium. Plates are incubated in BOD incubator and are observed periodically over 3- to 5-months periods. The media is prepared by suspending the ready-made powder media in distill water, heating to boiling, to dissolve the media completely. The media are sterilized by autoclaving at 15 lbs pressure (121 °C) for 15 min.

12.4.2 Thallus Fragment-Derived Culture

The thallus fragment-derived culture can be used to obtain mycobiont of lichen species that do not produce apothecia. The method was described by Yamamoto (1985) and popularly known as 'lichen tissue culture method'. Generally, the thallus fragment-derived cultures are liable to contamination with micro-organisms present in the thallus surface; therefore, small thallus fragment (in μm range) is recommended to use (Yoshimura et al. 2001).

A segment of about 1 cm length from the thallus is separated by scissor or knife and washed under tap water for 30 min to 1 h The thallus fragment is then homogenized with mortar–pestle in distil water. The homogenate is filtered through nylon sieve mesh 500 μm. The solution is again filtered through nylon sieve mesh 150 μm, and thallus fragments thus obtained on the mesh are picked up with the sterile needle under the dissecting microscope and inoculated onto the surface of slant malt–yeast medium. The slants are maintained at 15 °C in dark. Two weeks after the inoculation, the mycobiont hyphae and algal cells can be observed on slants. The symbionts can be isolated and further sub-cultured in desired culture medium that suits the growth and metabolite production in the cultured mycobiont.

Although the production of lichen compounds by a mycobiont sometimes stops or is quite meagre in amount suggested the mycobiont needs algal stimulation for the production of secondary metabolites. Enough quantities of desired molecules can be achieved by batch cultures and fermentation. Mycobiont cultures of some selected Indian species are presented in Figs. 12.1 and 12.2.

12.5 Culture Conditions and Mycobiont Extraction

Petri plates containing the cultured mycobiont should be incubated in the dark at 22 °C and 70–80 % relative humidity in BOD incubator and observed periodically over 3–5 months. When the mycobiont is fully grown, the cultured mycelial mat (150 days old or more) can be used for dye preparation. Mycobiont is removed from petri plate by keeping plates in the water bath at 80 °C and dried with lyophilizer for 12 h. The obtained biomass is grinded using a mortar and

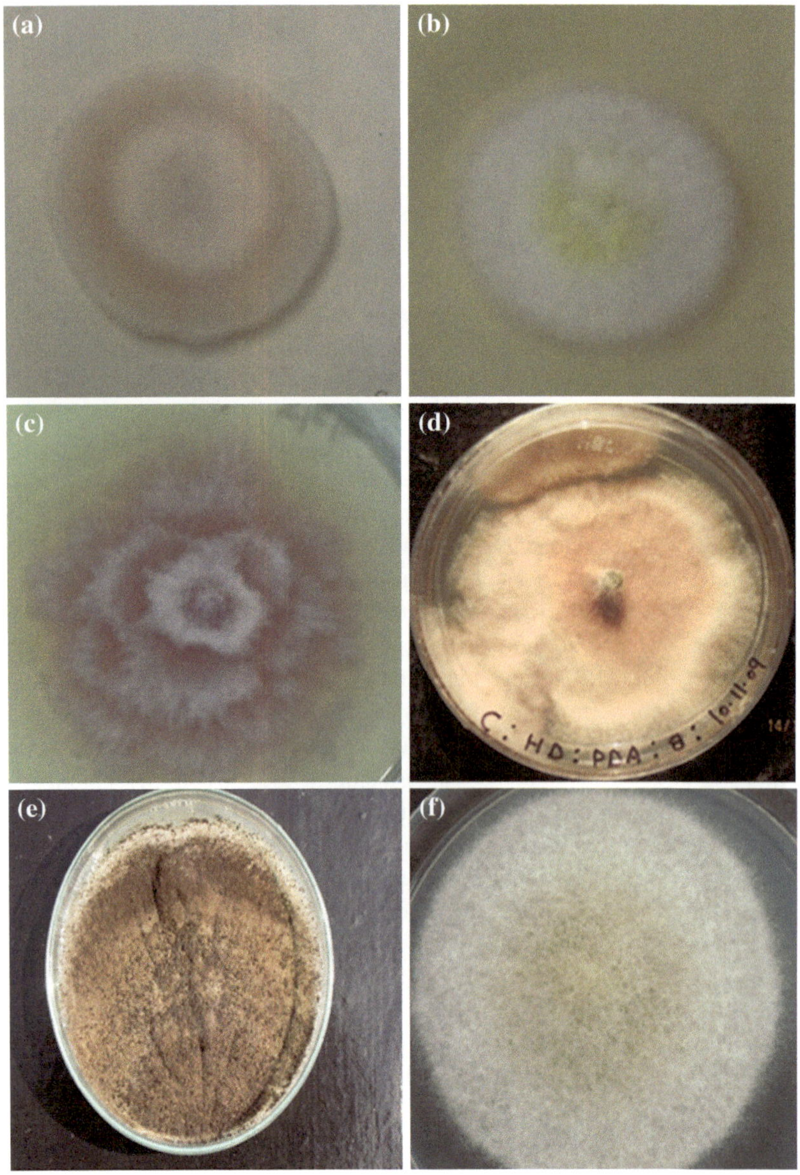

Fig. 12.1 Mycobiont culture of different Indian lichen species, **a** *Bulbothrix satschwansis* (Zahlbr.) Hale, **b** *Everniastrum cirrhatum* (Fr.) Hale, **c** *Everniastrum nepalense* (Taylor) Hale, **d** *Heterodermia diademata* (Taylor) D. D. Awasthi, **e** *Parmelaria subthomsonii* D.D. Awasthi, **f** *Parmelaria thomsonii* (Stirton) D. D. Awasthi

pestle, and the powdered mycobiont is used for dye preparation by the same methods as applied to extraction through natural thallus.

Lichens and the cultured mycobiont samples selected for extraction of dyes should be thoroughly washed under tap water. The samples are then shadow-dried in air or sun-dried within a temperature range of 37–40 °C so as to reduce the moisture content of the samples. The dried samples are crushed to break down the material into small units, preferably powder.

12.6 Extraction of Dyes from Lichen Thallus and Cultured Mycobiont

There are basically three most common extraction methods for lichen dyes that are known. In fermentation method, the lichens are allowed to ferment both in aerobic and anaerobic conditions to give rise to coloured products. In boiling method, the lichens are heated in specific temperature for

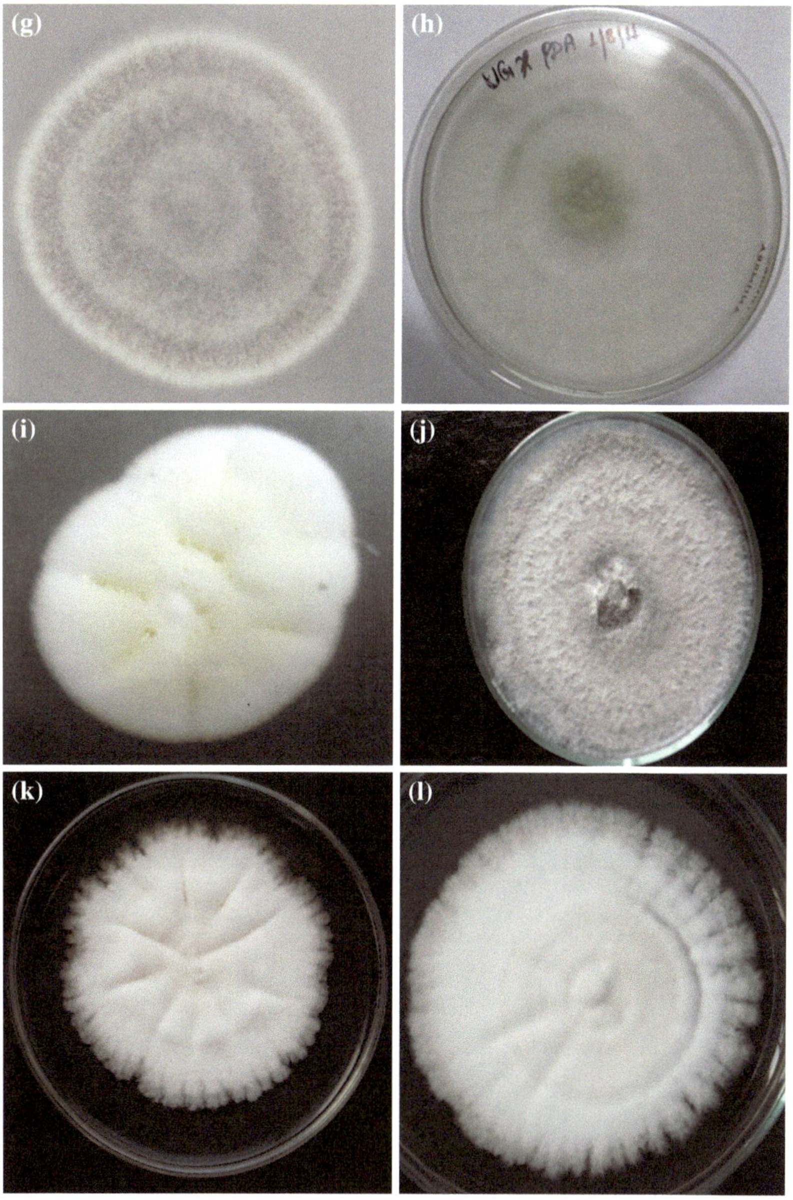

Fig. 12.2 Mycobiont culture of different Indian lichen species, **g** *Ramalina conduplicans* Vain., **h** *Usnea ghattensis* G. Awasthi, **i** *Xanthoria parietina,* (L.) Th. Fr., **j** *Stereocaulon foliolosum* Nyl., **k** *Sticta nylandriana* Zahlbr, **l** *Usnea pseudosinenesis* Asahina

hours to extract coloured compounds, while in the solvent extraction method, lichens are kept in polar and non-polar solvents for few weeks to generate coloured products. All the other additional methods are modified forms of the above-mentioned methods. Before the lichens are subjected to the different extraction methods, they are processed for dye preparation. Lichen dyes are unique as no mordant or intermediary agents are required to be taken up by fibres. The use of mordanting substances, acids or alkalies is sometimes used to vary the colour in lichen dyeing.

12.7 Methods for Extraction of Lichen Dyes

12.7.1 Cow Urine Method (CUM)

CUM is the oldest method of dye extraction from lichens, and it is vanished completely in modern era. In this method, the lichens are steeped in stale urine for about 3–4 weeks. The fibres are added to the filtrate and again left for few days to get dyed. In some modified methods, the mixture of cow urine and lichens is kept at moderate heat for long time and substance having a thick and strong texture like bread is taken out and made into small cakes. These pieces are wrapped in dock leaves and hung up to dry in peat smoke. The dried dye can be stored for years, and when needed for dyeing, it can be dissolved in warm water to dye fibres. In countries such as Sweden and Scotland, this method was used to produce red coloured dye from *Lecanora* sp. Later, cow urine was replaced with ammonia as it was found that diluted ammonia solution can also solve the purpose of cow urine in dyeing.

12.7.2 Ammonia Fermentation Method (AFM)

AFM is the most widely used method during eighteenth and nineteenth centuries. The simplest procedure involves extraction of lichens in diluted ammonia. The powdered lichen samples are added to diluted ammonium hydroxide solution (one part NH_4OH and 10 parts distilled water), and the content is mixed thoroughly and left for 3–4 weeks at room temperature. The extract is then filtered and fibres are added. The content is again left for 3–4 weeks. The fibres are then removed from the flasks, rinsed and dried.

In some modified versions of the AFM, the lichens are first boiled in the solution of ammonium carbonate. The mixture is then cooled, and ammonia is again added to make the mixture damp and is kept for 3–4 weeks. In another processes, the extracted lichens in ammonia are acidified so that the dissolved dye precipitates and is washed. Ammonia is again added, and the solution is heated in air until some colour is developed. The colour is again precipitated with calcium chloride, and resulting insoluble coloured solid is obtained that does not fade in light.

12.7.3 Boiling Water Method (BWM)

The powdered lichen samples are added to distilled water and heated till boiling. The mixture is maintained at simmer for 1 h. The content is filtered into a clean flask, and the filtrate is again maintained at simmer for at least 2 h until some colour is obtained. Pre-soaked fibres are then immersed in dye bath and are slowly heated at maximum 90 °C for 2 h. The dye bath is cooled after dyeing; the threads are rinsed in cold water and dried. Some dyers extract dyes and colour the fibres simultaneously. A layer of lichen sample and a layer of fibres are put until the dye bath is full. The dye bath is then filled with water and brings to boil till the colour develops. It is reported that small amounts of acetic acid put with the lichens assist in exhausting the colour.

12.7.4 DMSO Extraction Method (DEM)

The solvent extraction method was developed just before the dawn of twentieth century. In this method, the powdered lichen samples are added to 50 ml crude dimethyl sulphoxide solution. The content is stirred vigorously and left for 1 month at room temperature. After 1 month, the content is filtered into another clean flask and pre-soaked threads are added for dyeing. The threads are removed from the flask after 1 month, washed with distilled water and were left for drying. After dyeing, the fibres can be stored at room temperature. The colours can be named with those matching Ridgway colours (Ridgway 1912).

According to Casselman (2001), lichens are substantive dyes and do not require a mordant; however, the use of mordant increases uptake of colour, improves fastness and varies the colour. The lichen dyes obtained from natural thalli and cultured mycobiont can be quantified. The solvents used for extraction of dyes are evaporated using rotary evaporator to get dry weight of the dyes.

12.8 Chemistry of Lichen Dyes

Lichens produce variety of secondary metabolites, some of which are common in plants or in higher fungi, but about 80 % metabolites are specifically produced by lichens. These secondary metabolites undergo series of chemical reactions in the presence of air, water and solvents to produce coloured compounds used for dyeing fibres. The secondary metabolites, often called lichen acids, are of fungal origin. More than 1,000 secondary metabolites are so far known in lichens worldwide. Beside the externally visible crystallized and non-crystallized pigments that are deposited in the upper surface of the lichen thallus, also colourless substances are common, which are predominantly found in internal parts of the thalli. According to their chemical structures, most lichen substances are phenolic compounds (orcinol and b-orcinol derivatives), dibenzofuranes (usnic acid), depsides (barbatic acid), depsidones (salazinic acid), depsones (picrolichenic acid), lactones (protolichesterinic acid, nephrosterinic acid), quinones (parietin), and pulvinic acid derivatives (vulpinic acid).

The depsides and depsidones, common lichen compounds, are aromatic in nature formed by joining two or sometimes three phenolic units. Lichens have evolved diverse biosynthetic pathways mainly polymalonate, shikimic acid and mevalonic acid pathways to produce this diversity of compounds.

It has been proved that among the diverse group of secondary metabolites, esters, depsides and depsidones are precursors of orcein, the coloured compound formed in lichen dyes. These chemicals hydrolyse in aqueous solution to give orsellic acid. The orsellic acid undergoes decarboxylation reaction to yield orcein/orsinol. The orcein reacts by sequence of condensation reaction incorporating substituent from the solvents to give various derivatives of orcein. The mixture of orcein derivatives gives the actual colour from the lichens. Further, the colouring of the fibres is due to the chemical reaction taking place between the orcein derivatives and chemical constituents of fibres. The ortho–hydroxyl–aldehyde group present in the lichen dyes reacts with the free amino group present in natural protein fibres, such as silk and wool, and forms stable Schiff base (compounds having C=N function) by azomethine linkages.

$$\underset{\text{Depside}}{C_{16}H_{14}O_7} + \underset{\text{Water}}{H_2O} \rightarrow \underset{\text{Orsellicacid}}{2C_8H_8O_4}$$

$$\underset{\text{Orsellic acid}}{2C_8H_8O_4} \rightarrow \underset{\text{Orcein/Orsinol}}{2C_7H_8O_2} + \underset{\text{Carbon dioxide}}{2CO_2}$$

$$\underset{\text{Orcein}}{2C_7H_8O_2} + \underset{\text{Ammonium hydroxide}}{2NH_4OH} + \underset{\text{Oxygen}}{3O_2} \rightarrow \underset{\text{Orcein derivative}}{2C_7H_7NO_3} + \underset{\text{Water}}{6H_2O}$$

Pierre Robiquet in the year 1829 isolated orcein as large colourless crystals from lichens by extraction with ethanol. Unlike other dyes, the chemical structure of orcein was unknown until 1960s. Hans Musso during the year 1956–1965 worked out the chemistry of orcein and revealed that it contains a variety of phenazones. The actual orcein is a mixture of hydroxy-orecins, amino-orceins and amino-orceinimines.

12.9 Lichen Dyes Extracted from Indian Lichen Species

India is represented by the occurrence of more than 2,300 species of lichens belonging to 305 genera and 74 families of lichens. The lichens grow luxuriantly in the Himalaya and higher altitudes of Western Ghats. The rich lichen diversity in India can serve as raw material for making dyes due to their unique chemistry. Studies revealed that Indian Parmelioid lichens are potential source of natural dyes and provide brilliant colours in different solvents, and wide range of colour dyes such as brown, orange,

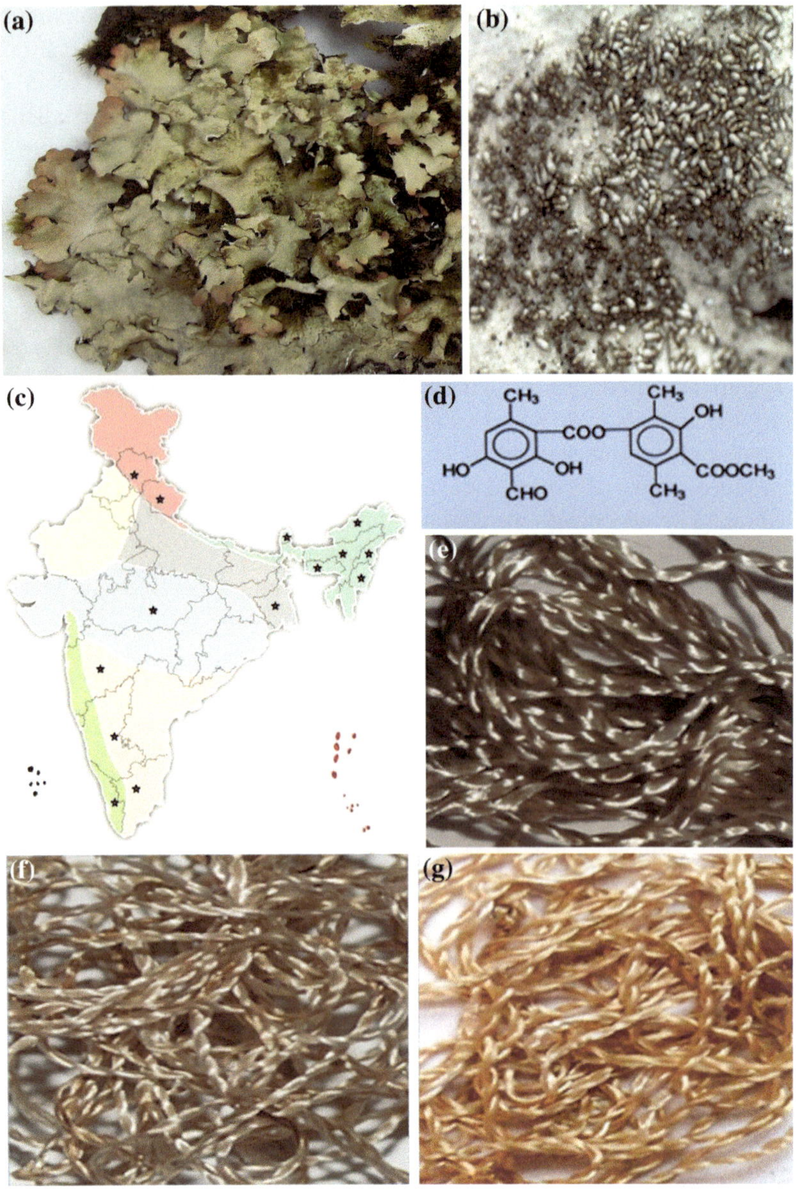

Fig. 12.3 **a** Natural thallus of *Parmelinella wallichiana* (Taylor) Elix and Hale, **b** isidia, **c** map showing distribution of *Parmelinella wallichiana* in India, **d** chemical structure of Atranorin, **e** silk thread dyed through AFM, **f** silk thread dyed through DEM, **g** silk thread dyed through BWM

yellow, purple, pink and grey can be obtained (Figs. 12.3, 12.4, 12.5, 12.6, 12.7 and 12.8). AFM, BWM and DEM methods are reported to be used for extraction of lichen dyes to colour different fibres such as silk, Tussar silk and cotton. Lichen dyes are environment friendly and give better quality of colour than synthetic dyes. Lichen dyes not only provide colours to the fibres but also imparts musky odour to the fibres. Furthermore, the dyed products are reputed to be insect-proof as the secondary metabolites render the fibres distasteful to the insects.

Out of the three methods, the dye colours extracted from AFM are reported to be much brighter as compared to BWM and dimethyl sulphoxide extraction method (DEM). AFM gives different shades of pink, violet, orange, grey, brown and yellow colours. The DEM gives shades of green, brown and yellow, while mostly shades of orange, brown and yellow colours are

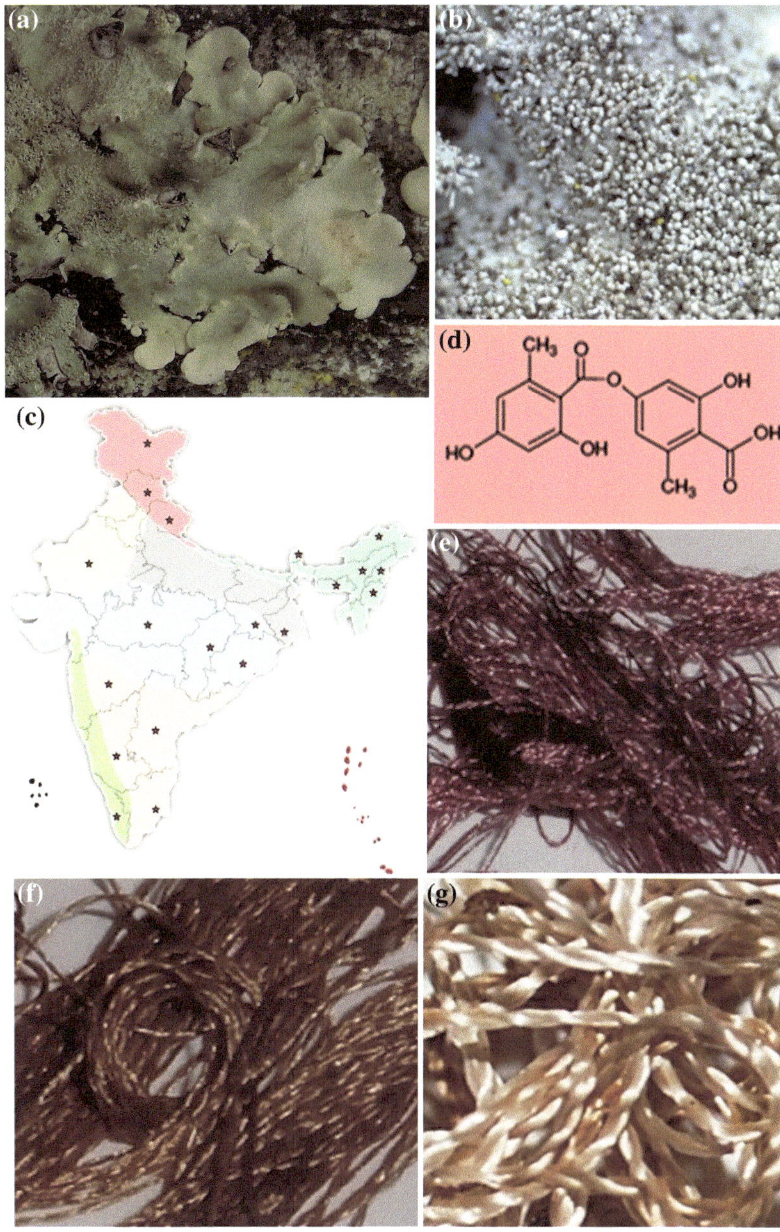

Fig. 12.4 **a** Natural thallus of *Parmotrema tinctorum* (Despr. ex Nyl.) Hale, **b** isidia, **c** map showing distribution of *Parmotrema tinctorum* in India, **d** chemical structure of lecanoric acid, **e** silk thread dyed through AFM, **f** silk thread dyed through DEM, **g** silk thread dyed through BWM

produced in BWM. Lichens can dye natural fibres of all kinds, such as wool, silk and synthetic fibres, and plastic buttons and animal hides substantially do not require any mordant. The colours produced are the fastest known.

India, being a mega-diversity region of the world, exhibits rich diversity of different plant groups including lichens. Indian lichen species and the colours observed through different extraction methods are listed in Table 12.2. Out of the different growth forms of lichens, the foliose and fruticose lichens, bigger in shape and size, were studied for their dyeing potential through BWM, AFM and DEM (Table 12.3).

The lichen secondary metabolites present and corresponding colour of dye produced by Indian lichens revealed that purple-colour-producing lichens such as *Parmotrema tinctorum* and

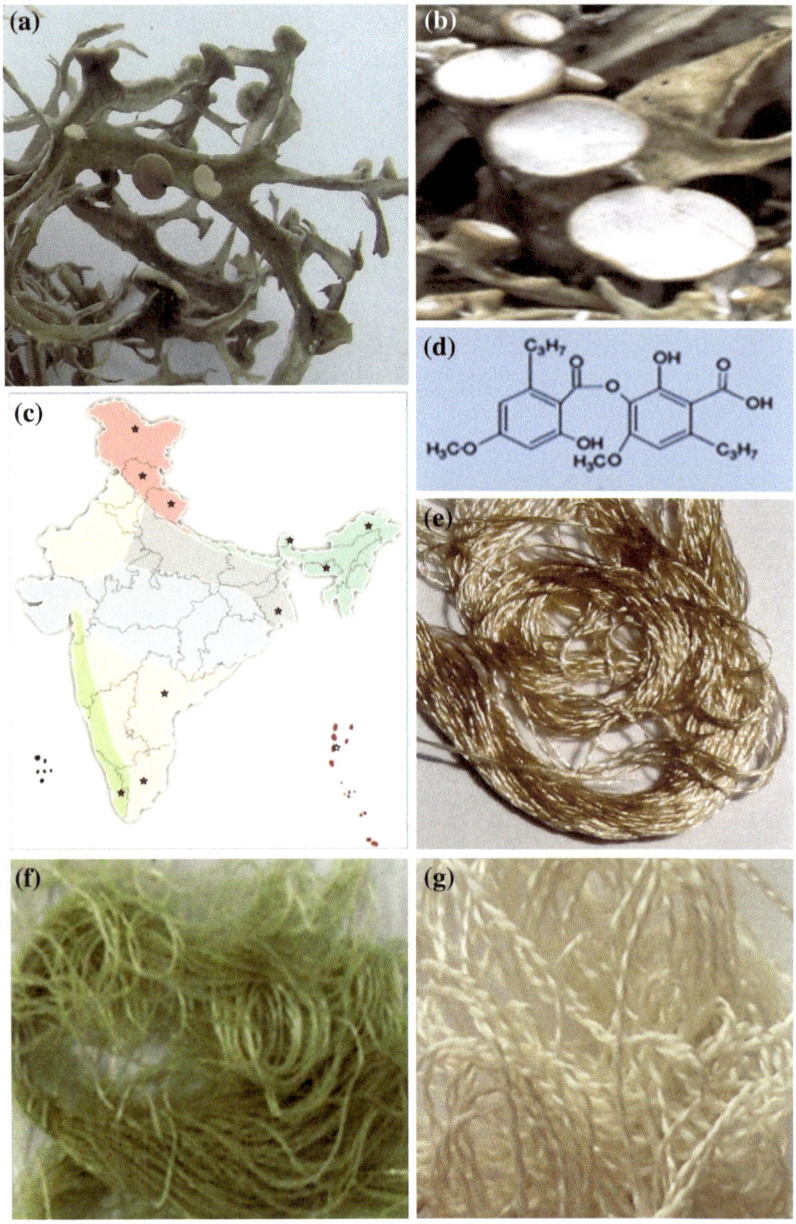

Fig. 12.5 **a** Natural thallus of *Ramalina conduplicans* Vain., **b** apothecia, **c** map showing distribution of *Ramalina conduplicans* in India, **d** chemical structure of sekikaic acid, **e** silk thread dyed through AFM, **f** silk thread dyed through DEM, **g** silk thread dyed through BWM

Punctelia reducta have lecanoric acid. The lecanoric acid is a *p*-depside that hydrolyses to orsellic acid and undergoes a series of chemical reaction to form colour-producing substance orcein. *Sticta nylandriana* containing gyrophoric acid and *Heterodermia diademata* containing zeorin also produce shades close to purple-pink. The Pink-purple dye is produced by *Evernia mesomorpha* representing diluted purple may be due to the presence of evernic acid and divaricatic acid, where both are structural homologue of lecanoric acid. The blue-purple dye colour is produced by *Usnea longissima* and *Usnea ghattensis* through AFM may be either due to usnic acid, *p*-depside called barbatic acid or depsidone fumaroprotocetraric acid or it may be due to the combined action of some of the secondary metabolites.

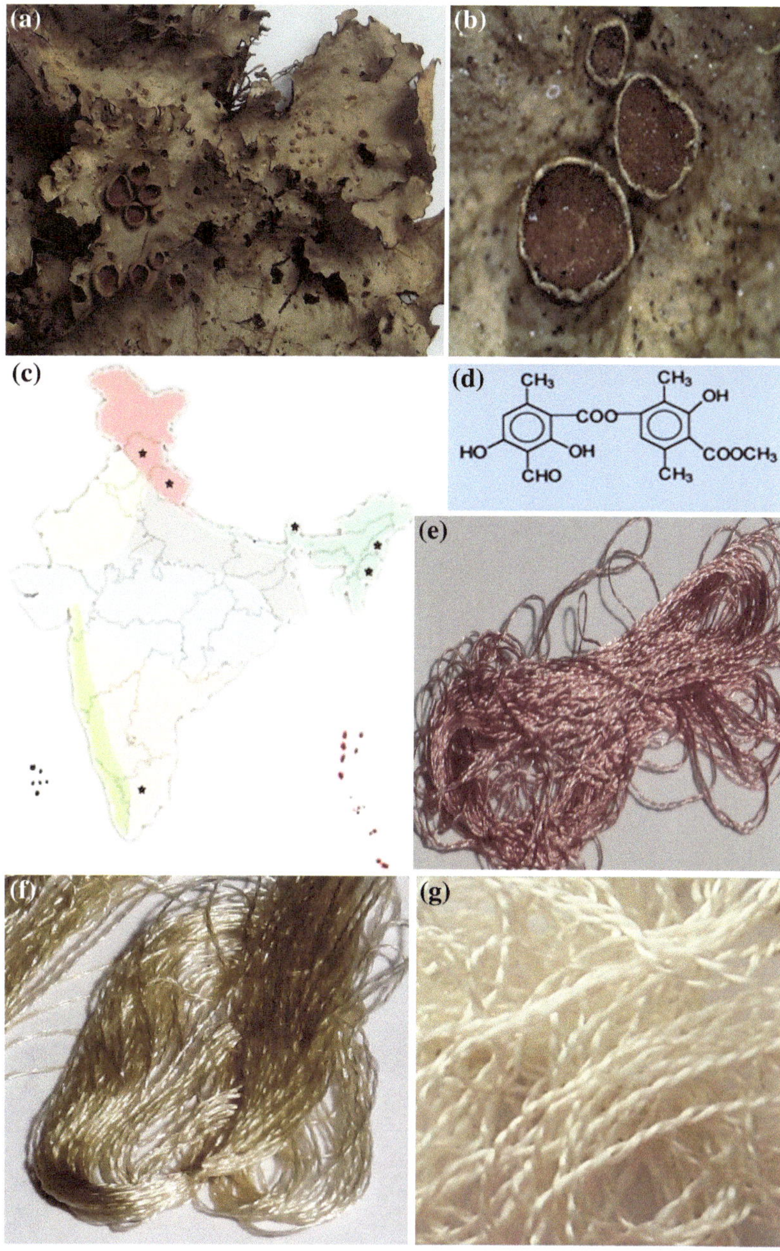

Fig. 12.6 **a** Natural thallus of *Sticta nylanderiana* Zahlbr., **b** apothecia, **c** map showing distribution of *Sticta nylanderiana* in India, **d** chemical structure of atranorin, **e** silk thread dyed through AFM, **f** silk thread dyed through DEM, **g** silk thread dyed through BWM

Bulbothrix satschwensis, *Everniastrum cirrhatum*, *Everniastrum nepalense*, *Parmelinella wallichiana*, *Parmotrema reticulatum*, *Usnea undulata* and *Xanthoparmelia somloensis* contain salazinic acid in their thalli and produce brown dye through AFM and orange dye with BWM. Salazinic acid is responsible for the production of orange shades of colour (Brough 1988), and due to increased fermentation time, the orange shades became darker brown. *Usnea stigmatoides* with stictic acid complex produces brown colour through AFM. The grey dye is produced from *Flavoparmelia caperata* and *Nephrmopsis nephromoids* through AFM. *F. caperata* contains ceparatic acid which may be responsible for the colour, while *N. nephromoids*

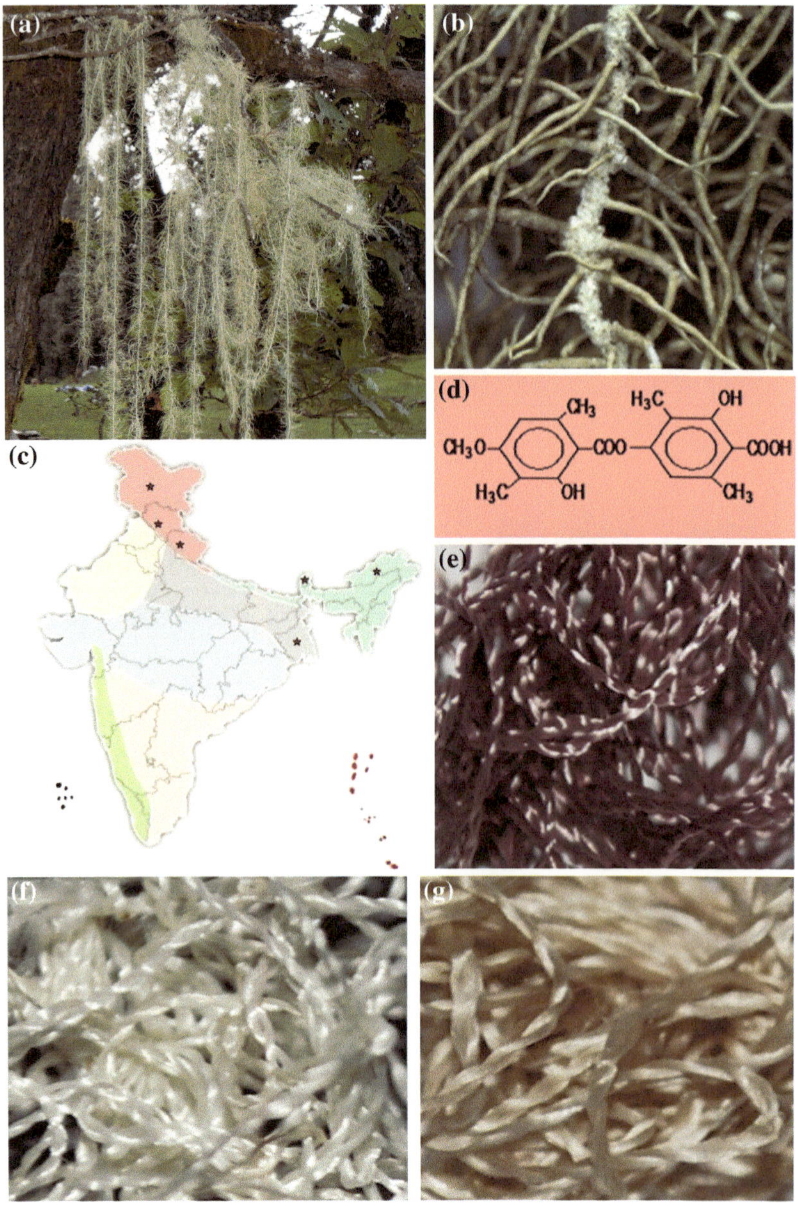

Fig. 12.7 **a** Natural thallus of *Usnea longissima* Ach., **b** decorticated central axis, **c** map showing distribution of *Usnea longissima* in India, **d** chemical structure of Usnic acid, **e** silk thread dyed through AFM, **f** silk thread dyed through DEM, **g** silk thread dyed through BWM

have ceparatic acid with usnic and protolichesterinic acids. *Parmotrema nilgherrensis* produces grey-brown colour through AFM. Lichens containing both atranorin and salazinic acid produce yellow colour, while salazinic acid is responsible for orange and brown dye colours. *Dermatocarpon vellerum*, *Loberia retigera*, *Peltigera rufescence* and *Ramalina conduplicans* produce shades of green colour through DEM, whereas *Nephromopsis nephromoides*, *Parmelinella wallichiana*, *Parmotrema reticulatum*, *P. tinctorum*, *Stereocaulon foliolosum*, *Sticta nylandriana*, *S. platyphylloides*, *Usnea undulata* and *Xanthoparmelia somloensis* produce shades of brown colours through the same method.

Out of several Indian lichen species screened for the production of dyes, *Usnea ghattensis* G. Awasthi and *Heterodermia diademata* (Taylor) D.D. Awasthi are promising lichens that produce dyes from both natural thalli and cultured

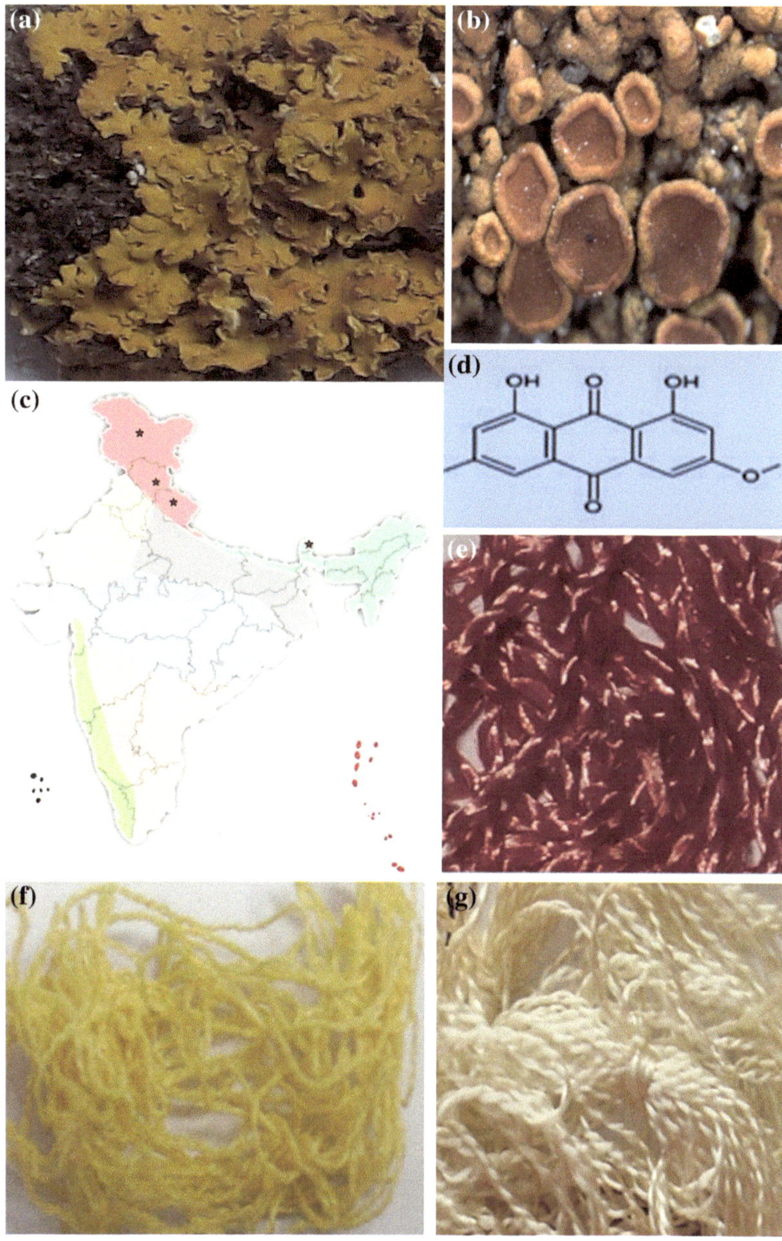

Fig. 12.8 **a** Natural thallus of *Xanthoria elegans* (L.) Th. Fr., **b** apothecia, **c** map showing distribution of *Xanthoria elegans* in India, **d** chemical structure of parietin, **e** silk thread dyed through AFM, **f** silk thread dyed through DEM, **g** silk thread dyed through BWM

mycobionts (Figs. 12.9 and 12.10). *Usnea ghattensis* can produce purple dye both from natural thalli and cultured mycobiont, whereas *Heterodermia diademata* can produce reddish purple colour through AFM. AFM yields the best colour for both natural thalli and cultured mycobiont, whereas colours obtained from DEM are light stable as compared to other solvents. The dye colours vary depending upon fermentation time and alteration of pH of the solvent and temperature. Fibres dipped for longer duration in dye solution yield darker colour, whereas fibres dipped for short period of time yield light colour. The mycobiont culture of the dye-yielding lichens will be helpful to get bulk biomass as a source for dyes and can be used for commercial

Table 12.2 List of lichen secondary metabolites in different Indian lichens and extracted lichen dye colours through different methods

S. No.	Lichens	Secondary metabolites	Colour obtained through BWM	Colours obtained through AFM	Colour obtained through DEM
1	*Bryoria lactinea* (Nyl.) Brodo & D. Hawksw.	Fumaroprotocetraric acid	Pinkish buff	Chamois	Marguerite yellow
2	*Bulbothrix setschwanensis* (Zahlbr.) Hale	Salazinic acid	Buffy brown	Mikado brown	Reed yellow
3	*Cetrelia braunsiana* (Mull. Arg.) W. Culb. & C. Culb.	Alectoronic and α-collatolic acid	Ivory yellow	Light yellowish olive	–
4	*Dermatocarpon vellerum* Zschacke.	No lichen substance present	Marguerite yellow	Buffy olive	Light turtle green
5	*Evernia mesomorpha* Nyl.	Divaricatic acid, Evrnic acid and Sekikaic acid	Pale pinkish buff	Cameo brown	–
6	*Everniastrum cirrhatum* (Fr.) Hale	Salazinic acid, Atranorin and Protolichesterinic acid	Sayal brown	Light yellowish olive	–
7	*Everniastrum nepalense* (Taylor) Hale	Salazinic acid and protolichesterinic acid	Chamois	Mikado brown	Reed yellow
8	*Flavoparmelia caperata* (L.) Hale	Usnic acid, Protocetraric acid, Caperatic acid	–	Light brownish olive	Cartridge buff
9	*Flavopunctelia soredica* (Nyl.) Hale	Lecanoric acid	Orange-cinnamon	Buffy brown	Marguerite yellow
10	*Heterodermia diademata* (Taylor) D. D. Awasthi	Zeorin	White	Vinaceous russet	Cartridge buff
11	*Heterodermia leucomelos* (L.) Poelt	Zeorin, norstictic, salazinic acid and triterpenoids	Pinkish buff	Tawny olive	Primrose yellow
12	*Loberia retigera* (Bory) Trev.	Triterpenoids and thelephoric acid	Cartridge buff	Vinaceous buff	Clear fluorite green
13	*Nephromopsis nephromoides* (Nyl.) Ahti & Randl.	Lichesterinic acid and Protolichesterinic acid	Ivory yellow	Eoru-olive	Olive buff
14	*Parmelaria subthomsonii* D. D. Awasthi	Atranorin, alectoronic and α-collatolic acid	Ivory yellow	Isabella color	Marguerite yellow
15	*Parmelaria thomsonii* (Stirton) D. D. Awasthi	Atranorin, alectoronic and α-collatolic acid	Ivory yellow	Isabella color	Marguerite yellow
16	*Parmelinella wallichiana* (Taylor) Elix & Hale	Salazinic and consalazinic acid	Clay color	Deep olive	Wood brown
17	*Parmotrema nilgharrensis* (Nyl.) Hale	Atranorin, Alectoronic and α-collatolic acid	Ivory yellow	Avellaneous	–
18	*Parmotrema reticulata* (Taylor) Choisy	Salazinic acid, Atranorin and Consalazinic acid	Sayal brown	Light brownish olive	Deep olive buff

(continued)

Table 12.2 (continued)

S. No.	Lichens	Secondary metabolites	Colour obtained through BWM	Colours obtained through AFM	Colour obtained through DEM
19	*Parmotrema tinctorum* (Despr. ex. Nyl.) Hale	Atranorin and Lecanoric acid	Chamois	Perilla purple	Mikado brown
20	*Peltigera rufescense* (Weiss) Humb.	No lichen substance present	White	Ivory yellow	Deep turtle green
21	*Punctelia reducta* (Ach.) Krog	Atranorin and Lecanoric acid	Pale pinkish buff	Hay's brown	Cartridge buff
22	*Ramalina conduplicans* Vain.	Usnic acid, sekikaic acid aggregate and salazinic acid	Cartridge buff	Isabella color	Turtle green
23	*Ramalina hossei* Vain.	Usnic acid and sekikaic acid aggregate	Pale pinkish buff	Olive yellow	White
24	*Ramalina sinensis* Jatta	No lichen substance present	White	Olive yellow	White
25	*Roccella montagnei* Bel. em. D. D. Awasthi	Erythrin	Ivory yellow	Naphthalene violet	Marguerite yellow
26	*Stereocaulon foliolosum* Nyl.	Atranorin and lobaric acid	Chamois	Isabella color	Colonial buff
27	*Sticta nylandriana* Zahlbr.	Atranorin, gyrophoric acid and unknown substances	Chamois	Dark vinaceous	Dark olive
28	*Sticta platyphylloides* Nyl.	No lichen substances	White	Isabella color	Olive buff
29	*Usnea ghattensis* G. Awasthi	Usnic acid	Pinkish buff	Dark dull violet blue	Reed yellow
30	Usnea longissima Ach	Usnic acid, barbatic acid and fumaroprotocetraric acid	Pinkish buff	Dark mineral red	Cartridge buff
31	*Usnea stigmatoides* G. Awasthi	Stictic acid complex	Pinkish buff	Light brownish olive	White
32	*Usnea undulata* Stirt.	Salazinic acid, Usnic acid	Mikado brown	Light yellowish olive	Deep colonial buff
33	*Xanthoparmelia stenophylla* (Ach.) Ahti & Hawksw.	Salazinic acid, consalazinic acid and usnic acid	Mikado brown	Buffy olive	Deep colonial buff
34	*Xanthoria elegans* (Link) Th. Fries	Parietin	Ivory yellow	Corinthian red	Olive ocher
35	*Xanthoria parietina* (L.) Th. Fries	Parietin	Marguerite yellow	Congo pink	Ivory yellow

production of herbal dyes. The mycobiont could be grown in large amounts in the fermenters and could be used in small-scale industries too for producing colour compounds. The extraction of naturaldyes from mycobiont culture is economic and eco-friendly approach. The vast natural resources of lichens in India can also lead to employment generation at the village level if

Table 12.3 Yield of different lichen dyes (dry weight in grams) in three solvents (mean ± Standard deviation)

S. No.	Lichens	Method					
		BWM	% Dye	AFM	% Dye	DEM	% Dye
1	*Bryoria lactinea* (Nyl.) Brodo & D. Hawksw.	0.71 ± 0.06	11.8	0.82 ± 0.03	13.7	0.74 ± 0.04	12.3
2	*Bulbothrix setschwanensis* (Zahlbr.) Hale	0.76 ± 0.03	12.7	0.94 ± 0.02	15.7	0.81 ± 0.03	13.5
3	*Cetrelia braunsiana* (Mull. Arg.) W. Culb. & C. Culb.	0.63 ± 0.04	10.5	0.74 ± 0.01	12.3	–	–
4	*Dermatocarpon vellerum* Zschacke.	0.54 ± 0.02	9.0	0.68 ± 0.02	11.3	0.64 ± 0.04	10.7
5	*Evernia mesomorpha* Nyl.	0.55 ± 0.05	9.2	0.70 ± 0.03	11.7	–	–
6	*Everniastrum cirrhatum* (Fr.) Hale	0.71 ± 0.05	11.8	0.81 ± 0.04	13.5	–	–
7	*Everniastrum nepalense* (Taylor) Hale	0.69 ± 0.07	11.5	0.78 ± 0.04	13.0	0.71 ± 0.06	11.8
8	*Flavoparmelia caperata* (L.) Hale	–	–	0.79 ± 0.04	13.2	0.72 ± 0.02	12.0
9	*Flavopunctelia soredica* (Nyl.) Hale	0.67 ± 0.06	11.2	0.72 ± 0.03	12.0	0.69 ± 0.05	11.5
10	*Heterodermia diademata* (Taylor) D. D. Awasthi	–	–	0.77 ± 0.03	12.8	0.72 ± 0.04	12.0
11	*Heterodermia leucomelos* (L.) Poelt	0.68 ± 0.04	11.3	0.75 ± 0.05	12.5	0.68 ± 0.02	11.3
12	*Loberia retigera* (Bory) Trev.	0.76 ± 0.07	12.7	0.99 ± 0.04	16.5	0.86 ± 0.05	14.3
13	*Nephromopsis nephromoides* (Nyl.) Ahti & Randl.	0.64 ± 0.05	10.7	0.78 ± 0.06	13.0	0.71 ± 0.03	11.8
14	*Parmelaria subthomsonii* D. D. Awasthi	0.68 ± 0.05	11.3	0.79 ± 0.01	13.2	0.75 ± 0.03	12.5
15	*Parmelaria thomsonii* (Stirton) D. D. Awasthi	0.66 ± 0.03	11.0	0.77 ± 0.02	12.8	0.74 ± 0.02	12.4
16	*Parmelinella wallichiana* (Taylor) Elix & Hale	0.79 ± 0.06	13.2	0.93 ± 0.06	15.5	0.82 ± 0.03	13.7
17	*Parmotrema nilgharrensis* (Nyl.) Hale	0.64 ± 0.04	10.7	0.81 ± 0.02	13.5	–	–
18	*Parmotrema reticulata* (Taylor) Choisy	0.61 ± 0.06	10.2	0.79 ± 0.03	13.7	0.72 ± 0.01	12.0
19	*Parmotrema tinctorum* (Despr. ex. Nyl.) Hale	0.65 ± 0.04	10.8	0.76 ± 0.02	12.7	0.69 ± 0.03	11.5
20	*Peltigera rufescense* (Weiss) Humb.	–	–	0.68 ± 0.02	11.3	0.64 ± 0.04	10.7
21	*Punctelia reducta* (Ach.) Krog	0.69 ± 0.02	11.5	0.79 ± 0.05	13.2	0.74 ± 0.03	12.4
22	*Ramalina conduplicans* Vain.	0.74 ± 0.02	12.3	0.91 ± 0.05	15.2	0.85 ± 0.05	14.2
23	*Ramalina hossei* Vain.	0.58 ± 0.03	9.7	0.69 ± 0.07	11.5	–	–
24	*Ramalina sinensis* Jatta	–	–	0.67 ± 0.04	11.2	–	–
25	*Roccella montagnei* Bel. em. D. D. Awasthi	0.63 ± 0.04	10.5	0.70 ± 0.03	11.7	0.68 ± 0.06	11.3
26	*Stereocaulon foliolosum* Nyl.	0.59 ± 0.05	9.8	0.68 ± 0.05	11.3	0.63 ± 0.04	10.5
27	*Sticta nylandriana* Zahlbr.	0.66 ± 0.03	11.0	0.78 ± 0.06	13.0	0.71 ± 0.01	11.8
28	*Sticta platyphylloides* Nyl.	–	–	0.76 ± 0.03	12.7	0.73 ± 0.03	12.2
29	*Usnea ghattensis* G. Awasthi	0.86 ± 0.04	14.3	1.26 ± 0.02	21.0	1.19 ± 0.05	19.8

(continued)

Table 12.3 (continued)

S. No.	Lichens	Method					
		BWM	% Dye	AFM	% Dye	DEM	% Dye
30	*Usnea longissima* Ach.	0.64 ± 0.03	10.7	0.79 ± 0.03	13.2	0.70 ± 0.03	11.7
31	*Usnea stigmatoides* G. Awasthi	0.73 ± 0.04	12.2	0.81 ± 0.06	13.5	–	–
32	*Usnea undulata* Stirt.	0.75 ± 0.02	12.5	0.92 ± 0.03	15.8	0.84 ± 0.03	14.0
33	*Xanthoparmelia stenophylla* (Ach.) Ahti & Hawksw.	0.60 ± 0.03	10	0.76 ± 0.01	12.7	0.72 ± 0.05	12.0
34	*Xanthoria elegans* (Link) Th. Fries	0.57 ± 0.06	9.5	0.62 ± 0.06	10.3	0.60 ± 0.04	10.0
35	*Xanthoria parietina* (L.) Th. Fries	0.53 ± 0.05	8.8	0.61 ± 0.04	10.2	0.57 ± 0.06	9.5

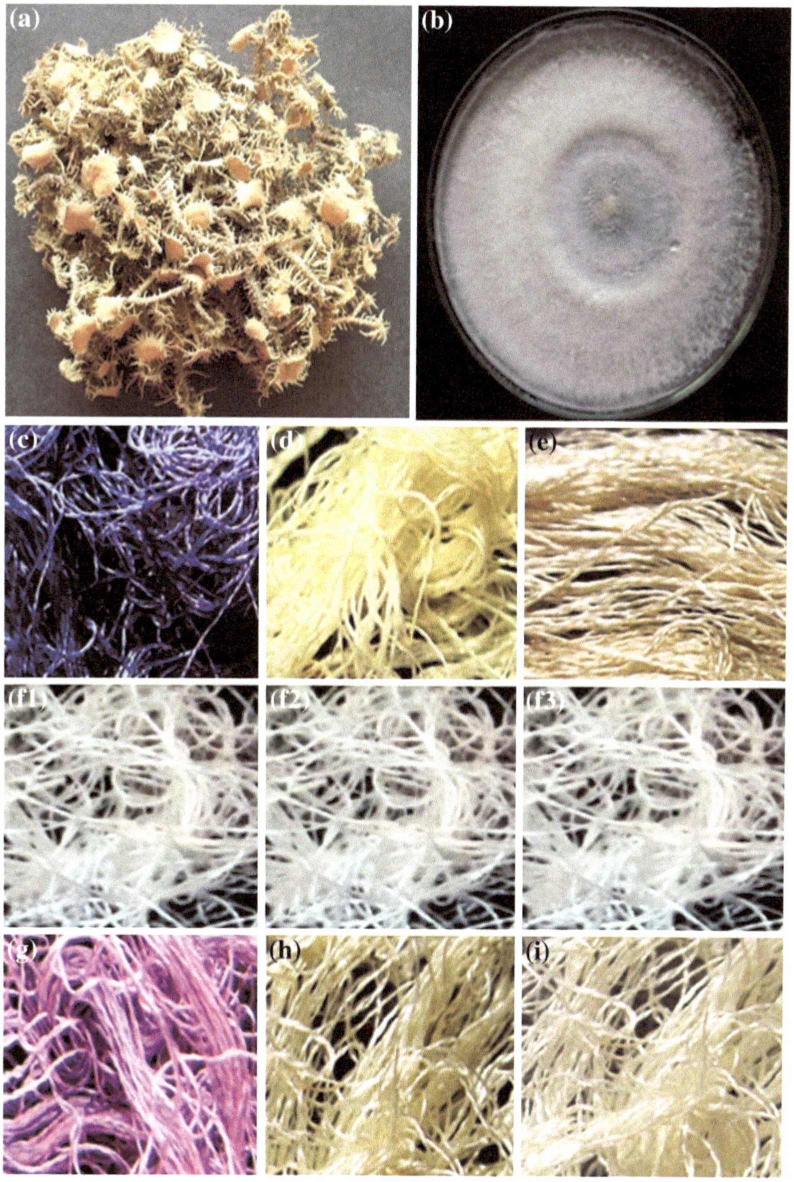

Fig. 12.9 **a** Natural thalli of *Usnea ghattensis* G. Awasthi, **b** cultured mycobiont of *U. ghattensis*, **c**, **d**, **e** silk dyed from natural thalli through ammonia fermentation method, boiling water method and cow urine method, respectively, **f1**, **f2**, **f3** control threads, **g**, **h**, **i** silk dyed from mycobiont through ammonia fermentation, boiling water method and cow urine method, respectively

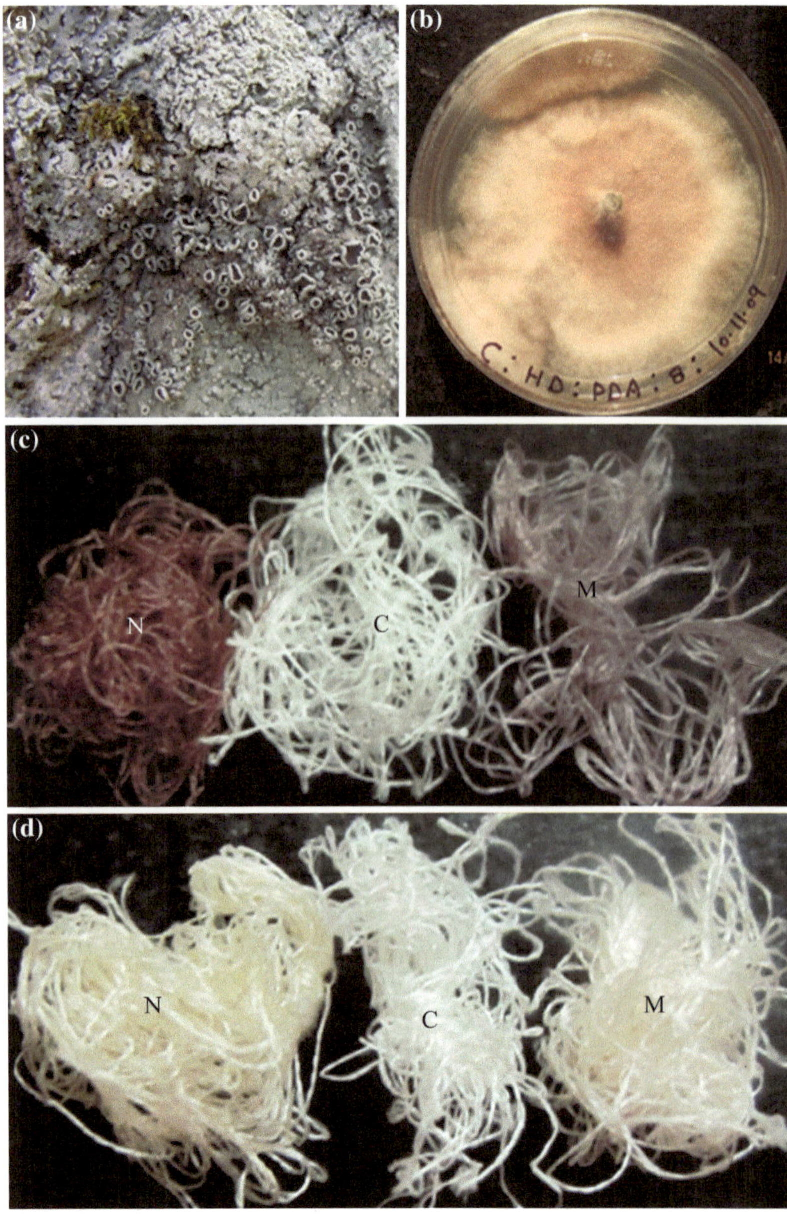

Fig. 12.10 **a** Natural thalli of *Heterodermia diademata*, **b** Mycobiont culture of *Heterodermia diademata*, **c** dye prepared by ammonia fermentation method, **d** dye prepared from cow urine method, (*N*) dye prepared from natural lichen thalli, (*M*) dye prepared from mycobiont, (*C*) control (Natural Tussar silk)

sustainably utilized for production of dyes, and the lichen dyes can prove to be a boon to the textile industry in India and abroad.

Acknowledgments Authors are thankful to Director, CSIR-National Botanical Research Institute, Lucknow, for providing necessary laboratory facilities and Department of science and Technology, New Delhi, for financial support under project no. GAP - 215625. PS is also thankful to Miss Sugam Gupta for providing photograph of *Xanthoria elegans* (L.) Th. Fries.

References

Ahmadjian V (1973) Methods of isolation and culturing lichen symbionts and thalli. In: Ahmadjian V, Hale ME (eds) The Lichens. Academic Press, New York, pp 653–660

Ahmadjian V (1993) The mycobiont (fungal symbiont). The lichen symbiosis. Wiley, New York, pp 8–15

Brough SG (1988) Navajo lichen dyes. The Lichenologist 20(3):279–290

Casselman KD (1994) Historical and Modern Lichen dyes: Some Ethical Considerations. Norw Breakfast Club Newslett I(1):1–13

Casselman KD (2001) Lichen dyes: the new source book. Dover Publications, Mineola, New York

Dean G, Brown D, DeSouza D, McClintok D (2012) Adventure with lichen dyes. Turk Red J 17:1–10

Hart KH, Cox P (2000) A cladistic approach to comparative ethnobotany: dye plants of South-Western United States. J Ethnobiol 20(2):303–325

Hodge K (2006) Dyeing with lichens & mushrooms. http://blog.mycology.crnell.edu/2006/12/12/dyeing-with-lichens-mushrooms/. Accessed on 17 July 2014

http://www.chriscooksey.demon.co.uk/lichen/orcein.htm. Accessed on 17 July 2014

Kok A (1966) A short history of the orchil dyes. The Lichenologist 3:248–272

Margareta SF (1981) The chemistry of plants animal dyes. J Chem Educ 58(4):301–305

Perkins P (1986) Ecology, beauty, profits: trade in lichen based dyestuffs through western history. J Soc Dyers Colorists 102:221–227

Richardson DHS (1988) Medicinal and other economic aspects of lichens. In: Galun M (ed) Handbook of lichenology, vol 3. CRC Press, Boca Raton, pp. 93–108

Ridgway R (1912) Color standards and color nomenclature. R. Ridgway, Washington, DC

Shukla P, Upreti DK (2014a) Natural dyes from Himalayan lichens. Ind J Tradi Knowl 13(1):195–201

Shukla P, Upreti DK (2014b) Assessment of dye yielding potential of Indian lichens. Ind J Plant Sci 3(1):57–63

Upreti DK, Joshi S, Nayaka S (2010) Chemistry of common dye yielding lichens of India. ENVIS Forest Bull 10(1):122–133

Upreti DK, Tiwari P, Shukla P, Dwivedi A (2012) Natural thalli and cultured mycobiont of *U. ghattensis* G. Awasthi: a potential source of purple dye yielding lichen from India. Ind J Nat Prod Resour 3(4):489–492

Yamamoto Y, Mizuguchi R, Yamada Y (1985) Tissue cultures of *Usnea rubescens* and *Ramalina yasudae* and production of usnic acid in their cultures. Agric Biol Chem 49:3347–3348

Yoshimura I, Yamamoto Y, Nakano T, Finnie J (2001) Isolation and culture of lichen photobionts and mycobionts. In: Kranner I, Becknett R, Varma A (eds) Methods in lichenology. Springer, Heiderberg, pp 3–33

Index

D.K. Upreti et al. (eds.), *Recent Advances in Lichenology*,
DOI 10.1007/978-81-322-2235-4